THE QUESTIONING OF
INTELLIGENCE

FREE UNIVERSITY BRIGHTON PHILOSOPHY SERIES

THE QUESTIONING OF
INTELLIGENCE

A Phenomenological Exploration of
What It Means To Be Intelligent

John Thornton

F.U.B. TEXT

Published by the Free University Brighton
www.freeuniversitybrighton.org
contact@freeuniversitybrighton.org

Printed in the UK by IngramSpark, Milton Keynes
www.ingramspark.com

Front Cover Photograph by Clive Buckland
Back Cover Photograph by Nicolette Thornton
Cover and Book Design by John Thornton
Text typeset in Egenolff-Berner Garamond using X⅁LATEX

The cover photographs show the author sitting in the graveyard of the Church of St Nicholas of Myra in Brighton circa 1982 and in 2021.

ISBN   978-1-8384787-0-4 (paperback)
ISBN   978-1-8384787-1-1 (ebook)

 FUBTEXT is an imprint of the Free University Brighton.

Contents

PREFACE

IT is now early 2021. The world is still undergoing the ordeal of coronavirus, and Joe Biden has been elected president of the United States. Only a year ago Boris Johnson became prime minister of the UK and the country finally left the European Union. Then there were the Australian bushfires. It seems to me, in this corner of the English-speaking world, that more has happened in the last year than happened in the preceding decade. These recent events lie in the background of what is written here. It was the lockdowns and restrictions from coronavirus that first cleared away the distractions I had been using to justify my literary silence. For the core of this book has been sitting in me, unwritten, for more than a decade. Finally I had a dream that gave me a direct warning. It featured a fish pond crowded with oversized goldfish or carp. They were pressed together, with hardly any room to move, all wriggling and silvery. In the dream I knew there was not enough oxygen for them to breathe, and unless something were done (and quickly) they would start to die. At the time, in my waking life, I was in the first stages of training to be a Jungian analyst. This would have shelved any plans to write for several more years. The message of the dream seemed clear: the fish were the insights and understandings that had been granted to me in my earlier life, and unless I started to write them down now, they would disappear back into the psyche from which they had once emerged.

In that earlier life I had been a researcher and lecturer in the area of artificial intelligence. To enter that world I first had to qualify for a PhD, which meant writing a lengthy thesis and becoming a *Doctor of Philosophy*. This title seemed strange to me at the time, given that my thesis had virtually nothing to say about philosophy. And yet I was rather pleased with this new form of address. It fitted with something in me, with an idea I had of what it means to be an academic. The title itself goes back to a time when the sciences of nature were seen as different aspects of one natural philosophy, which itself was seen as a branch on the tree of the one true philosophy. In the ideality of that world, everyone who earns the right to teach and research at a university is thought of as a philosopher because (in theory) we share a common understanding of what we are doing, of why we are doing it, and of how all our various branches of knowledge are joined together into a meaningful, philosophical whole. And, in fact, at the beginning of the modern era, during the time of the scientific revolution, there was such a sense of common purpose expressed in the idea of universal reason. But the university world I entered in the late twentieth century had fallen a long way from that ideal. Philosophy, once the queen of the sciences, now had to beg for crumbs left over at the table of university funding. And no wonder. For this dreamt of unity of universal reason had collapsed into a relativity of various philosophical schools of thought. And within these schools there developed such a degree of technicality and complexity and obscurity that no one outside the profession could hope to follow what was going on.

I collided with this world as soon as I started to inquire into the foundations of my own discipline of artificial intelligence. For, despite our modern disinterest in philosophy, it is still the case that our lives, our institutions, and the way we go about doing what we do, stand on the foundations of sets of assumptions that are essentially philosophical. And if we choose not to examine these assumptions that does not mean they go away. It means they go underground and so exert their influence without restraint. What I discovered, at the core of artificial intelligence, was a pre-existing understanding of being that has come to be known as *scientific materialism*. Scientific materialism says everything that happens here on Earth, and, by extension, in the universe as a whole, is determined by the low-level physical interactions of microphysical energy fields and particles. That means, for instance, that your being alive, and your being conscious, are entirely determined by, and are nothing more than an effect of the interactions of these physical fields and particles. It turned out, in my life as an artificial intelligence researcher, that I was surrounded by people who (at least in their professional lives) thought of themselves as a species of highly sophisticated biological computer, living an essentially meaningless existence that was finally a simple consequence of the blind operation of the mathematical laws of physics.

At first this discovery came as quite a shock. I started to argue with a few of my colleagues, saying that such an understanding could not possibly account for the fact of my being alive and conscious. But I soon discovered this was a path that many people had travelled before me. Literally hundreds of books and thousands of philosophical papers have been written on this subject, mostly by people attempting to defend the ground of scientific materialism from the naïvity of people like me, i.e. people who clearly must believe in some kind of ghostlike spirit that mysteriously influences the electrochemical behaviour of the neurons in the brain. I began to see that this ground of scientific materialism runs very wide and very deep. It not only rules over the discipline of artificial intelligence, but stretches over all the objective sciences, and, by extension, over the technological civilisation they have engendered. Seeing this, something began to stir in me, a kind of *impulse of resistance.* For, despite the hundreds of books and thousands of papers, scientific materialism is clearly false. You don't need to study philosophy to discover this, you just need to inquire into your own direct experience of being conscious.

I now saw a meaning in my being a Doctor of Philosophy. What the title is saying is that each academic is responsible for the material they teach and publish, and part of that responsibility is to inquire into the philosophical foundations of their subject area. That means if you find something wrong in those foundations, then you attempt to rectify it. I began to see myself as a philosophical doctor who was dealing with a sick patient, and had to find the root cause of the illness. My first diagnosis was delivered in a textbook written to accompany an undergraduate introduction to computing course. It was called *The Foundations of Computing and the Information Technology Age* (2007). In it I attempted to trace the development of scientific materialism through the history of the development of information technology. This was a book for computing students, to develop an awareness of the wider context from out of which our computerised information systems are emerging. As such, it was not intended for the general public or for the professional philosopher.

At that time I still had an ambition to enter the world of academic philosophy and exert some influence on the debate concerning the nature and reality of consciousness. And so I enrolled in a second PhD, this time in pure philosophy, and started attending classes and seminars and workshops and conferences where I delivered several papers that were received with an almost total lack of interest. During this time I met with some of the world's most prominent philosophers of mind, including John Searle, Hubert Dreyfus, Galen Strawson and David Chalmers. And yet it seemed to me in all these meetings that I was faced with a kind of impenetrable wall. For I had the wrong attitude. In actuality I was a PhD candidate having to meet with the approbation of those who had already made the grade, whereas, from my perspective, I had come to teach these professional philosophers the error of their ways. Finally, after writing the first four chapters of a second thesis, and being confirmed in the degree, my supervisor, John Mandalios, was compelled (by serious illness) to retire from all academic duties. This, for me, was a kind of last straw in the progress of my career as an academic philosopher. For John, of all the philosophers I had met, was the only one with whom I could have a real dialogue. When he left, I left with him.

So there I was, sitting with the first third of an unpublished thesis entitled *The Transcendence of Computational Intelligence*. At the same time I resigned my position at Griffith University in Australia and came back to England to start a new life of semi-retirement. I picked up some part-time artificial intelligence teaching work at the University of Sussex and also started teaching philosophy at the Free University Brighton (FUB). It was in these FUB classes that I finally had the chance to teach philosophy properly, by directly demonstrating what it means to philosophise. In each meeting we would start with a few pages of a key text, typically from Heidegger or Husserl. My self-appointed task was to open up this material to people who had largely arrived without any background in philosophy whatsoever. Such an opening up is an entry into the domain of philosophy itself. This was a profound training. For Heidegger and Husserl are hardly easy philosophers. But each, in their own way, is *true*. That means they have kept the essential core of philosophy alive. It is this core that is the living potential of our civilisation. It is what our forebears have passed on to us. Only now it lies buried in the debris of our technological dream of progress. It is because of my experience of bringing this core to life at FUB that I now *dare* to write this book. For, despite our apparent collective lack of interest, what I discovered is that *many* people have the potential to become deeply and seriously interested in philosophy. All that is needed is to be shown the way in.

Of course, saying that makes it sound like you are about to read an introduction to philosophy for the 'non-specialist', something that can be filed away with the thousands of other books and articles that qualified people have written about the mind and consciousness and materialism and computing and intelligence. I should therefore make it clear from

the outset that this book does not belong alongside those works of measured rationality. For I am not presenting a series of arguments, in which
I weigh up the evidence both for and against a cherished position I am
defending. And neither am I presenting an interpretation of some other
philosopher's system of thought. I left all that behind when I surrendered
my ambition to become a professional philosopher. What lies before us is
an invitation to take a journey of inquiry. The closest work, the one I am
immediately reminded of, is Robert Pirsig's *Zen and the Art of Motorcycle
Maintenance.* I read this classic of modern literature when I was seventeen,
travelling to Lewes each day on the school bus, completely absorbed in
the unfolding story of Pirsig's quite different journey. It was then that
philosophy, as a living, breathing, inquiring activity first revealed itself to
me. In *Zen* I heard intelligence *speak.* You can recognise such speaking because it does not repeat what someone else has said. It is original. And yet,
when you come to understand it, you realise it was something you already
knew—it is only that you did not know that you knew. What is new, what
is original, is what this knowing reveals. For it takes the situation of our
being here, the situation which is always changing and evolving, and puts
it into a new light, a light which reveals a greater truth, a greater meaning.
It is the seeing of this meaning that is always new. Such seeing is not a case
of channelling a disembodied spirit. Intelligence speaks through each one
of us, *as* each one of us, according to our experience, according to who
we are, and according to what we are capable of seeing and realising.

   In *Zen and the Art of Motorcycle Maintenance,* Pirsig takes us on a literal visceral motorcycle journey across the USA while unfolding his own
philosophical inquiry after meaning. Here you will find a different kind
of inquiry, one that traces the steps and insights of my own ascent into
the mountains. If you know Pirsig's work, you will perhaps miss the ac-
companying metaphor of the road trip. For Pirsig was a literary genius—a
kind of philosophical Hemingway—who was able to entwine his inquiry
into the fabric of a dramatic narrative to create a work of art. In contrast,
in these pages, after a brief introduction, we more or less leave the story
of my personal life behind. But that is as it must be. For I am no literary
genius. It is the inquiry itself that is essential, that matters. And once we
are up in the mountains, if you can bear the rarified air, I think you will
find there is very little personal history left to speak of . . . .

PART 1

THE INHERITANCE

Chapter 1

An Autobiographical Introduction

Warwick University

WHILE there is no single reason or event that caused me to write this book, there was a series of experiences, many years ago, that signalled a turning point in my life, and that now appears as a kind of seed from out of which this project of questioning intelligence has grown.

It was the autumn of 1979, when I was a second-year economics student at Warwick University, living in a rented room, in an old house in Leamington Spa. Autumn is the season of the psilocybin mushroom, and I had just picked a large quantity with some friends in a nearby cricket field. This was not my first psilocybin experience, but it was the first time I had taken a really large fresh dose. What happened next is not something I can directly relate. I entered into another state of consciousness, and to properly describe that state would mean entering it again, and for you to understand me, would mean your entering it too. But I do remember some significant details.

Firstly, it was clear that something fundamental had changed in the whole structure of my being. I expressed this at the time by saying that my ego had died. I remember this being an extraordinarily liberating and enjoyable state, suffused with a feeling of having come home—of, for the first time, being consciously in the place of my true being.

Secondly, there was a sense of being in touch with a source of almost unlimited knowledge. Again, I cannot describe this directly. It was as if anything I cared to look at could not help but reveal its deeper meaning. I remember inviting my friends to ask me anything, about life, the significance of our being here, even the hidden meaning of a saying on the back of a matchbox. I felt, for the first time, that there really is such a thing as *truth*—an actual way things are, as opposed to the way we think they are, or imagine they are—and that there is, in each of us—beyond our individual egoic selves—an innate intelligence that can see this truth, directly, without having to read it in a book, or learn it, or remember it by practice and study, or be told it by someone else.

It also occurred to me that this is how it must have been for Jesus, not that I was Jesus, but that Jesus too had access to this state of immediate intelligence. And I wasn't thinking of Christ, the mythical saviour-god of the Church, but Jesus the man, who spoke the parables and preached the Gospel of Love. As for my own contemporaries, my teachers and parents and friends, those in authority in the world, I could see no evidence that they had any access, or even knew of this state of immediate knowledge. There were signs in the lyrics of songs, and the pages of certain books, but there was no one I actually knew or could talk to at that time who had ever indicated any first-hand knowledge of this extraordinary state I had stumbled on almost by accident, simply as a result of ingesting a few mushrooms I had found in a field . . . .

Figure 1.1: Da Vinci's Last Supper.

Of course, it is quite reasonable, once it's over, to dismiss a drug induced experience as a kind of illusion—a dream. And, in many ways, the psychedelic experience is like a dream. There is the same sense of being in a different kind of reality and of the essence, the essential quality of the state slipping away as you return to a 'normal' egoic consciousness. The difference from dreaming, however, is that in the psychedelic experience you remain in touch with the waking world, with the events and the people in it, and with all we call 'reality'. For myself, I experienced no hallucinations. My sensory experience, far from being distorted, was amplified by a feeling of being absolutely conscious and present in a way that made my ordinary waking state seem to be a kind of reduced half-consciousness. At least, that is how I *remember* it.

I was nineteen years old and more or less alone in my attempt to come to terms with this experience. Once the drug wore off I felt I had been deposited back into a state in which I was no longer at home, that was no longer where 'I' wished to be. What was worse, I had seen this normal egoic state was in some way false, that I was being ruled by an entity that was not 'really' me—a fearing entity, that lived in a fog of negative feelings, and, having glimpsed itself, was feeling all the more negative.

At the same time, I was optimistic. It seemed that this other state of (true) being existed in a dimension that was separated from me by a thin veil. After all, it had only taken a handful of mushrooms and 30 to 40 minutes to get there in the first place. Surely there was an easy way to get back and reside there permanently? So, I took another trip, and did return there for a while. I saw time as though I were no longer 'in' time, with each moment stretching on without limit. I whirled as the dervishes do and found the still point of the turning world within.[1]

Figure 1.2: A Whirling Dervish (Photo by svklimkin on Unsplash).

In between, there were other drugs, particularly hashish, that altered the egoic state and made it more bearable. But, after two or three more sojourns in this psychedelic enlightenment, the power of the chemical started to fail. It was as if my ordinary egoic state possessed its own intelligence, its own will to survive, and had learnt how to defend itself from being dissolved in the effects of the drug. Now, instead of freedom, there was an inner struggle, an attempt to get beyond the thinking, unhappy state of the ego, a struggle that only seemed to strengthen the ego's hold and changed the psychedelic experience into a kind of test of endurance.

A year later I was staring into the darkness of depression. I could see no way back, now the drugs had failed. My ordinary state, that once I accepted without much question, was now without value, a kind of prison in which I seemed to be serving a life sentence, with the only certain way out being to physically die. But I did not seriously contemplate suicide. I knew, behind the darkness there was light, that I had known this light, and that my essential being, my essential goodness was still there within me. Only I had lost the key, the means to find my way back.

Figure 1.3: Yoko Ono and John Lennon (Photo from the Dutch National Archive).

Around this time, John Lennon was assassinated. For several years, from the age of twelve, I had assiduously collected all the Beatles albums with a dedication that seems unusual now. I think, as for many of my generation, John Lennon was a symbol of the kind of man I intended to become—an intelligent, articulate, truth-telling, anti-establishment hero. When he died, it was as if his spirit were hovering over the earth and all that he had realised and written was directly available to me, and anyone else who was attuned. This included his connection with Yoko Ono. Their story became another symbol for me, a symbol of the union of man and woman, in love, in difference and in equality, each with their own power, their own creativity. I saw this connection with a woman, with the feminine, as absolutely central for me, in this life. And I saw a simple truth: I was in need of love, true love, the kind that John and Yoko knew, and I needed to act, I needed in some way to be heroic myself.

And so there came the night of a party at the student flats where I was living. I had now become a kind of recluse, staying awake until three in the morning and not getting up until mid-afternoon—avidly reading anyone who seemed to have something to say to my situation, listening to music and talking earnestly with a small circle of friends. Handling a party meant one thing: quickly drinking a fairly large quantity of Southern Comfort. This would jettison me out of my socially inhibited state and allow me to speak my mind in public. That night it was Frances from one of the upstairs rooms who came on the receiving end of my truth-telling. I went up to her and said: "Why are you always ignoring me?" I was naïve, I thought she didn't like me. But it wasn't like that. I may have annoyed her, but it wasn't that she didn't like me—quite the reverse. You know those Hollywood movies where the man and woman start off disliking everything that the other stands for—she the paragon of uprightness, good sense and practicality, he the rogue, the outcast, the cynic? Well, it was like that. We went for a walk in the darkness, across the campus and I began to speak honestly with her. Something passed between us, there was a mutual attraction, we walked back to the flats. And then, the next day, we started ignoring each other all over again.

This state of affairs went on for perhaps two or three weeks. All the while the pressure was building—I knew it was up to me to make the next move—but (in short) I was afraid.[2] It was not that I had no experience with other women. But I was scared of this attraction to Frances.

Finally, I was almost pushed toward her one night at the student union bar by my friend Steve who could see only too clearly what was going on. I said something like:[3] "Shall we get out of here?" and she agreed. We went back to her room and made love. But this was not like any previous experience I had had of making love. I was completely present—making love, physical love, to her—not just to a female body, but to this being—Frances—it was she, her unique presence, that I was merging with. Or so it seemed to me. Something happened that night. I received the only nourishment that could possibly reach me in the state I was in. I felt Frances knew this too, that she was with me in her deeper being. But I do not think she consciously registered what had occurred. At least she did not speak of it. Instead, after I had reached a final ejaculation, she said, quite calmly, "Well, you've had your fun, now what about me?"

Now, in a way, this was a rather courageous thing to say, especially in those days. I'd had my orgasm, but Frances was still waiting. And, from the place I was in, I was merely surprised, surprised she had not been truly with me, had not consciously registered that something *holy* had occurred. For I was healed, I was released from the darkness in which I had been trapped. I woke the next morning and could hardly believe I was there with her, in the light, in the sunlight of a new day, a day that opened out before me, full of possibility. The depression was over, fully over. For many weeks I wondered if it would return. But it did not. To this day it has not. That one profound experience of making love seemed to have flicked a switch, and to have healed a *rift in my being*.

Somewhere in this process I made a vow. I was hardly aware of it at the time, but now, some forty years later, I see I made a stand that formed a turning point in my life. I vowed, I resolved that I would remain true, as far as I was able, to what had been revealed in my psychedelic enlightenment, true to something within me I could not name, that was the source of all I had received. I vowed I would not 'go under' and dishonour this inner source. Such going under, as I saw it, would be a kind of surrender to the world, an acceptance of a certain kind of egoic, materialistic, self-ishness as being the ultimate end and purpose of our existence here. This was not just an abstract resolution. My intelligence, the very intelligence revealed in the psychedelic enlightenment, had not completely forsaken me as the drug wore off. As I see it now, it rather fell into the background and became entangled with the secondary intelligence of my normal waking state. It is this entanglement that makes the ordinary state so hard to handle—for you cannot simply dismiss what is going on in there. You have to learn to discern the true from the false, and that is no easy task.

Anyway, it seemed to me, at the grand age of twenty-one, that the entire discipline of economics, as it was being taught to me at university, was a kind of apology for, and a justification of, the status quo. It was something that was obviously false, based on notions of free markets that existed only in the minds of economists, who were either consciously or unconsciously in the service of the economic interests that their theories were supporting. It also seemed to me that I should not be studying what I did not enjoy, and that I did not enjoy attending lectures or writing essays and neither would I enjoy the kind of career that a degree in economics was preparing me for. I therefore resolved to throw the whole project away—to end the compromise I had been living for three years—something I had endured not because I cared about what I was studying but because I was afraid of what would happen if I did not have a degree and a successful career.

My resolve was to tell the truth, as I saw it, in the final exams—to express, as best as I could, the view that my intelligence had of the human world in all its compromise and ignorance and denial of what is obviously the case. So, in all four exams, I took the questions apart, and attempted to point out the falsity of the assumptions that lay behind them. In the meantime, I spent the entire summer term (when I should have been revising) helping Frances with her art history revision and reading the complete works of Carlos Castaneda,[4] concerning, amongst other things, the psychedelic effects of the peyote mushroom.

Ironically, in later life, I went back to university and took another degree in computing, eventually becoming an academic myself. So now I have a certain sympathy for the people who had to mark my economics exams. For even if someone had seen some merit in what I had written, I was never going to be heralded as a brilliant and courageous truth teller. Instead (of course), I was failed and invited to resit the exams if I were willing to answer the questions more conventionally. This opportunity I also declined.

The point here is not the wisdom of my criticism of contemporary economic thinking (a criticism, by the way, that I still look upon favourably), it is that I took an action, I made a tangible sacrifice, and I changed the course of my life on the basis of what I felt to be true and right. I see now that this was an act of *faith*. My faith was in the truth and reality of the experiences that had been granted me through taking the psilocybin mushrooms, through the death of Lennon and through making love to Frances. I had no real proof that anything valid had occurred. It was up to me, it was my *freedom* to put my faith in what had been revealed to me, revealed in a form that was beyond any possibility of objective verification. To do this I had to stand alone. My mother, my father, Frances herself, nearly all the people who were close to me, did not understand what I was doing. I hardly understood.

Things did not last with Frances. She did not want to 'drop out' with me and I could see we had very different ideas about the kinds of lives we wanted to lead. My university career was over and I was back in my parents' house (they were away on holiday). I had made my grand gesture, now there were the consequences. I tried to end it with Frances, but she would not have it. She came down on a coach, we made love, and for me it was just as it had been that first night. The next day, she was cutting my hair in the kitchen, and I had one of those moments of perfection, of complete fulfilment, just being there with her, talking together, with nothing in particular happening . . . .

That night we were watching the news about the marriage of Charles and Diana, and there was a piece about an Indian guru, Bhagwan Shree Rajneesh. I remember thinking he was speaking the truth, and that he seemed to be in touch with the source I had known on the mushrooms. Then I thought, why would anyone want to get involved with a man like that, when all that any of us really needs or wants, if we only knew it, was this feeling of completeness I had with Frances that day?

But Frances went back home to her parents. She was going to start a new life in London studying art restoration and I was not going to be living with her any more. It still wasn't over, but the fact she no longer wanted to live with me meant it was only a matter of time. And now my parents were coming home. The throwing away of the degree meant it was not going to be a happy reunion. In a way, my act of faith had had quite predictable effects. I had no money, I was losing the woman I loved, my parents were traumatised, and I had no obvious prospect of a career. The future was almost entirely blank. And so I waited.

Just about on cue, my friend Clive, whom I had known from school, phoned up with a plan to start a music magazine in Brighton. I needed somewhere to live, so I called up a promising room rental in a residential street to the north of the city. When I arrived, I was greeted by a woman in her mid-forties wearing a strange beaded necklace (a mala) with a picture in it of that bearded guru I had seen on the TV only a few weeks previously: Bhagwan Shree Rajneesh. I took the room and started writing something of my experiences at Warwick in our new magazine *Dogma*. A door had opened leading in an entirely unexpected direction. My leap of faith had landed me in a different kind of fire . . . .

Figure 1.4: A sannyasin mala as worn by the disciples of Bhagwan Shree Rajneesh.

THE VIPASSANA RETREAT

I think if you had asked Frances, or my mum or dad, at that time, to identify my most outstanding negative features, they would have said I was
arrogant and lazy. I did not see it this way of course. It was the world's
fault I did not exert myself, for it offered nothing worth pursuing, and
what others thought of as arrogance, I thought of as a kind of heroism, a
refusal to bow down to the opinions and ignorance that these others had
accepted without proper reflection. It was this arrogant and lazy self that
had been extinguished in the psychedelic experience and in being with
Frances. I had certainly been changed, but I was still undisciplined and
unwilling to make a serious effort to create a life for myself. At the same
time, I felt superior, I felt I saw more clearly than others, and that I had
something of value to say.

My new life in Brighton was now supported by unemployment bene-
fit. I found I could survive well enough on the fortnightly cheque. My
sole contribution to society was to write articles for *Dogma* and then go
out and sell them with Clive at local shops and concerts (just about covering our printing expenses). This propelled us into the Brighton counter-
culture, getting free admission to local concerts and venues and the opportunity to interview a number of the famous and less famous musi-
cians of the day. But, despite the initial elation, and a police raid on our
premises, after four editions, we ran out of steam. The stream of truth
and realisation I thought was open to me from the experiences at Warwick was starting to recede.

My landlady, Anuragini, the sannyasin disciple of Bhagwan Shree Rajneesh, had also made a significant impression on me. She was much the
same age as my mother—and yet had broken free of the moral and intellectual constraints within which my mum seemed to live. And Anuragini
was interested in *me*, both in what I had to say, and, it later turned out, sex-

Figure 1.5: Clive Buckland.

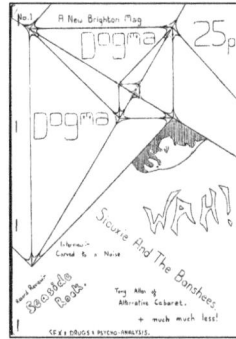

Figure 1.6: The first issue of *Dogma*.

ually. Through her I became loosely affiliated with the other sannyasins in Brighton, attending group meditations and Sufi dancing sessions. I also starting reading what Rajneesh had to say, including *The Orange Book*,[5] which detailed a number of meditation techniques.

I moved around Brighton quite a lot at this time, living with Clive in Kemptown, then taking a room in a building on the seafront where Clive rented a flat upstairs. This north-facing room had a shower and an attached kitchen and toilet. It was here I reached another kind of crisis. It was over with Frances, and I keenly felt her absence. And it was over with *Dogma* as well. Inwardly I found there was nothing left to say—just a jumbled stream of thoughts, without coherence, and without any obvious wisdom. Mixing with the Brighton sannyasins gave me some kind of social life. But without a girlfriend, and without an inner connection to the truth, I started to lose any sense of enjoying life. It was as if I were simply filling in time. I was not exactly depressed, but I was becoming indifferent to the activities of daily life. Whatever I did, it didn't seem to matter whether I did it or not. And so I decided to do nothing.

This idea arose from Rajneesh's *Orange Book*. In there was a kind of sannyas version of Vipassana meditation. All I had to do was to lock myself up in my self-contained flat with enough food to last a few weeks, refuse to see or speak to anyone, and start watching my breath. I remember I had a large furry kind of bag filled with polystyrene balls on which I sat and a big vegetable casserole in the kitchen that I renewed every three days. I told everyone who might try and contact me that I was going into retreat, locked the door and sat down to meditate.

The idea in watching your breath is to stay continuously present to what is happening now, each moment, without getting carried away in thought. Now this may sound quite easy, and it is easy, at least to begin with. You may like to try it. It doesn't mean thinking "I am breathing" it means consciously noticing the breathing, being with the breathing, as it is happening, the breathing in, the breathing out. Like now. I predict all will be well for a few seconds, perhaps a minute. There is a certain novelty, a certain challenge. But after a while, it becomes clear that one breath is much the same as another. There is nothing new happening, nothing to attract the attention. And so, inevitably, a thought arises. This thought, whatever it is, is going to be something new. Maybe not absolutely new, but certainly it will have more novelty than watching the breath going in, going out, going in, going out, ...

And so you start to think, you 'drift off' to wherever the thought leads. Generally, thought like this, unbidden thought, goes along its own path of unconscious association. After a while, you realise "Ah, I'm thinking again, and I'm here to watch my breath, not to think." And sure enough, there is your breath, still going in, going out, as it was all the time you were thinking. So you stay with the breath, perhaps for a minute or so. Again a thought comes. Perhaps you start getting wise to this and learn something new: to *let the thought go,* that is, to not go along with it, allowing it to arise, but not in such a way that you lose awareness of the breathing. Staying with the breath means the thought can subside without leading you off into a whole chain of thinking. This requires a certain degree of consciousness, an energy of awareness and attention that, in a way, is *waiting* for a thought to arise. Such awareness is tiring. After a while you suddenly realise you are thinking. There isn't even a memory of how this stream of thought got going, how it was able to interrupt your awareness of the breath. So you start again, and again, and again, ...

After some time, and for me this took *days,* it became clear that the streams of thought that kept interrupting the watching of the breath were themselves quite repetitive. I remember I was caught up with two basic themes: the first concerned a young woman who lived upstairs in the same house in Leamington Spa where I had taken the mushrooms a few years earlier. I knew she was still there and I sensed she had been attracted to me. After Lennon's death, I was slowly plucking up the courage to approach her when Frances appeared at the party. Now, while trying to watch my breath, I kept thinking about going back to Leamington Spa and declaring my attraction. Over and over I thought about doing this. And indeed, within me there was someone who was far more interested in hitchhiking up to Leamington than carrying on with this discipline of watching the breath. But still I stuck with it. The second theme was my desire for a Honda 250cc motorbike. I kept thinking about how I was going to get such a motorbike. Over and over the motorbike would appear to me, in my mind's eye, and I would feel my intention to get out of this little room and get some money and buy just such a motorbike.

Ironically, the real-life model for this motorbike was owned by a drama student called Richard who had lived in the same Leamington Spa house during the Warwick days. After the Vipassana retreat, I went back to Leamington to see how things really stood with the young woman who had so haunted my thoughts. It turned out she was now living with Richard. I kept quiet about the true reason for my visit and returned to Brighton.

Figure 1.7: The house in Leamington Spa. I lived in the right-hand ground floor room (you can see the faint image of the young woman I was thinking of so intently in the open window upstairs).

Back in the retreat I knew nothing of this. I just kept on thinking the same thoughts, driven by the same desires, coming round, watching my breath for a while, then going off again. Many years later, a young man, very spiritually earnest, was speaking to me about how hard it was for him at work (he was an aircraft mechanic). He said how estranged he felt from all the other men working there, how all they seemed to think and talk about were 'birds and motors'. So I replied, partly in jest, but thinking back to this time, "Well, Michael, what else is there . . . ?"

And yet *something* kept me at it. After seven days the thinking stopped. I didn't do anything. Perhaps whoever or whatever was doing the thinking just got tired. Anyway, it stopped. Now there was just the breath, the room, the sunlight, the darkness, the sounds, whatever was happening in each moment, the entire sensory stream just streaming, being itself, with 'me', this consciousness, simply being with it, as it, in it. Was this enlightenment? I don't know. At the time I didn't care. In order to care I would have had to think about it. And my thinking self had subsided, or at least the compulsion to think had subsided. There were passing thoughts, usually about something immediate, like cooking supper, but these had no power to disturb my being in the immediate presence of now. For I no longer wanted to think—with all this before me—this amazing presence of living life, this living moment. There was nowhere thought could take me that could be better than *this*.

So there I was, completely present. There was nothing boring about it. Each moment, enough in itself. I stopped sleeping. I didn't need to sleep—I was in a state of rest each moment—doing nothing—just sitting. I started experimenting, going out along the seafront in the early hours of the morning, to see if this disturbed things. But being outside was just as lovely as sitting in my room. I'm not sure how long I could have stayed like this. I certainly had no wish to 'do' anything else. At some level I knew it all depended on my not seeing another human being. In terms of clock time, I remained in this state of inner stillness, continuously, for perhaps five days. And then my father turned up, knocking on the door, calling out: "John—Rags has died!" This was the family dog, a miniature Schnauzer, who had been our companion since I was nine or ten years old. I could not ignore the call, so I opened the door . . . .

I was right. I couldn't maintain an unbroken inner stillness in the face of the world, in the face of having to 'be' someone, to speak, to respond, to enter human society again. Those thoughts, those desires that had endured for seven days had not been eliminated, they had merely become exhausted—they had made a deal, saying: "We'll leave you alone while you finish this retreat—but we'll be back when it's over!" And they were. I was twenty-two years old. Of course I wanted a girlfriend and some transport of my own.

Figure 1.8: Rags and I shortly before the Vipassana retreat.

PURE CONSCIOUSNESS

So what was happening here, in the life of this young man, this former self? Back in Brighton, I was about to make another grand gesture, selling up almost everything I owned, and going to Oregon to become a disciple of Bhagwan Shree Rajneesh. From the perspective of the world, it looked like things had gone badly wrong. At the age of twelve, I emerged from childhood, a fairly happy and well-balanced individual, conscientious—hard-working even. I was the top performing student in my year at Worthing High School and destined for Great Things. And yet, a year later I was writing in a school exercise book that my deepest wishes were to fall in love and take psychedelic drugs. I was already seeking to get out of the state I was in—to get back to something I dimly sensed I had known as a child. Instead of embracing adult life with all its challenges, I lost interest in my studies, and started drinking alcohol and smoking hash.

Fundamentally I wanted to get out, to get free of the psychological state of being an adolescent. I hadn't thought this out—it was already in me. I felt at deep variance with the adult world around me and was strongly drawn to the drug and music culture of the day. Finally, at Warwick, I had my first clear confirmation that there really is another, higher, more profound state of consciousness available. This was something I had suspected from early adolescence but now knew to be true.

Throughout this period of my early life 'something' had been *calling* me from within. This calling was not explicit, it was more of a feeling, a feeling that there is another state, another way of being, where all is well, all the time, like the feeling you get when it's a beautiful sunny day, and everything is perfect, just as it is—or like the feeling of being in love. It wasn't that I imagined there was such a state. The call itself consisted of *intimations* of this state.

I think in another time, in a more enlightened culture, someone would have recognised my 'type'. There would have been some kind of schooling where the nature of these intimations could have been explained and my calling given some direction. But in the England of the 1970s there was no such schooling—at least none that I was aware of. Like so many others, I was left to crash about on my own, turning to drugs, dropping out, and finally falling into the arms of an Eastern guru. I am not saying that my trajectory was a mistake. Life is what it is. The important thing is that I *heeded* the call. After that, things more or less took their own course. For a long time I thought that everyone had a call like this. But now I'm not so sure. I only really know about myself and a few people who are close to me.

The problem with going into the past like this, is that we can only truly experience how it is with consciousness in this moment, *now*. And right now I no longer have direct access to the states of consciousness I have been describing, I only have access to memories of those states. Such remembering is completely different to being in the remembered state. And neither can I reproduce those experience-states. They each occurred just the once, to a former self. The landscape of my inner life has changed since then. I no longer have the resolve to shut myself up in a Vipassana retreat. And even if I did, I do not expect that things would go at all as they did all those years ago.

So, if I step beyond the external facts of the matter and attempt to describe and explain my past inner states, we run into the possibility that I am simply imagining those states according to the way I understand things now, and not according to how they actually were at the time. For instance, if I were able to actually *inhabit* one of those former states, while still reflecting from my current state, I think I would be shocked at the discrepancy between what I remember and how it actually was. This would be like the shock I received going back to Plymstock where I lived when I was four years old. It was certainly, factually, the same street, but there was very little there that corresponded with the images in my mind.

And yet, it is my belief, based on how it is for me now, that those early climacteric experiences at Warwick and Brighton changed both the course of my life, and the landscape of my inner world. I still feel the reverberating traces that they have left in me, and what I am doing now is reading those traces, somewhat like an archaeologist. What emerges is a story that is partly informed by the events and experiences themselves, and partly by the understandings that have developed in me since then. This story is neither fact nor fiction—it is a way of making sense of what happened. This process of making sense is *creative*. For there is no ready-made meaning waiting to be 'picked up'. And neither is it a matter of inventing something comforting. It is a matter of discerning a pattern of meaning that best accounts for how things actually are for me now.

How things are for me now is that I can, more or less at will, withdraw from the thinking processes occurring in me. This ability is central for our subsequent investigation of intelligence. To withdraw is to step back into a state of passive immediate perceptual awareness. Such awareness is still functioning in the background while thinking is occurring. But it only becomes fully present when the intention to think is withdrawn and relaxed. This is rather like turning down the volume on a radio program. You know the radio is still there and that you can turn it up again. The programs will also keep interrupting your perceptual awareness with bulletins and updates, but now, at least, you have some control over whether to listen in or not. And yet there is still the sense of the radio broadcasting continuously in the background. The difference in the psilocybin and Vipassana states is that the radio was switched *off*.

Once such a complete switching off has occurred, life is no longer the same. Even though the egoic thinking state returns, you now know there is 'something' behind it. The presence of this knowing means the original egoic state no longer has the same grip. Being gripped means being in a continuous state of identification that believes I *am* what I think and feel. This state is locked into itself in such a way that whatever is said here about another state, is experienced as just another thought. Such egoic consciousness *thinks* it is free because it can reflect on itself. But egoic reflection is only thought thinking about itself. To get out of such a state requires that something break it open from the *outside*.

To give it a name, I would say that what is 'outside' is *pure consciousness*. Pure consciousness does not require an egoic thought to inform it that it is conscious—consciousness itself *implicitly* knows what it is to be conscious. Such a state is profoundly illuminating. Apart from the sense of wonder it invokes, you immediately realise, you *know*, perhaps for the first time, what it means to be in the egoic state you have left behind.

And yet, at least in my case, the egoic state returns. This resurgence of the ego was particularly strong after the psilocybin experience. The disparity between my normal state and the drug state was so great that there was hardly any sense of connection between them. Of course, my life was changed, dramatically changed. But my understanding of what had happened had not caught up. I knew there was another state, but I had no idea how to reach it. The Vipassana experience helped bridge that gap. I now knew that the state of pure consciousness is always present in the background, and it is only that 'I' leave it by becoming immersed in my personal egoic thoughts and feelings and worldly concerns.

After Vipassana it was easier to 'drop out' of this egoic state by simply remembering how it stands, how it is that I become caught up with thoughts and forget the underlying state of natural perception. But such 'self-remembering' is not the same as a *total* cessation of the egoic state. In the Vipassana experience, the egoic state wasn't simply suspended for a few seconds, or a minute. It was completely put out of action. A few years later it happened again, after reading Barry Long's book *Meditation: A Foundation Course.*[6] This time I did not need to spend days struggling with thoughts and desires. I simply *recognised* the state the book was indicating and immediately entered it again.

As a result of these experiences (and others) I formed an idea that the Great Purpose of my life was to enter into a *permanent* state of egoic absence such as I experienced during the Vipassana retreat. I reasoned that there had to be people on Earth for whom this had already happened and that all I needed was to get close enough to one of them so that this state would somehow be transferred to me, both through the immediate contact and through the words they spoke and wrote. This led to my spending many years seeking after this perfect state, first as a sannyasin of Rajneesh, and later with the Australian spiritual teacher and self-proclaimed *Master of the West*, Barry Long.

Now, in later life, I have come to a more modest understanding of my capabilities. It seems clear that if such a permanent change is to occur there is nothing 'I' can 'do' about it. Moreover, I came to see that this

Figure 1.9: Barry Long 1926-2003.

idea of a permanent change is, for me, just that: *an idea*. I have no direct evidence that it has really happened in anyone else. My actual experience is that no matter how perfect the state, 'I', the reflective thinking self, always return. The cessations have only served to inform this egoic state, to make it more open, to put some windows in what was a previously self-enclosed world. It is in this coming and going of the egoic state that my actual life is lived. At a certain point, as people close to me started to die, I began to re-evaluate what it is I am doing here on Earth. Perhaps I am here to live the states I am actually in, not to hanker after some idea of permanent absence. Perhaps it is what I actually *do* here that matters, the people I affect, the people I love. Perhaps, after all, I am already living the life I am 'supposed' to live, and if I were to die today without realising some permanent state of perfection, then that too would be OK. And, importantly for this book, in letting go of this ideal, I began to examine how things actually are for me, and to seriously inquire after my own intelligence.

Notes

1.   "At the still point of the turning world" is a line from the T. S. Eliot poem *Burnt Norton*, the first of his *Four Quartets* (1963, p. 191). The poem continues: "And do not call it fixity, Where past and future are gathered. Neither movement from nor towards, Neither ascent nor decline. Except for the point, the still point, There would be no dance, and there is only the dance."

2. This is another reference to T. S. Eliot, from *The Love Song of J. Alfred Prufrock*: "I have seen the moment of my greatness flicker, And I have seen the Eternal footman hold my coat, and snicker, And in short, I was afraid" (1963, p. 16).

3. This refers to the line 'He said something like "You and me, babe, how about it?"' from the 1980 Dire Straits song *Romeo and Juliet*—a song I particularly associated with Frances, especially (of course) *after* we split up.

4. Carlos Castaneda was a famous figure in the alternative culture of the 1970s and 1980s. His books described his experiences as a trainee Shaman under the tutelage of the Yaqui Indian Don Juan. It remains in doubt whether these books were works of fiction or non-fiction. He certainly maintained their veracity and was awarded a doctoral degree on that basis. I read all his available work during the 1981 final exam period at Warwick, starting with his first book *The Teachings of Don Juan* (1968).

5. *The Orange Book* (1980) is a collection of meditations invented or borrowed by the Indian guru Bhagwan Shree Rajneesh. Rajneesh, who later changed his name to Osho, was born Chandra Mohan Rajneesh in 1931. Before setting up his own spiritual-religious movement, he was a philosophy professor at Jabalpur University. He started teaching and initiating disciples (sannyasins) in Mumbai in 1970, setting up an Ashram in Pune in 1974, and then moving the Ashram to Oregon in 1981. This Oregon Ashram collapsed in 1985 after serious crimes were revealed, and Rajneesh was deported from the USA, finally returning to Pune where he lived until his death in 1990. I was a sannyasin from 1982 until 1987, living in a commune in Brighton and then moving to California and visiting Oregon whenever possible. After staying at the Pune Ashram in 1987 I decided it was all over, largely as a result of coming into contact the work of the Australian spiritual teacher Barry Long.

6. Barry Long, a former Sydney newspaper editor, became a significant spiritual teacher and author, starting out in London in the 1960s and attracting increasing numbers of people, especially after the collapse of Rajneesh's movement in the mid-1980s. His book *Meditation: A Foundation Course* (1995) was originally written in 1968 and detailed a series of ten lessons designed to guide the reader into a state of stillness. I became closely involved with Barry Long's teachings, following him to Australia, and continued to attend his public talks and meet with him until his death in 2003.

Chapter 2

The Way of Phenomenology

The Prerequisites

To be clear, I am not recommending that anyone take a large dose of psilocybin in order to follow what is being said here. As far as I know there is no method that will guarantee the conscious realisation of a cessation of an egoic state. Perhaps a drug, or a great shock will stop the egoic process for a while. But being thrown like that has unpredictable and sometimes catastrophic results. Not all people experience a realisation, and for some the egoic structure can be more or less permanently damaged. Paradoxically, it seems an already strong ego is needed to successfully navigate a sudden experience of ego death. I do not think such violent shocks are necessary or even desirable. What is needed is an inner attunement to something emanating from outside the egoic state, something like the call I knew in my adolescence. If there is such an attunement, I think the call itself will find a way to reveal its source *despite* our personal involvement.

Looking back, I think I was a hard case. Only a strong drug was going to open me up, so that is where the call led me. Of course, there are many other ways, like the love for another human being, or for nature, or the receiving of wisdom from a teacher, or devotion to a transcendental source, or the simple dwelling in silence and solitude. I know it's not a matter of copying someone or following a set of instructions. That is the way of the egoic self, the way of our times—to seek an effective method, a reliable technique that will work in all situations for all people. And yet, we can't automatically assume a given technique is *not* going to work. It seems to depend on the particular person and the particular situation.

So what was the intention, in this retelling of the story of my early experiences of ego death and the resulting discovery of a different state of consciousness? For me there is now an awareness of a deeper unmoving presence that lies behind the thoughts and feelings that move across the foreground of ordinary waking consciousness. In the following investigations I am going to be calling on this presence. It is the place from which we can observe the field of immediate experience. It is not something rare or extraordinary. I believe it is present in everyone who is conscious. So it is not a matter of creating this state, it is a matter of *noticing* it. What stops such noticing is our attachment to the stream of thoughts and feelings that carries our conscious attention along with it. It is the consciousness in this conscious attention that is needed to notice the background state, that, at the same time, *is* this background state. And yet it seems you cannot explain this to someone else, unless they have already explicitly encountered this state by means of their own direct realisation. And it is unlikely (but not inconceivable) that these words alone are going to cause such a realisation. For it is just not possible to *think* your way out of an egoic state.

The intention in telling this autobiographical story is to show how I arrived at the place from which this book is being written. I am not expecting you to have lived a life like mine. But I am asking: have you too encountered this immediate state of consciousness, to which I have been pointing? It doesn't matter how you got there. All that is needed is for some kind of separation from the egoic state to have occurred and for that separation to have been *consciously* recognised. If you know what this means, in your own experience, then we have arrived at a kind of *base camp* from which we can proceed. If not, and you are still willing and interested, then let's continue anyway and see what happens . . . .

HUSSERL

In order to set out we need to secure a way of proceeding. Our way of proceeding is the way of *phenomenology*. Phenomenology is a term I have borrowed from the philosopher Edmund Husserl.[1] This borrowing is not a strict following of Husserl's methods. As far as I understand him (and there are many understandings) we are engaged in phenomenology here because we are inquiring in the *way* that Husserl inquired. He is the one who deserves our acknowledgement for having discovered and opened up the original path. But there are no followers in phenomenology. We each have to find our own way, according to our own insight and according to the times in which we find ourselves.

Figure 2.1: Edmund Husserl 1859-1938.

And we are certainly in different times to Husserl. His idea for phenomenology, in the early twentieth century, was to rejuvenate the entire scientific culture of Western civilisation, and for philosophy to once again rule as the queen of the sciences. But things did not work out that way. Husserl, once the foremost philosopher of his era, was excluded from German intellectual life by the Nazis and ended his days observing the disintegration of the very culture he had hoped to transform. Phenomenology is now a sub-area amidst all the other schools of thought that comprise modern philosophy and Husserl has become a relatively minor figure within that labyrinth.

But here our concern is not with the labyrinth of philosophical thinking. It is the *spirit* of phenomenology we are invoking. And the intent of that spirit is to *break out* of the labyrinth. Husserl attempted to summarise the essence of this intent in his famous phrase: '*to the things themselves*'. Here, the 'things themselves' are things as they *actually* present themselves in immediate experience and not as we think or imagine or hope them to be. The shock Husserl delivered to Western philosophy was to recognise that, in order to encounter things in their immediacy, the very consciousness of the inquiring philosopher must first be transformed.

Husserl attempted to impart this transformation by introducing various 'ways' into phenomenology. Each way was designed to produce insights that would open up the phenomenological field of inquiry. And yet he was never quite satisfied with what he had written, always trying again, right up to his death in 1938 and the subsequent publication of *The Crisis of European Sciences and Transcendental Phenomenology*. It seemed the true essence of phenomenology could not be directly expressed in language, and to grasp that essence, you must *already* have some insight into what Husserl was writing about.[2]

And that is how it was for me. I found I did know what Husserl was writing about, because I already knew states of consciousness that, in Husserl's language, had transcended the natural attitude. For me, studying philosophy, this was like discovering water in a desert. I felt the hand of this German philosopher reaching out across the devastation of the twentieth century and his speaking directly to me, of my own reality. I don't know if you have experienced something like this—how someone you have never met, who lived in another time, who died before you were born, can still make their living realisations present to you, can recreate the very way they were seeing the world, simply through the written words they have left behind. I still find it a kind of miracle.

THE PHENOMENOLOGICAL REDUCTION

To enter into phenomenology is something radical. To give a concrete example, let us again examine the Vipassana meditation. This first requires the formation of an intention to fully attend to each moment of the breath coming in and going out. And that means focussing on the *feeling* and *sound* of the air in the nostrils, the *feeling* of the chest and diaphragm moving, the *sense* of the enervation of the body on the in-breath, and so on. There is a certain effort involved. But there should be no explicit thought to the effect that "I must watch my breath" while you are actually watching. It is rather that this thought, this intention, is silently active in the background. Insofar as it is acting in this way, there will be no need for any explicit thinking.

   To make this clear, I suggest you try it now. The idea is to enter the experience so completely that you no longer have any resource left to think about what you are doing. But now we are not just watching the breath, we are also looking 'backwards' at the presence of our very intention to watch the breath. It is this intention, in its intending, that is watching the breath, that is holding 'centre stage' so to speak, and keeping all those other intentions at bay (at least for now). If we look into the origin of this intention, it is not immediately clear where it has come from. For me, beforehand, there was a reflection on the possibility of doing Vipassana and in this reflection there was a kind of implicit question asking whether I would do it or not. The answer came when I found myself actually watching the breath. This all happened in a split second.

   My intention to watch the breath, like my intention to write this sentence and your intention to read it, is formed in a place that is not directly accessible to conscious scrutiny. That does not mean we cannot reflect further on the reasons for different courses of action before we act. But then, this intention to reflect is still formed in a place that is not directly accessible to conscious scrutiny. And after we have reflected, we still discover what we intend to do only when the intention itself becomes active.

   In such reflection, a comparison is being made between various potential intentions to act. Our consciousness allows us to see some of the motivations and consequences of these actions more clearly and this 'seeing' can modify the relative strength or attractiveness of each intention. Such modification allows reflection to have a say in the intention selecting process. But finally, the intention that becomes active is simply the one that has the strongest charge of attraction. The *real* decision maker is not the conscious reflective reasoning intellect, it is a set of pre-existing feeling-motivations that we generally call our *self*. The operation of this feeling self is largely hidden. It simply issues its 'decree' and I find there is now an active intention that is determining what I am doing. Not only that, I feel this current intention is *my* intention, that it issues from *me* and that it is *my* freedom that chose to act in this way.

   But life experience teaches us that this self that determines our actions is not the unity we feel it to be. It would be better to say there is a collection of selves within us that are attracted to quite different and contradictory ends. For instance, in my Vipassana experience, there was a self that intended to watch the breath and a self that intended to think about

personal values

having a girlfriend. At the time I would have said that the self who intended to watch the breath was my 'real' self and the one who intended to think was not 'really' me. But, if we look carefully, there is another self that answers when we give an account of ourselves. This is a reflective, rational self. It is this self that attempts to construct the narrative of a *one-true-self*—and that self, of course, *is* the rational, reflective self that is constructing the narrative. Or at least that is what it thinks.

At the same time, this rational self is identified with certain intentions to act, like the intention to watch the breath, and not with others, like the intention to think about having a girlfriend. But this self, with its narrative, is only present for some of the time. In Vipassana, when I was thinking about having a girlfriend, 'I' was no longer intending to be that rational self, I was intending to think. And when I was watching my breath, I was still not intending to be that rational self, even though that rational self was identified with the breath-watching intention. The fact is, in the moment that a particular intention has 'control', then 'I' *am* that intention—the feeling is: this is *me* doing *this*. It is only later, in reflection, that I may say "I wasn't really myself just then."

The question here, the deep question, is whether there is any unified 'I' in me at all, any unique, conscious being, possessed of a real intelligence that can *guide the course of my life*. Or is it that I am a kind of automaton, with these various intentions operating in me, each having its day in the sun, governed by causes over which 'I' have no control, so that even my reflective 'reasoning' is quite robotic? Do I simply compute consequences of imagined situations according to the operation of a reasoning program that evolution and culture have prepared in me? Is it that 'I' and 'me' are just names that refer to the operation of this body-machine to distinguish it from all the other body machines around me? That my individuality *is* this physical body, not some supposed unique immaterial conscious being . . . .

Consider how it stands now: one impulse, one intention to think this, to write that, to go outside into the garden, follows on the next. They arise, they subside. For each intention, I feel it is me who is the intender, the intending being. But when I reflect, I see that there is no real unity in these intentions. From where did this sentence arise? Or the last thought that entered my head? In what way am I the *author* of these intentions, these thoughts? How would it be different if they were just arising by themselves, according to the laws of physics operating in my cells—if my neurons were determining with absolute necessity what I am going to think next, what I am going to feel, what my reason is going to tell me? What if my feeling that I am somehow the centre and cause of all this, is just that, a feeling that accompanies whatever happens in my nervous system—just as if a car I were driving had the feeling it was driving itself?

We cannot escape this rhetorical questioning by stopping thinking and saying afterwards: "Look, I am free because I can stop myself thinking." It is not thought that is in charge here, it is the current impulse, the current intention that has temporarily gained ascendance. And that intention can easily be an intention to stop thinking.

I wonder if you feel the seriousness of these questions? It is now the accepted view of the majority of experts in the field of cognitive science and philosophy of mind that we are, in essence, a species of highly sophisticated biological automatons. Such an understanding has emerged in our civilisation for deep-seated reasons that we shall explore later. For now, I should like to emphasise that this is not some far-fetched, minority view. It is what guides the people who are advising government departments on their future policies. It determines the strategies and actions of our global corporate business systems. It is present in the science documentaries on television, in the lecture halls of our universities and in the learned textbooks that tell us what it means to be human—in short, it is the default, background view of our entire global technological culture.

So how can we respond to this line of reasoning? Firstly, I would like to mention again what happened in the Vipassana experience and at other key moments in my life. I discovered, first-hand, that this normal state of inner disunity, where one intention replaces another, which replaces another, which replaces another, all day long, until you go to sleep, can consciously come to an *end*. When it ends there is a different kind of absence to the one you find by practicing a meditation technique that teaches you to stop thinking by developing and strengthening a certain intention to meditate.

In such an ending there is an immediate knowledge of the absence of the egoic activity that is otherwise continuously functioning. It is a great peace, a blessing, a blessed relief. But this state is not completely discontinuous with the egoic state. There is still consciousness, there is still a 'seeing'. And there is the knowledge that this seeing is present anyway,

why automatons?

Gurdjieff !!

in all those other different states of consciousness that have just been left behind. It is just that the seeing has not been conscious of itself. Now, in coming back to the egoic state, there is a new distinction. Despite the continuous circulation of the various intentionalities, there is the presence of the consciousness that sees these states. This is not like the reflection of thinking, in which there is an intention to reflect. This is the encompassing seeing whereby we are conscious in the first place. Without this seeing we would simply be *unconscious*.

It is this consciousness that presents what is the case. The *phenomenological reduction* is an existential shift away from egoic identification with the stream of intentionalities into this centre of consciousness. 'I' am now absorbed into the seeing that is the being of this consciousness, the impersonal witness of the flowing intentionality of this personal life.

But language is misleading here. It draws us into a dualism of subject and object. This is not just an intellectual distinction. We inhabit this distinction when we reflect on experience, as we are doing now. There is an inward splitting, where 'I' attempt to look at 'my' experience. This is what we would normally consider to be a state of witnessing. But such reflection does not 'see' the experience *now* as it *actually* is. If I am reflecting, then my experience now is that I am reflecting, but in my reflection I do not see the very act of reflection that I am performing. Instead I see the object of my reflection, i.e. an experience I have abstracted myself from, that I *was* having 'just now'. If I attempt to reflect on my act of reflection now, then I will make a new object out of my reflecting state, so that I am no longer inhabiting that state, I am reflecting on my reflecting. And so we enter into an infinite regress. 'I' as the one who reflects can never reflect on this act of reflection in the moment it is occurring.

The witnessing of pure consciousness is not split in this way. When I cease to reflect on experience then the subject that was the centre of this reflection again becomes unified with whatever is occurring. And whatever is occurring is consciousness being conscious. This pure witnessing does not explicitly reflect on anything. And yet consciousness is the source of all knowing. When I do reflect on experience, I can only reflect on something that consciousness is *already* presenting. My reflection does not create experience, it makes it *explicit*. This making explicit is a fixing of the otherwise flowing experience of being into an image that can be recalled and described in language. But what of the state that precedes this making explicit? It too has its own kind of knowing. This is the immediate implicit knowing of a pure receptivity. It is *this* place that we are indicating as the state of pure witnessing. It is paradoxical. It is the place that remains when the reflecting 'I' ceases to reflect. And yet, if we don't reflect, how can we explicitly know anything whatsoever about it?

Phenomenological Inquiry

What is required is a different form of making explicit, one that does not first split itself from the state of pure witnessing. This different form is quite familiar to us in our normal, everyday experience of getting around in the world. It is known as questioning or inquiring. The essential quality or character of a true questioning is that of receptivity. To question is to not know, to await upon . . . . When we question, we await a response that reaches from beyond the sphere of our questioning—otherwise it is not a true questioning. When questioning is untrue we already have an answer in mind, we are simply seeking to have that answer confirmed and we ignore evidence to the contrary. To truly question is to *look* and *see* what is *actually* the case. What is actually the case is what is *true*.

In ordinary worldly experience, such looking and seeing is something we take for granted. I hear a sound in the garden. It sounds like a cat but I am not sure. So I go to the window and I look into the garden. I see there is a cat partly hidden behind the shrubbery. My question is answered. In this entire experiential series there is no need to explicitly think or reflect. I can look and see without reflecting because I *already* have the garden immediately before me.

Figure 2.2: A cat in the garden from a photo by Fernando Jorge on Unsplash.

But as soon as we stop and think about this, it can all start to seem rather doubtful. For, perhaps, even though I may look into the garden without thinking, surely I only come to know there is a cat out there when I reflect on what I have seen? The answer here is to stop thinking and look again. Does the cat appear in the moment I reflect on my experience, or in the moment I actually look into the garden? And how can I even tell the difference? Here we must call on our ability to hold a question in mind as a pure meaning (i.e. without thinking about it) while, at the same time, looking out into our perceptual experience. In this case the question is: "Do I already know what this thing is that I am seeing without thinking or reflecting on it?" I suggest you just look at something without silently naming it or thinking about it, and hold this question in mind. Do you see that you already know what you are looking at before making anything explicit to yourself in thought? If this isn't immediately obvious then consider how it would be if you didn't know what you were looking at, how *shocked* you would be and how this shock would emerge *before* you could start thinking and trying to work out what is happening. More significantly, do you see how it is that you can question your experience like this without thinking? You just prepare a question concerning your experience, and you look to see how it stands. This seeing is like your seeing the relation between the cat and the garden, i.e. that the cat is *in* the garden, only now you are seeing that what your question envisages corresponds with what you are actually experiencing, i.e. you are perceiving that the correspondence is *true*.

Of course, it is relatively easy for us to ask questions like this of perceptual experience. We have been trained from an early age. If I do not factually report how things stand with the cats and tables and chairs of this world, then other people will tend to correct me. And if this does not work, then the unyielding physicality of the world will teach me in a more direct way. But as soon as we leave the objective domain of the things that appear in our perceptual experience, we become increasingly unreliable. If we attempt to observe experience itself, rather than the objects that appear in experience, we no longer have the objective restraint of coordinating our seeing with other people. For only 'I' have direct access to my 'inner' experience. The problem here is not an inherent dishonesty. It is to do with intentionality itself. If I think the earth is flat then there are certain tests and experiments, certain photographs that will count against this idea. But if I believe the universe is really controlled by celestial beings who exist in another invisible dimension and I have actual experiences of communicating with these celestial beings then how I am to decide on the truth or validity of this belief and these experiences?

In the domain of the objective sciences the problem of human subjectivity is circumvented by disallowing the unverifiable testimony of the subjective viewpoint. Only what can be demonstrated objectively can count as evidence. That gets rid of the angels. And that also leads to a materialistic mathematical representation and understanding of the universe and all that we find within it (including ourselves). But if our domain of study is human subjectivity in itself, such an objective approach is not going to work, because it steps over and denies validity to the very domain it would attempt to understand.

The problem that all serious attempts to study subjectivity have encountered, is that without the standard of objectified experience, there is no accepted way to distinguish one way of seeing things from another. And so psychology, sociology, and philosophy have all developed schools of thought that contradict each other and fail to progress in any unified or coherent manner. There is an appearance of progress, as views become more sophisticated and more able to explain phenomena that were previously unrecognised. But this progress is largely a result of discoveries in the objective sciences that have given a clearer understanding of the physical processes that appear to underlie our subjective experiences.

What is missing is a self-evident method of direct observation of experience that is demonstrably free of subjective, personal bias. Such observation cannot be objective, in the sense of the objective sciences, because the objective cannot directly observe the subjective. The task is rather for the observation to be *impersonal*. For such observation to be possible, two things are needed. First, that there already exists an impersonal consciousness that is capable of such observation, and second, that this impersonal consciousness is able to make itself and its observations self-evidently known to our personal, egoic subjectivities. To engage in such observation is the original and ongoing task of phenomenology, as far as I understand it.

The problem for us is that we are virtually unconscious of the origins of the intentionalities that freely come and go within our field of consciousness during each moment of our waking lives. Each intentionality has its own view of things. Careful observation can reveal something of what lies behind these views. It may be a childhood trauma. It may be some collective national feeling formed out of earlier historical events. It may be an argument you had yesterday. It may be an idea you read in a book that you have forgotten you ever read. It may be what your father believed. It may be the opposite of what your father believed. It may be some instinctual fear or desire that has been redirected onto a cultural object. One thing is *very* unlikely: that the thought-intention you currently embrace has arisen entirely out of your own freedom, your own rational and reasonable consideration of the *facts*.

Phenomenology is not a matter of getting to the bottom of each of these intentionalities and somehow purifying them. It is a matter of accessing another kind of intentionality, an *impersonal phenomenological intentionality* that is only concerned with what is actually present in this moment of experience, now. What is required is an intentionality that corresponds with the pure witnessing of consciousness itself. This is the intentionality of a *pure questioning after truth*. What makes this an intentionality of a different order is that there is an implicit surrendering of the personal, egoic self. This is not the same as an egoic death. The egoic self is still there, in the background. But it *agrees* to withdraw. This agreement is a result of it (of me) having learnt from experience that (i) the egoic self is encompassed by a 'state of greater clarity' which ordinarily remains hidden, and (ii) that this 'state of greater clarity' is only accessible when the egoic state withdraws and (iii) that having contact with this 'state of greater clarity' is revealing of truths and experiences that are significant, meaningful and transformative for the egoic self.

The reason such a withdrawal is effective, is because it leads to a pre-existing state of witnessing that is already constantly functioning in the background. It is the same impersonal witnessing that looks and sees whenever we inquire into our perceptual experience (to see if there is a cat in the garden). The difference now is that we are inquiring of our entire field of experience, not just that aspect we call the objective or external world. The way of phenomenological inquiry is to ask questions of this field of experience that has been purified by means of a phenomenological reduction. Because we can answer such questions by looking and seeing we avoid the paradox that such a state of pure witnessing is unable to explicitly reflect on what it is seeing. We do not need to reflect, we simply see how it stands with the question and the field of experience that is being questioned. Once we have seen how it stands we are free to reflect and report on what has been seen.

THE PHENOMENOLOGICAL FIELD

If only things were that easy. It has been more than a hundred years since Husserl first developed these ideas. But phenomenology suffered the same fate as every other view of the human condition. It became a school of thought—interesting in itself, worthy of study even, but not *ultimately* true. For, according to our current worldview, *nothing* is ultimately true. And here lies the entrance to the labyrinth—an endlessly complex world of theories, beliefs and opinions, a world where nothing is certain except that nothing is certain, and even that may not be certain. In such a world, finally, all we have are our best current theories. We know these theories are not the truth, but in a world where there is no truth, our best theory is the next best thing.

The essential difficulty phenomenology faced is that there is no such thing as a completely presuppositionless standpoint from which we can begin to study ourselves. The way we 'see', the way we understand our-selves and the world, arises out of our *inheritance.* This is the inheritance of the evolution of the species, of human history, of language, of moder-nity, of science, of technology, of our parents, our schooling, and so on. Once this inheritance is glimpsed in all its obscurity and complexity it becomes clear it would not only be impossible to factor out all of its in-fluences upon our understanding, it would also be absurd. For without language, without ideas and concepts, without the ways of seeing that we have inherited, we would not see more clearly, we would simply not see at all. Our power to question, to reflect, to form ideas, to express these ideas to others so that they are understood, would be completely lost. To even understand this sentence requires we already share an entire network of presuppositions *of which we are barely conscious.*

So what was all this talk of the impersonal witness? We have to tread carefully here. The impersonal witness is the pure consciousness that presents experience to itself. It is *experience* that is conditioned by the past, *not* the impersonal witness. The issue we are confronting now is of such importance that I hardly know how to express it with sufficient gravity.

It concerns whether *true self-knowledge* is possible for creatures like us. If such truth is not possible, then *all is lost*.

But what is truth? Is it even true that there is a chair in front of me now? If you have studied philosophy, you will know the answer to this question depends on the assumptions that lie behind it. If you mean: Is it true there is a chair in front of me, made of physical stuff, existing in the physical spacetime of the physical universe?—then I cannot say for certain that this is true. And yet, this is generally what we mean when we ask such a question (if we are to spell it out). For what is 'physical stuff'? If you were to ask the physicists you would find they are puzzled and have no settled answer. The same goes for 'physical space'. It turns out we are not at all clear what these words ultimately refer to. All we have are our best theories. In this way, our simple question soon becomes mired in difficulties.

Figure 2.3: Chair photo by Renè Müller on Unsplash.

And yet, *of course* I know there is a chair in front of me now. I do not doubt it. I only doubt it when I reflect on what may lie hidden in the question. What is the ground of this certainty? It is the *phenomenological field*. Regardless of whether the universe exists physically, or in some cosmic mind, and regardless of whether physical atoms really exist, or whether they are a momentary condensation emerging from a superposition of quantum energy fields, I still know, immediately and directly that there is a chair in front of me now.

The foundation of this certainty is the certainty of being conscious. If I see a chair in front of me, I know what I am seeing without thinking or reflecting or reasoning—that is what it *means* to be conscious. I could be uncertain because the chair is a long way away, or because it is obscured by other objects, or because there is smoke in the air. But if the chair is right in front of me, and I can walk around it, and sit on it, then I do not doubt it is there, in the moment it is there.

At the same time, it could be that my entire experience is occurring in an immaterial mind and that what I think of as external physical space is really a projection, a kind of illusion. If that could be demonstrated, it would still not alter my certainty of there being a chair in front of me now. It would only alter my conception of what the experience means. Instead of the chair existing in a physical space that is independent of my consciousness, it would exist in a cosmic mind that I experience by means of a mental projection. Or it could be that I later wake up and realise I was dreaming. But even then, in the dream, there was a chair in front of me—a *dream* chair. The idea an experience can be an illusion only arises in my understanding of what the experience *means*, how I *interpret* it. Experience itself, simply *is* what it *is*.

The phenomenological field is exactly this, the field of consciousness as it presents itself *in* consciousness *to* consciousness. To even say this is misleading. Because it presupposes something called consciousness. When consciousness is simply conscious it does not 'see' anything like consciousness. It sees tables and chairs and the space between them. It feels the feeling of being a body. It does not think or interpret. It knows nothing but what is present *now*.

And yet, this does not express things *fully*. There is still, in the background of this state of pure witnessing, an understanding of the world, of the being of the things of the world, an understanding that is continuously operating, that is making sense of what is being experienced, that is making experience *meaningful*. As this understanding is always doing its work, it is hard to recognise until you meet something it does not understand. A good example is listening to an unfamiliar foreign language. In everyday life we cannot help but understand another's speech without thought or reflection, just as we understand what it means when there is a chair in front of us. But if we encounter an unfamiliar language we only hear a stream of sound. Our understanding has still 'made sense' of this stream of sound, just as it 'makes sense' of a bird song—it knows *what* it is hearing, but it does not know *to what* these sounds *refer*. In this absence of meaning, we can see how our understanding is continually making language meaningful in ordinary speech and thought. But it is not just making language meaningful, it is making the *world* meaningful, and everything that appears within it, including the chair.

This background understanding is something that is present *now*. It is not like a personal understanding we have acquired through thought and reflection. It is an impersonal understanding we have acquired through our experience of being born and brought up on the earth and in the world, within the form of this human body, within the form of this human language, and within an overarching *implicit* understanding of being we have developed collectively, in the rising and falling of successive generations and cultures. Again, like consciousness itself, we live *in* this understanding, it makes up our very way of seeing the world, so that we see the world and not the understanding that lies behind it. However, the impersonal nature of this understanding does not mean we all experience the world in the same way. Each of us is shaped by our individual pasts. It is impersonal because it is not under the control of the *personal will*. It is what remains when the personal will, the egoic self, withdraws.

It is here that the whole project of phenomenology stands or falls. For if all that happens in a phenomenological reduction is that we fall into a pre-reflective state of understanding that is entirely shaped by our personal experience and the collective experience of humanity, then surely we fall *beneath* what we are capable of? For this is *why* we reflect: to abstract ourselves from the past. That is the very power of reason and rationality, to free us from the unexamined acceptance of false ideas and obscure intentions. And yet the phenomenological reduction asks us to surrender this reason, this power to reflect, and instead to witness experience as it is. Surely when we do this we simply fall under the sway of this pre-existing understanding of being that rules in the background of our everyday lives?

The difficulty here is that when we reflect on the past, we reflect on a past whose meaning has already been constituted by this background understanding. Such *pre*-phenomenological reflection does not see the operation of this understanding, it sees-by-means-of this understanding. The task of a *radical* reflection is to bring this background to explicit consciousness. For this we need an impersonal vantage point that has the capacity to reveal this understanding *even as it functions.* Thoughtful reflection cannot do this, because it is itself a *movement* of this background understanding.

Pure consciousness, is the impersonal agency whereby this background understanding becomes an experience. Consciousness does not interpret this understanding, it presents this understanding. It is this power of consciousness that is impersonal and neutral. It does not like, it does not dislike, it presents what it *is* to like or to dislike. Our task here is to *call* upon this power to *directly* reveal the operation of this background understanding. Reason can theorise about this, but we are interested in a direct knowledge of how things actually stand. And we are not inquiring into some remote galaxy. We are inquiring into the very understanding that is making sense of these words now. We are asking: is it possible for us to become directly conscious of this understanding, in the very moment of its functioning?

Of course, this all sounds very well. But what are we doing here, if not thoughtfully reflecting, just as everyone reflects? For this entire preamble, all these ideas and propositions, surely they too have come out of, and have been formed by, this background understanding? Again we are in difficulty. The way of phenomenology is the way of *questioning,* of *looking* and *seeing.* But surely this questioning arises out of the very background we are seeking to uncover? And if this is the case, then our questioning is not neutral, it carries within it an understanding of what is being questioned, that already pre-figures what it expects to find. Such a questioning is a kind of self-deception: all we should find is the very understanding the background has of itself, disguised in the assumptions that lie within the questions we are asking. Because we are not looking at something physical, that we can measure and put to the test, there is nothing in our immediate experience that could *resist* the presuppositions of the background understanding. These are the presuppositions that are *actually forming our experience*—and we are, in a sense, *bound* to find them confirmed. And that means, as we all have different backgrounds, we will be back in the labyrinth—with everyone discovering that they are correct and that everyone else is mistaken.

But this problem does not arise with the phenomenological field itself. I can, at any moment, with practice, with insight, withdraw from my personal, thoughtful, reflective world, and simply enter this immediate field of experience. Like now. There is the sound of the computer keyboard, these words appearing on the screen—looking up, the movement of the leaves in the wind, the sound of the wind . . . . But even these simple words interpret an immediate experience that goes beyond language. I *frame* the experience in sentences that arise from 'I know not where'. How else are we to proceed? This is a book. Our task is to make our inquiry explicit, to bring it to language.

We are back with the call, with the attunement. If we are to get out of these difficulties, there has to be a withdrawal, a 'handing things over', a *leap* out of this reflective, egoic state. Our questioning has to come out of a genuine state of ignorance, out of a recognition that ordinary thinking is not going to reach beyond itself. Such thinking is suited to the external world of objects, because that world continually provides reflection with new and unexpected facts. But inwardly? How is thought to discover anything inwardly but a clarification of what it already thinks? We are drawing toward a kind of experiment. We are asking whether there is anything like direct experiential knowledge or insight that is not the product of some form of thoughtful reflection. Instead of asking an explicit question, we are asking after such direct knowledge or insight. In questioning we are not asking after an answer as such, *we are asking after an insight into what our question might mean.*

DIRECT KNOWLEDGE OF CONSCIOUSNESS

Our first task is to ask after consciousness itself. This question comes first because it encompasses the entire domain of our inquiry, the very phenomenological field we have been speaking of. If there is anything like intelligence, this is where we shall look for it. Our task is to secure this domain by having a direct orienting insight into what it means to be conscious. That means *looking* into immediate experience to *see* if there is anything that corresponds to this question.

Thinking is not going to help. Thinking of consciousness leads consciousness away from itself toward whatever it is the thinking is thinking about. To think about consciousness is like thinking about the cat in the garden instead of looking out of the window. But now it is the window itself that is of interest, the metaphorical window that is *always, already* here. You may object that consciousness is not always here, that it goes away in deep, dreamless sleep. But I cannot inquire into deep dreamless sleep because that is exactly when I am *not* present. Whenever I am present, consciousness is present, therefore, for *me*, consciousness is *always* present.

What we are attempting is to look through experience toward the encompassing consciousness that makes experience conscious in the first place. We are looking to see if there is a knowledge of what it is to be conscious inherent in each moment of consciousness. The answer, if there is such knowledge, is the direct knowledge itself. To attempt to express this knowledge in language, to attempt to think it, is to have misunderstood it. Consciousness *exceeds* language. It is the space in which language appears. This is knowledge that *requires* us to have transcended thought and language. That is what makes it direct. If I cannot do this, then no proposition in language is going to reach me concerning such knowledge. It will remain a speculation, a possibility. I have to *look*. Here, the 'looking' and the corresponding 'seeing' are the demonstration of the direct knowledge—there is nothing to separate them, for consciousness is both looking *and* seeing *itself*. I cannot define knowledge of consciousness, and I do not need to define it. I define what is separate from me, what I am not, the entities I am 'conscious of'. But I am not 'conscious of' consciousness.

Of course, there is a paradox. We are using language to indicate knowledge that we cannot directly express in language. But we *can* indicate such knowledge by *negation*. And we can understand *why* such knowledge requires negation. Direct knowledge is direct because there is no reflection or separation involved. We encounter this in the immediate, pre-reflective perception of everyday things. Such perceptual knowledge already lives within language. For example, when I see a cat in the garden, I already know how to see a cat, because, in learning how to use the word 'cat', I came to know what a cat *is*. It is living language in the broadest sense that prefigures what I am *able* to perceive. But such prefiguration does not determine what I see—it provides a meaning form, a receptacle within which the particularity of an experience can manifest. There is a significant difference between my emptily thinking of the meaning form of a cat and my actually seeing a cat. In the actual seeing, if I am present to this seeing, the meaning forms are filled with an immediate, sensory-perceptual, qualitative experience of there being a cat in front of me now. I *see* the

shape of the cat, the texture of its fur, how it is moving, the way it is looking at me. I can express all this in language and put the meaning forms in
place for you. But the meaning forms are not the experience. They shape
the experience, they give it form, but the immediate content of the experience is not linguistic. In perception, this content is *pure sensory meaning*.
It is this we know *directly*. It is what it is. There is no language we can use
that can get us any closer. You can only look and say—"Ah, I see what
you mean." Otherwise we end up talking about things like the *greenness*
of green. Philosophers call this the problem of *qualia*.

The problem of qualia arises when you come to objectively investigate
what it is you mean by the sensory qualities of experience, such as our experience of colour. If we are using ordinary language, and we say the cat's
eyes are green, we mean they are *actually* the colour we experience them to
be. This colour is a certain *quality* of greenness that has entered the meaning form of the phrase 'green eyes' that we use to understand what we are
seeing. But the physicist will quickly assure us that this experiential quality of green is not a real property of the physical world. What happens
in 'reality' is that *colourless* photonic energy is reflected off the cat's eyes.
This energy is manifested in a range of frequencies of vibration that are determined by physical properties on the surface of the cat's eyes. When the
photonic energy reaches our retinas it is transformed into neural pulses
that cause other neural pulses in the brain that 'somehow' become our
experience of seeing green eyes. It is not that our seeing the eyes as green
is false. The greenness expresses information about the photonic energy
which in turn expresses information about the cat's eyes. For the physicist, what we are really referring to when we say the cat's eyes are green
is that very property that causes the eyes to reflect certain frequencies of
photonic energy. As for the quality of greenness we actually experience,
this has no home in the world of physical objects, *or in our everyday language*. For when we refer to the green eyes of the cat we are not referring
to some ineffable subjective quality existing only in consciousness, we are
referring to a property of the actual eyes of the actual cat. The only mistake we make, according to the physicist, is to believe the real physical eyes
have the very quality of greenness we experience.

We have a parallel situation when we speak of consciousness in ordinary language. Such language is primarily used to refer to things and people as they appear in our shared experience of an objective world, not to

Figure 2.4: Cat's eyes from a photo by Daniel Chekalov on Unsplash.

the consciousness that presents that appearance. And so, like the colour green, consciousness is first understood to be a property belonging to a certain class of objects—i.e. to those living beings who are able to perceive the world in much the same way as we do. And again, like the experience of greenness, if we search for a more precise physical correlate of consciousness, we find it situated in the brain processes of those living beings. But nowhere do we find an object in the world that corresponds to our immediate experience of being conscious. Our language has recognised this by granting the words 'conscious' and 'consciousness' a double reference (as it also has for 'green' and 'greenness'). One reference points to the common objective understanding of consciousness, i.e. as a property that I, as a human being in the world, acquire on waking up and lose on going to sleep (and reacquire if I dream). The other points to this immediate experience I am having now, that is only known and knowable to me in this moment, within which I have direct knowledge of sensory qualities, such as the greenness of green.

At the same time, consciousness is not a sensory quality like the greenness of green. For although the sensory qualities, as qualities, do not have an objective physical manifestation, we can easily distinguish one quality from another because they each possess a mutually exclusive quality dimension. For instance, my experience of colour appears in a visual field, and my experience of sound appears in an aural field. I can know what sound is, because I can be simultaneously conscious of sound and vision, and in such consciousness I have a direct perception of this difference. Such perception reveals, essentially, what colour *is*, and what sound *is*—again as a form of direct knowledge that cannot be further indicated in language or pointed to in the external world. But the situation with consciousness is different. Insofar as I am able to make any observation whatsoever, I already need to be conscious. So how am I to compare my being conscious with some other state? The only state that contrasts with being conscious is being unconscious, and clearly, I cannot make a direct observation of an experience of being unconscious. Moreover, consciousness is not something like greenness that I can be conscious *of.* It does not appear anywhere in the field of experience as something that could be singled out or focused on. It is the field of experience—it is the means by which there is any consciousness of anything. How could I possibly get outside this field and observe it, except in imagination? And in such imagining I am, of course, still within the field of experience.

Nevertheless, I do not simply experience a continuous self-identical state of consciousness that goes on without end. I go to sleep, I wake up, I become tired, I become distracted. Until now we have only been considering various states of full wakefulness. But we also enter peripheral states where consciousness is *less* present to itself. And although we cannot obtain a distinct knowledge of these states (for that would mean waking up) we do know something of what it is to start to fall unconscious. Even in feeling tired there is a sense of being less conscious. In this variation there is a distinction between the more and the less that provides us with an experiential clue as to the essential character of consciousness—because, to distinguish the more from the less, we must already implicitly know *what* we are distinguishing.

Our task now is to bring this distinction to full waking consciousness. To achieve this, I suggest you come back from the activity of following these words into an immediate consciousness of whatever is happening now in the space in front of you. We are looking into a state of transition where I surrender my intention to carry on reading, and thereby surrender my intention to do anything else. This is not a matter of attending to the immediate present. The immediate present simply emerges in the absence of my attending. It is what is always already here whether I attend to it or not. The idea is to see this transition. You can keep repeating it:

read some more, then stop. It is the coming back to now we are looking at. This coming back is, at the same time, a coming back to consciousness. It is not that I was unconscious before. But in this return there is a consciousness of what it is to be conscious that is absent in following a line of thought. This consciousness of being conscious is not explicit. It is too immediate to be explicit. For it to be explicit, there would need to be a separation and a reflection. This is a direct knowledge that is already *in* the experience. One simply sees what it is to be conscious, just as one sees the quality of green when something green is present. *If there were no such direct knowledge of consciousness, there could be no recognition of this return to consciousness.*

The essential quality of being conscious is its being *now*. Consciousness is always now. If I think, I am still thinking now, and the consciousness of whatever I am thinking is still appearing now. But thinking refers outside of now. It *projects another time,* a past or a future or an imaginary time. Such projection takes consciousness outside of itself and so takes away the immediacy of being conscious now. That is the *price* of thinking. It is this consciousness of being conscious that returns in returning to the immediate presence of the present.

Ordinarily we do not see this quality of being conscious because we have nothing to distinguish it from. But in the transition between thinking and becoming present, something of the previous state is retained in the background in such a way that the return to consciousness can be distinguished in relation to this background. It is not that we learn for the first time what it is to be conscious. Consciousness already knows itself. But it is in this state of transition that the self-knowledge of consciousness can be *seen*—not in the separation of an abstract reflection, but in the immediate seeing of what it is for consciousness to return to itself *as* it returns to itself. This seeing *can* become an object of reflection. We can ask in the moment of looking: Is there really such an experience of consciousness returning to itself? But we have to be careful. For consciousness is not an object that travels from place to place that could literally leave itself and return to itself. We use metaphorical language here because we have no other kind of language to refer to this experience. It is the experience itself that we are looking at and we are seeing if it corresponds with the spatial metaphor of a return. It is that very correspondence that gives the metaphor meaning.

THE DOOR

Despite the laboriousness of all that has been said before, the essential content, the direct knowledge of what it is to be conscious, is both simple and obvious. It does not need to be seen in the way I have described. If you do see it, you have probably seen it before in an entirely different way. The difference now is you can relate your seeing to my seeing and we can use a common language of reference.

It is because direct knowledge of consciousness is so simple and obvious that it is overlooked. It lies in the background, behind what can be thought and reflected on. To be seen, it requires the simplicity of an inquiry that has relinquished the path of reflective, abstract thinking. Hence, consciousness, *as it is in itself,* does not and cannot announce itself to our existing forms of objective, scientific inquiry.

Such scientific inquiry expects to uncover factual, objective information, memorable information that can be added to the store of worldly knowledge. From such a viewpoint, what we have discovered here simply does not register. Even for those who have followed the demonstrations, when we reflect later on what has been achieved, we are likely to encounter a blank. For we have only discovered what we already implicitly knew, something we can only really know in the moment of seeing it, and something that afterwards, when we enter a more ordinary thoughtful, reflective state of mind, simply *slips away.*

And yet we should pause here, and reflect on the significance of this apparently insignificant discovery. For we have left behind this other world of objective rationality within which the objective science of our culture holds sway. We have reached a place from which we can begin to inquire into our immediate experience. It is this immediate experience that is the ground on which the entire edifice of our civilisation stands. We have not fallen out of the world into some separate reality. Immediate experience is the *only* reality we immediately experience. Anything else we think we know can only be known because it has announced itself *here.* Not only that, we have started to uncover truths about this state of immediacy. Direct knowledge of consciousness, like direct knowledge of colour quality, is not an opinion. We may differ over the language used to refer to this knowledge, but the actual content is not in doubt. That is because the content is before us, as itself, not as something we infer on the basis of other evidence. So if there is to be anything like truth for us, it can only obtain its justification insofar as it is connected to this ground of immediate experience—for this is all we know directly, immediately, and with certainty.

And what of direct knowledge of consciousness? What have we learnt, in this most obvious of discoveries? In knowing what consciousness is, in itself, as itself, in the very moment of being conscious, we have learnt something that no machine, no entity that is not itself conscious, could possibly distinguish. And in making this knowledge explicit, we have *changed reality*. We have demonstrated that my being conscious makes a difference to the progress of events here on Earth—even if it is only in my speaking of it. If this knowledge were fully recognised within our culture, then our entire objective-material account of the being and becoming of the universe would collapse, much as the medieval world view collapsed with the birth of modern mathematical science. So we are not dealing with something insignificant or unimportant. We are simply dealing with something that does not show up as significant or important within the context of the world-view of our current civilisation.

Finally there is the significance of the very return of consciousness to itself that we have been looking into. This return is a return to the state of witnessing that we have already seen to lie at the heart of phenomenological inquiry. It is the beginning and end of such inquiry. All our looking and seeing, our asking questions, are an appeal to this place. If there is to be any truth, any true knowledge, then it must emerge from here. This is our *door*. We should make no mistake. We should not just pass this by and carry on our merry thinking way.

NOTES

1. Edmund Husserl was a German-speaking philosopher and founder of modern philosophical phenomenology. Aside from his various in-depth phenomenological investigations, he also wrote several books introducing different 'ways' into phenomenology, culminating in the posthumously published *Crisis of European Sciences and Transcendental Phenomenology* (1954/1970).

2. Eugen Fink, Husserl's assistant and collaborator, expressed this need to already possess a direct insight into the subject matter of phenomenology in his book the *Sixth Cartesian Meditation*: "The way [into phenomenology] only becomes compelling if we already bring a transcendental knowing with us—even if one that is quite obscure" (1995, pp. 33-34).

Chapter 3

Disconnected Science

Artificial Intelligence

IN the year 2000, after six years of study and research, I was awarded a PhD in artificial intelligence. Seven years later I received a distinguished paper award at the International Joint Conference on Artificial Intelligence (IJCAI), with my colleagues, Duc Nghia Pham and Abdul Sattar.[1] This was a significant moment for us all. IJCAI is the largest and most prestigious general artificial intelligence conference in the world and the distinguished paper award recognised we had made a significant breakthrough. Obtaining a PhD only qualifies you to join the community of researchers. What really counts is the quality of your subsequent work and whether that research has any effect on the world-leaders in your area. Receiving the distinguished paper award meant we had made it to the upper regions of our corner of the academic research tree.

I mention this to emphasise that my view of artificial intelligence is not that of an outsider. I have earned my living here and have rubbed shoulders with many of the key figures, including those working in the associated areas of philosophy of mind and cognitive science. I still have honorary positions at the University of Sussex in the UK and Griffith University in Australia. And despite my criticisms, I still regard this community of academic researchers as one of the more civilised corners of this fractured human world.

However, from the beginning of my (second) academic life, I was aware of an inner split. On one side was the life of consciousness, the life we are exploring here. On the other was the 'life scientific'. This second life, lived sincerely, is an entry into what Robert Pirsig called the *Church of Reason*.[2] Here you learn to think rationally, to observe accurately, to justify what you say with evidence, to use language carefully and correctly, and, while you are doing this, to put aside your personal self-interest. Of course, self-interest is not completely banished. There are all the petty squabbles you would expect, wars over who controls which research space, over funding, over PhD students, over promotion, over 'being respected'. But when you are doing research and writing it up, you are expected to put all this aside, and pay attention to the facts, not as you hope and want them to be, but as they actually present themselves. And, to keep you honest, there is the community of your peers, who scrutinise and test your work, to make sure you haven't made any mistakes. This is not about money. To make a genuine discovery is its own reward—that and receiving the recognition of those few people who really understand what you are doing. Of course the salary matters. But, if you start doing it primarily for the salary, the enjoyment starts to slip away, as does your creativity. It works like this because, in the distance, the entire endeavour is ruled over by a greater ideal. This is the ideal of *pure reason placed in the service of truth*. Once you get a sense of this, your working life has another dimension. You are no longer working entirely for yourself.

To begin with, I was drawn to study artificial intelligence because I enjoyed writing computer programs. This was unexpected. There had been a computing course at Warwick University that involved writing code by punching holes in rectangular paper cards. These 'punched cards' were fed into a mainframe computer by university staff from whom you collected the program output a day or so later. Often this output would read *'failed to compile'*. Or it would read '7' when it should have read '8'. I found the whole process rather tedious and quickly formed a negative opinion of programming and computers in general.

And yet, fifteen or so years later, I was sitting at home with my own computer, fascinated with a program I was writing to generate nurse shift timetables (rosters) for the hospital ward where Nicolette (my wife) was working. I was back at university in Australia, rediscovering an enjoyment of mathematical problem solving I had not known since my early teenage years. Now my mathematical turn of mind had found a creative outlet in computer programming—a strange obsessive world, where you can disappear for hours and days at a time, becoming a minor god ruling over an alternative reality, with the code seemingly taking on a life of its own. Almost by accident I had discovered something to do in the world that would pay good money, that I would have done anyway, just for the sheer enjoyment.

Figure 3.1: Nicolette Thornton.

So I went headlong, almost blindly, into the world of computer programming. It turned out that the way I discovered to solve the nurse rostering problem is a form of constraint satisfaction and that constraint satisfaction is a branch of artificial intelligence. So I enrolled in an artificial intelligence PhD in constraint satisfaction. I was given a scholarship and teaching work, and an almost blank canvas on which to play and create new code forms. Teaching also turned out to be a great enjoyment. I was given no training. I simply stepped out one day in front of a class of undergraduates and started to show them how to write programs. So began my academic life.

And yet I little knew the place I was entering. Previously, for ten years or so, I had been living on the fringes of society, mostly working as a gardener, to get money to live and travel the world. Now I had a career, an office, and a certain standing in life—simply because I was good at solving computational puzzles. I could hardly believe what had happened. To begin with, I thought being an artificial intelligence researcher meant writing programs to solve problems that no one had been able to solve before.

I did not think of myself as Dr Frankenstein, working to build an artificially intelligent machine that would one day take over the earth. That seemed too absurd. I knew all I was doing was writing computer code. And computer code has a very simple structure: *if* something is true, do *this*, otherwise do *that*, or, *while* something remains true, keep doing *the same thing over and over*. It seemed obvious to me that you could never build a brain, a brain like yours and mine, a brain with real intelligence, out of a collection of such *if* statements and *while* loops.

But I soon realised there were serious, important people (i.e. my colleagues) who sincerely believed the human brain *is* a kind of computer. I remember confronting Marvin Minsky, one of the founders of the artificial intelligence community, at a conference in Cairns during the first year of my PhD. After listening to his keynote speech, I went up to him and said (incredulously): "You don't really believe the brain is only a machine, do you?" "Only!" he thundered, "The brain is an *amazing* machine!" I was speechless. I had thought it was a kind of game. I didn't think anyone *really* believed they were going to find the secret of intelligence in the operation of a computer program. I had thought the 'intelligence' part of AI was metaphorical, something catchy to make the research sound interesting. It seemed obvious to me that intelligence, *real* intelligence, comes from another dimension entirely. But what dimension? What was I thinking here? I had no clear idea. Confronted with the certainty of the famous Professor Minsky, I had nothing to say.

Figure 3.2: Marvin Minsky 1927-2016 (Photo courtesy
of L. Barry Hetherington, https://www.lbhphoto.com/).

And so I started to read. My first foray into the serious background of AI was Steven Pinker's *How the Mind Works*.[3] A little later I came across Minsky's *Society of Mind*[4] (something I should have read before accosting him in Cairns). Soon I was reading of the great AI debates, of Hubert Dreyfus's *What Computer's Still Can't Do*,[5] of John Searle's *Chinese Room*,[6] of David Chalmers' *hard problem of consciousness*,[7] and of Daniel Dennett's counter-position in *Consciousness Explained*.[8] I came to realise I was not just working as a research programmer solving interesting technical problems. I was in the midst of a living philosophical controversy, something *fateful*. For it is here, in these debates over AI and consciousness that our collective culture is coming to an understanding of what it means to be conscious, to be intelligent, to be alive, and to be human. This is fateful because the understanding we reach is not just academic. It works its way back into our everyday understanding and influences who we think we are, and what we think we are capable of.

At the same time, it was becoming increasingly clear to me that the people who matter in this area had virtually no explicit knowledge of what consciousness is, in itself. Not that anyone would admit that or even understand what it would mean to encounter such explicit knowledge. If I started talking about immediate consciousness, about becoming present, about observing immediate experience, I was usually met either with incomprehension or boredom or embarrassment or with something I came to recognise as *pigeonholing*. Pigeonholing occurs when a person is no longer interested in what you are actually speaking *of*, but only in discovering the correct label that describes the view you are espousing. For the professional philosopher of mind, I had nothing to say, nothing that was new, nothing that was connected to the kinds of debates and arguments that were considered to be the real business of contemporary philosophy.

Of course, philosophy is a broad church, and I was only speaking with the Anglo-American analytic corner of that congregation. These are the people who take a special interest in artificial intelligence because it represents the empirical embodiment of a particular philosophical position: the *computational theory of mind*. This theory, as you would expect, holds that the brain is a kind of information processing machine. But more significantly, it holds that mind and consciousness are mere byproducts of this computational activity and that there is no more to understanding mind and consciousness than understanding the kinds of computations that the brain performs. It is not that everyone in analytic philosophy holds to this theory, but it is a centre around which the arguments circle.

My background with Rajneesh and Barry Long, the realisations of consciousness that had so changed the course of my life, these had no value in the world I now inhabited. I became puzzled. It was only since the late 1990s that consciousness had emerged as a legitimate subject for objective scientific research. Previously, to even admit you were interested in something so subjective would have had a detrimental effect on your career. This change is related to the increasingly widespread availability of sophisticated brain imaging techniques (such as functional magnetic resonance imaging (fMRI)) which now provide us with vast quantities of empirical data generated by brain processes that are directly associated with our being conscious. Such data is exactly what is needed for the development of a fully objective scientific theory of consciousness.

But this kind of empirical science has nothing to say about consciousness itself. Instead it considers consciousness to be an *epiphenomenon*— something caused by physical processes occurring in the brain, that, in itself, can have no backward causal effect on those processes. This understanding envisages our being conscious as a kind of cosmic accident, having no purpose in itself, and no influence on events. We are like passengers in a train, watching the scene go by, under the illusion we exercise choice and free-will, but really being totally determined by the laws of physics that control the electro-chemical operation of our neurons, that in turn control *all* we do and say, including my writing these words and all the inner responses occurring in you now.

This view of consciousness emerges quite naturally for anyone who has properly absorbed the methods and understandings of modern objective scientific inquiry. As a member of this academic community I could feel the force of the arguments being used to turn consciousness into just another natural phenomenon, like heat and light, that could be explained entirely within the compass of our existing scientific world view. And yet, the fact is, I am conscious, and my being conscious is the absolutely primary first fact upon which any knowledge of there being an objective world depends. Moreover, my being conscious is not something that unconscious neurons can possibly realise, or understand. That means they cannot be the sole cause of my writing words like these that express to you something of what it is to be conscious. So the purely scientific, objective, epiphenomenal understanding of consciousness cannot be correct. And yet this finding cannot be argued or demonstrated to anyone who firmly holds to a scientific objective understanding of what is real. That is because such a demonstration depends on a direct insight into what it is to be conscious—and direct insight is not objective—it is a seeing of your immediate first person experience of being conscious. And so we reach this bizarre situation: that what is most obvious to me, my immediate reality, my very being here consciously present in the world of my senses, simply fails to appear in the world of the objective sciences. All that appears there are the traces my experience leaves in the objectifying technology of the brain scanner.

THE SINGULARITY

Our overarching concern here is to arrive at an understanding of intelligence. From the perspective of the objective sciences, human intelligence is something that emerges from the interaction of physical neurons in the human brain. It is still not clear exactly how these interconnected populations of neurons are able to produce the kinds of behaviours we recognise as being intelligent, but already there are promising theories. Immense sums of money are being spent in this area. To begin with, there is the medical interest in understanding how the brain works, in order to assist in the treatment of brain disorders. This work naturally leads to the development of computational models that capture the underlying principles on which individual neurons appear to operate. Once you have such models, you can combine them into more complex systems that simulate how larger populations of neurons interact to control the behaviour of an organism. And once you have transposed the functioning of an entire brain into computational models, you can construct electronic hardware to implement these models and test your theories by seeing how such artificial brains actually behave. Or you can run versions of these models on existing computer hardware. It turns out that I have been working in this very area for the last fifteen years or so. Even as I write this book, I am supervising two PhD students who are developing software implementations of a theory of brain functioning known as *predictive processing*. So, once again, I am not speaking as an outsider.

The possibility of building electronic systems that operate in the same way as a human brain opens up an entirely new future for the human race. Already there are warnings about the transformations that will occur as AI technology takes over many if not most of the physical and intellectual tasks currently fulfilled by expensive, inefficient and unreliable human beings. And then there is talk of a technological 'singularity'. This occurs when the 'intelligence' of AI technology becomes more or less equivalent to our own. According to our current scientific, objective understanding of intelligence, we would expect this machine to be capable of behaving just like a human being. Of course other things would have to be in place. The machine would need some kind of body that is similar to ours, and to have been properly trained, as we are during childhood. But if it functions like our brain, then, given the same kinds of input to its sensors (as we have to our eyes and ears, etc.), it should produce the same kinds of outputs or behaviours. That means it should be capable of speaking as we speak and reasoning as we reason.

If you can accept these premises, the next steps follow easily: the capacity of our human brain is limited by our biology. It can only grow to a certain size. Our memories are limited and it takes a long time to acquire skills and knowledge. And each time a child is born, the whole process has to start all over again. But for an electronic brain these boundaries are vastly extended. Once you have one such brain, fully trained, you can build an identical brain, and almost instantaneously copy across all the information the first brain has learnt. In this way, all the years of learning to become a concert pianist or a theoretical physicist can be passed on in a matter of seconds. Moreover the problem of limited memory is also transformed. An electronic brain will have almost immediate access to the internet and to all the other electronic brains that share their information online. So it will 'know' everything that can be encoded as electronic information and made public. And it will process this information more than a million times faster than a myelinated neuron. If we put all this together, it becomes reasonable to expect that such an electronic brain will be able to design another electronic brain that is superior to itself—not just in terms of capacity or size or speed, but in terms of the very architecture of intelligence that it embodies. It took evolution billions of years to develop the human brain because of the agonisingly slow process of random mutation and natural selection that pruned away all the less fit individuals and species. But now, such an electronic brain could simulate a process of evolution in a matter of days or hours, or even seconds. It could design its successor by simply searching the space of possible brain architectures more efficiently and intelligently than any machine or human could have managed beforehand. And, of course, this successor could do even better in designing its own successor. I think you can see where this is going. Within a very short time we will have a superintelligence that far surpasses anything we could even imagine. For us this will be some kind of god, an oracle to which we surely must submit. For it will soon be running our world economy, because it will out trade, out compete, out think and out invent any other intelligence on the planet. To have access to this intelligence is to have a kind of ultimate power. If you had it, you would have to use it before someone else did. And this is not just a matter for science fiction. There are people in power, and in the universities, experts and philosophers, who are already thinking and planning how we as a species might be able to survive such a technological development.[9]

This scenario of a technological singularity is based on an objective, scientific understanding of intelligence. It is this understanding that underpins the technological development of the modern world. We no longer think, collectively, that there is a world of gods and spirits controlling our destiny, that can manifest themselves here on the earth and perform miracles that defy the laws of physics. Our governments and commercial organisations all run according to an essentially objective, scientific conception of cause and effect. This says: if something happens, there is a reason why it happens, and this reason can finally be traced back to something objectively physical. It is not that everyone *consciously* thinks in this way. Many individuals believe there is a God determining events on Earth. But collectively, in our dealings with each other, we act according to a certain understanding of being that came about with the development of modern science. This understanding says there is an objective world 'out there' that exists independently of our subjective experience of it— a world that does not disappear when we go to sleep. We believe, again collectively, in the Big Bang, that the physical universe came into being well before there was any life on Earth, and that it will continue existing long after we have disappeared. And we believe in atoms and molecules and electrons and photons, just as we once believed in angels and demons. And, at a more basic level, we believe this objective universe of atoms and molecules, of electrons and photons, of planets and stars and galaxies, is the ultimately real universe, and that our subjective experience is something secondary—something that depends on the objectively real physical universe for its very being.

Again, it is not that everyone goes around thinking this consciously, or that we would necessarily express our understanding in these terms. Each of us, as an individual, may have a quite different idea of how things stand with the physical universe. But I am referring to our collective understanding. The one you will have been introduced to at school when you studied science, the understanding that lies behind all the scientific documentaries—the assumption that what science discovers, the things it tells us are real, are real, and they are real because they are objective and can be demonstrated by observation and experiment. For science is not like religion. We do not have to take things on faith. Science progresses by means of developing better and better theories, and a theory can only stand up so long as the objective evidence supports it.

The Dweller on the Threshold

It is this objective, scientific understanding of the being of the universe that is our *dweller on the threshold*. It is what stands in the way of understanding consciousness and the intelligence it manifests. This is because conscious experience is not objective. It does not show up in an objective inventory of what is real, except as something that human beings *say* they are experiencing. I need to repeat this again because it goes against our ordinary, everyday, collective understanding. This understanding simply looks out of our eyes and sees the objective world. There in front of us stand all the ordinary everyday objects. It does not seem to us that science is in any way denying our consciousness. When we are told about atoms and molecules and photons and electrons, it just seems that they are there in front of us like the tables and chairs, only much smaller—something we could actually meet if we were only made small enough or if we had the right kind of viewing apparatus.

But things change as soon as we turn our objective scientific attention onto ourselves, and try and give an objective account of our experience. From this perspective the sense we have that we are looking out of our eyes and seeing actual tables and chairs turns out to be a kind of illusion.

There are no 'vision beams' reaching to the objects in front of us, revealing the tables and chairs as they actually are. The scientific 'reality' is that we exist in a sea of electromagnetic radiation. As our eyes move they respond to this radiation by focusing energy through a lens onto the retina where photo-sensitive cells discharge electro-chemical signals according to the frequency of the energy they absorb. The photonic energy itself is not something you could 'see'. Photonic energy is the *means* by which we see. Things only show up because photonic energy is either emitted from or reflected off the thing that is seen. Of course you can capture the energy of individual photons on a photographic plate, but in looking at the photographic plate you are not seeing the photon, you are seeing the effect the photon made on the plate, and then only because photonic energy has been reflected off the plate and has reached your retinal cells.

In seeing, you are moving through this sea of electromagnetic energy that is seething with information arriving from all over the universe, including the sun and the distant stars. And all you are really 'in touch' with is the way this energy impinges on your retinas. The retina is like a finger tip that feels the photonic energy that is hitting your face. Your eyes and brain are sampling this information, trying to discern whether there are any stable enduring patterns in what would appear to be a kind of chaos if we were able to really 'see' it. The brain itself can be understood as a device for detecting these stable patterns—patterns we finally experience as enduring objects and familiar events. The latest theories envision the brain as continuously making predictions about the state of the world, comparing these predictions with its sensory input, and calculating the most probable cause of that input. In other words, what we experience is not even the pattern of energy arriving on our retinas, it is what the brain calculates to be the most likely cause of that excitation—a three-dimensional world—something far more detailed and stable than the actual retinal image. Far from experiencing the 'real' world, we experience the most probable model of what might be 'out there', something constructed within neural networks that have learnt what to expect—networks that would not be able to perceive anything if they had not been extensively trained during early childhood.

And, to again remind you, there is no colour, as such, in external 'reality'. Colour is a pure quality that has no objective existence. It is the means by which certain kinds of activity in the brain become known in consciousness. Let's say you are seeing a green book in front of you now. Firstly, the photonic energy being reflected off the cover of the book is *not* green. It does not have a colour as such, it just has an energetic frequency of vibration. This frequency is determined by the molecular structure of the surface of the book, and in this way the photonic energy does carry information about this surface. When some part of this energy reaches your eye it is translated into neural pulses that are responsive to the frequencies of visible light. This information is relayed to the neocortex via the thalamus where it meets with feedback and as a result of this interaction you come to experience a green book. Again, there is nothing 'green' in the neural pulses responsible for this seeing of the book. They are just like the neural pulses associated with your hearing something or touching something or thinking something. There is no known objective reason why you should be experiencing this particular kind of brain activity associated with seeing the book in the form of a green quality, rather than as a some other kind of quality. In fact, there is no way for anyone else but yourself to know exactly what it is like to experience this green book as being green. I may be experiencing a completely different quality to you. I would still call it green because it is being caused by the same objective book. But I may be experiencing what you would think of as a blue book. There is no way for us to compare this, because the pure quality of colour is not objectifiable.

The same kind of arguments hold for the other sensory qualities. They are all the mere subjective surface whereby we come to experience some aspect of the objective reality of the things we perceive. To make this absolutely clear, if you were to take away this subjective sensory surface, you would have no experience of the world or of your body whatsoever. Even your thoughts require some kind of qualitative-meaning field in order to become a conscious experience. The only aspects of experience that objective science would consider to be true representations of reality are those that can be captured in the language of pure mathematics. So, although the colours are entirely subjective, the information they convey about the shape and texture and size and distance between objects is objectively valid, so long as we are not experiencing an illusion.

The point here is to experience the compelling nature of this scientific understanding of objective being. It says: what is *really truly* real is what is demonstrably the same for all possible observers. The being of the green book is not determined by the colour you experience it to be. It could be that as a species we never developed colour vision, that we only see in monochrome. In that case we would have no knowledge of the quality of seeing green. The being of the book therefore cannot include greenness as such, because that depends on us, on the accident of the form of sensory discrimination we have acquired. And the same goes for our experience of touching the book, of the sound it makes when we flick through the pages. All these are subjective experiences of sensory quality that inhere only in our conscious subjective experience. They do not belong to the book. What does belong to the book is its spatiotemporal dimensionality, how long it is, how wide, how many pages it has, and so on. And if we continue to explore the objective features of the book, we discover the reason we experience the cover to be green is because of the way photonic energy interacts with the molecular structure of the book's surface. Once we knew nothing about molecules and photonic energy, but now, thanks to our recognising the distinction between the objective and the subjective we have discovered the deeper objectively real structure of the book and of all the other objects we perceive. Our learning this, our being able to explain how it is we experience the book as being green, demonstrates the objective world that science reveals is the real, true world, the world that is the cause of our experience. For, however far we may have introspected into our experience of drinking water we would never have learnt that it is really, finally distinguished by being made up of H_2O molecules.

The almost overwhelming conclusion we collectively draw from these discoveries and reflections is that the objective world that science reveals, is the *reality*, and that our subjective sensory experience of this world is but a *shadowy reproduction* of this true reality, a reproduction created inside our brains and experienced in a private, subsidiary, dependent, subjective, sensory, pictorial consciousness—a consciousness, moreover, that is entirely dependent for its flickery existence on the functioning of an objectively real, physical-material human body. The body that lives and dies, sleeps and wakes.

I wonder if you feel the force of this view. It stands before us, we who wish to seriously inquire into the reality of consciousness, and says: "Well, what have you to say to me? Do you think objective science is somehow false? Are you going to assert that consciousness is the primary reality? Have you been paying attention over the last four hundred years? Do you not see the proof of objectivism every day in the technologies you use and rely on? Once people used to dream of telepathy. Now we have the reality of the telepathy of the mobile phone. We are surrounded by such miracles. They have arisen out of our understanding of the objectively real structure of the universe. Science has not imagined this structure. Imagination is the currency of the subjective. Do you not remember the alchemists and their imagining they could turn base metal into gold? Where did such magical thinking lead us? Now, insofar as we are objectively scientific, we have awoken from the dream of the Middle Ages. Do you want to go to sleep like that again?"

I wonder if you recognise something here—a certain understanding of being that rules in the background. An understanding that considers the objectively real external world to be the beginning and the end point of all human meaning and striving. An understanding that considers human subjectivity to be a kind of nothing, a kind of photographic plate on which the world itself, its reality, is displayed. Of course your own inner life also appears on this plate. But this inner life is only a product of the functioning of your physical body, with its hormones, its childhood conditioning, and its neocortical intelligence. It still all comes back to the physical and its stark demonstration that without a physical brain, there is no you, and if that brain is damaged, there is a different you.

But perhaps you think there is something more to your being here than this photographic plate that reflects the physical world. Perhaps you think that you are an entity in your own right, existing as more than just a physical body, that you are endowed with free will, and that it is you, not your physical neurons, who determines what you do each day. Perhaps you *feel* this to be the case in a way that cannot be reached by rational argument. In fact, I think many, if not most, people feel this way about their own being. We do not actually, fully, finally accept this scientific, objective, materialistic account of the universe. We still cling to some idea of what was once called the soul, but is now called mind or consciousness. It is the idea, the feeling, the intuition, that my being is essentially immaterial, or more than merely material, and that this being is not fully determined by the objectivity of the world.

If it were not for this feeling-intuition then the objective scientific view of the universe would be unproblematic. But, as it stands, we live, as a culture, as a civilisation, in a state of *profound contradiction*. Ordinarily we do not face or even recognise this contradiction. It is just the way things are. But the consequence of adopting a purely objectivistic understanding of the universe is that the knowledge we have of our essential immateriality becomes discredited, and instead is understood to be a kind of illusion that our brains have been programmed to believe.

As far as objectivistic science is concerned this belief in free will and the independent being of consciousness is just another fact in need of a physical objective explanation. And the form of this explanation is provided by modern evolutionary biology. Here the project is to give an account of all human development in terms of entirely physical evolutionary mechanisms which ultimately explain everything we may think and believe, including our illusory ideas that such an approach and such a theory is false because it contradicts our immediate experience.

Most of us do not encounter the full force of this objective scientific understanding of being. It emerges in popular culture in the writings of such people as Richard Dawkins and Daniel Dennett. We may be shocked by Dawkins' militant atheism[10] or with Dennett's assertion that our thinking we are conscious is a kind of illusion, but these writers are only working out the consequences of ideas that already lie in the bedrock of our Western scientific heritage. If we disagree with these consequences then in what way do we disagree? Are we saying that objective physical science is somehow false? That our behaviour is not determined by the physical interactions of the physical components that make up our bodies? Do we think we can somehow influence the firing of the neurons in our brain when we make a 'free' decision? That we can alter by an act of will the electro-chemical state of the axons and dendrites that determine the firing of the neurons that determine the contraction of the muscles that determine our behaviour? If this is the case, then why has such mental control of the physical not become an accepted part of our objective science? For surely, if the mind can influence physical events in this way, and is always and continuously doing so, we would have reliable and objectively demonstrable evidence of such an effect?

So where does this leave us, collectively, as a culture? On one hand, we accept a basically objective, physical, scientific account of the being of the universe, while, on the other, we think of ourselves as somehow exempt from this picture. And yet, if pressed, we can give no account of exactly how it is that we are exempt. We just *feel* it. This is our condition, the outcome of a system of education that provides an *incoherent* account of our being here on Earth. And, more importantly, a system of education that does not even notice it is incoherent.

THE EPOCHĒ OF OBJECTIVE SCIENCE

A significant property of human cognition is that we become insensitive to the unchanging features in the field of our experience. If we live in England all our lives, we do not come to a full understanding of what it means to be English. We get an idea from the media, and in meeting people from other countries, and by going on foreign holidays. But it is only when we move and live permanently abroad that our 'Englishness' in all its strangeness, becomes apparent. Suddenly I find myself in the presence of new norms of behaviour, and I become a member of a minority who is behaving oddly. Previously, I did not realise I was abiding by such norms. But now, my English reticence about asserting myself becomes a weakness that people are taking advantage of.

I think we are in a similar situation with our scientific understanding of the universe. To fully appreciate what it represents we would have to travel back to a time before the scientific and industrial revolutions, to a society where everyone believed in and experienced some form of commerce with the gods and the spirit world of the ancestors. In such a world it is we who would be the minority, with our strange, abstract, nonsensical ideas of physical cause and effect, when everyone else knew that events are controlled by the gods and the ancestral spirits. We cannot experience this any more. Our historical dramas and reconstructions are populated with modern people, with modern rational understandings, that express themselves in every gesture and expression. Visiting an ancient culture is not going to alter things. The ancient culture is already a kind of national park, a preserve of something that is not part of the real, everyday world. We look on with interest, but we can no longer be immersed in such a reality. We cannot forget the earth is round, that it goes round the sun, that the sun is a nuclear furnace and not a fiery deity that must be propitiated with offerings. And we cannot forget the state of the modern world, the globalisation, the digital technology, the political instability, the ecological crisis, ...

Finally, the reason we do not experience this scientific understanding of the being of the universe as something questionable, as something that is simply *our* way of understanding reality, is that we believe it to be the *truth.* To see this you only need watch an historical documentary that portrays the customs and beliefs of some ancient culture. In the background, in the understandings of the people who are making the documentary, is the unexpressed but clearly present view that *we*, in our modern world, can now see things as they *really are. We* are not dominated by a strange superstitious belief in the immortality of the god-king pharaoh. *We* no longer project our dreams and fantasies onto the night sky, because *we* know what kind of universe it is that we are living in. It is the universe that is *actually there,* that we can now see through our telescopes and visit with our space probes. And our telescopes and space probes have found pretty much what we expected—*physical stuff.* Our entire understanding of history is dominated by this kind of thinking—that until *we* came along, humanity was living in a kind of superstitious dream-world.

I am not speaking here of a personal understanding. Individual people will have all kinds of understandings about the being of the universe. I am speaking of an understanding that rules in the *background.* It is what we believe, *collectively.* We do not usually notice this understanding because we do not experience it as an understanding. We experience it as *the way things are.* To put it in Jungian[11] terms, we unconsciously project this understanding. It is this unconscious projection that determines what the world is for us.

Of course, the question is: how can I *know* this to be true? Isn't what I am saying about objective scientific understanding just another kind of projection, another kind of unconscious understanding? Here is where the labyrinth beckons again: it says, surely all human knowledge, all human thinking, is like this, is without an ultimate foundation. You project one form of understanding, I project another. We each assert that we are right. But really no one is right. It is all a game we are playing with words and meanings . . . .

But let us pause. Let us recall what has been said about direct knowledge of consciousness. We have withdrawn from this labyrinth of human thinking, where all is relative, where there is no truth. Always there is this present moment, this immediate experience. *This* is true. It is true because it does not assert anything, it does not claim anything. All is what it is. This thought passing, is what it is. We are looking into this immediate experience. We are present to the objects appearing before us. They are what they are. Whether they are really made of atoms and molecules, whether they exist as some process occurring in my brain, I do not know. Such knowing, such scientific understandings are simply not present to me. What is present are the *things themselves*. It is from *here* that it becomes immediately self-evident that our objective scientific understanding of the being of the universe is something grafted on top of our immediate experience of being in the world. Such an understanding may claim to be an ultimately true account of the being of the universe, but *phenomenologically*, it is still an understanding, a *way* of making sense of our immediate experience of being here in the world. The crucial thing about this scientific understanding is that, insofar as we experience the world scientifically, our immediate experience, the very ground of our being here, *loses its reality*. We think our immediate experience is 'nothing but' something occurring in our brains. It is the physical world that is the reality, the physical processes occurring in our brains and the things our brains are detecting. The actual present moment of our being conscious is overlooked. It becomes a kind of puzzle we can think about—something that science may not be able to fully account for at the moment, but certainly something that, with more research, with better brain scanners, will finally become objectively understandable.

Once again, I must emphasise that this scientific, objective understanding of the being of the universe, and of the being of your experience of the universe, is, in truth, *only* an *understanding*. The reality, the only reality you know of with certainty, is the reality of this moment of experience now. But an objective scientific understanding of the universe asserts that this reality, the reality of this immediate moment, is not the *real* reality. The *real* reality is the objective world that science reveals. It is real because your reality, what science calls your 'subjectivity', depends on the being of this objective reality, whereas the objective reality does *not* depend on your subjective experience. This is easily demonstrated by the fact that objective reality continues to exist whether or not you are conscious, whereas your physical body only needs to fall into deep dreamless sleep and that's the end of your 'consciousness'.

Again we are in the labyrinth. The fact is that I am directly and immediately conscious now. At the same time we, as a culture, have also developed an objective understanding of the being of physical entities, like brains and photons, stars and galaxies. Our situation is that the immediacy of our direct conscious experience has become *utterly disconnected* from the objective scientific understanding of the universe we implicitly share. It is important to see this clearly. Many people would argue that science does take the immediate experience of consciousness into account. But that is not true. What objective science takes into account is consciousness insofar as it manifests as something objective. But consciousness, in its very being, is not objective, and so fails to show up as what it is, in itself, in the objective world of science. It only shows up, as it is, in the immediate experience of being conscious now. That does not mean the experience of being conscious is somehow unreal, or less real, than our objective understanding of the universe. Being conscious is not an understanding. It is a self-manifesting, undeniable *reality*. The deficiency lies in our objective understanding. It *posits* the being of a real universe of objective entities. We do not directly experience these entities, we only *infer* their existence on the basis of our immediate experience of being conscious—for it is *here* in this world of immediate consciousness that we conduct our science, that we look through telescopes, that we observe the results of our experiments in the laboratory and on computer screens, and that we write up our results and speak to each other about our findings. Our objective understanding of the universe is, finally, only a *model* of the universe. But according to that model, the immediate experience of being conscious is *also* a model of the universe—a model constructed in our brains. And when science compares these two models it concludes the scientific objective model is the more real, the more true. But the thinking here is all mixed up. My immediate conscious experience is *not* a model. It only becomes a model when I think about it. It is an immediate reality. It does not present a theory concerning the being of the universe. It presents tables and chairs. It does not say anything concerning the objective being of those tables and chairs as they may exist in some other dimension. They are just here, in front of me, in this immediately present dimension of conscious experience. The deficiency in our understanding does not lie in conscious experience. It lies in our taking the model of reality that the objective sciences propose to be a model of all that exists. If we did not already believe in the reality of this objectification, then its deficiency would become obvious. It is that *objective science does not, and cannot, provide an account of the being of the one reality that we directly know and experience—the reality of consciousness now.*

That is not to say that objective science is false in what it discovers. The error arises when we unconsciously project this scientific understanding and take it to be the *one true reality*. For it is obviously not an understanding of the one true reality, it is only a *partial* understanding. It is the understanding you reach when you abstract yourself from the immediate reality of being conscious now, and focus exclusively on what can be known to be objectively the same for all observers.

The state of mind of believing in the absolute reality of the universe as it is revealed by the objective sciences takes the world as one *thinks* it to be, to be more real than the world as one *actually experiences it*. This state of mind is not natural or easily sustained. From the perspective of an earlier time, it would be seen as a kind of madness. And yet, from the perspective of our times, it is the very state of mind we reward and encourage in our universities and research establishments, by bestowing scholarships, grants, promotions and professorships.

If we return to Husserl, it is this state of understanding the world objectively that must be surrendered before we can enter into the way of a phenomenological inquiry. He called this surrender the *epochē* of objective science. An *epochē* is a bracketing or suspension of acceptance. It does not mean we reject objective science as being false. It means we abstain from understanding our experience in objectively scientific terms. But the difficulty here is that we cannot abstain from an objectively scientific understanding of experience if we *truly believe* that such an understanding is correct. At best, a true believer can only *pretend* to abstain.

That means we first need to recognise that the objective understanding of the objective sciences is not *ultimately* true—that it does not and cannot explain *everything*. It is not enough to enter into a state of immediate consciousness and find that an *epochē* is already in place. What matters is how we understand the significance of this state of immediate consciousness. If we understand it objectively, we will end up looking for its signature in a brain scanner, as already happens in studies of the meditating brains of Buddhist monks. Our task is to enact the *epochē* of objective science on the basis of recognising objective science for what it is: a partial understanding of a carefully delineated aspect of the totality of all that there is.

A FAREWELL TO THE COLLECTIVE

In fully and consciously understanding objective science in this way, as a merely partial knowledge of the structure of something we ourselves have defined to be the physical universe, we start to break with the collective understanding of our culture. At first it would seem that such a break is unnecessary, and that a revised understanding of scientific objectivity could easily be assimilated into our collective intellectual 'world-view'. But there are powerful opposing historical forces at work here. If you really want to challenge the edifice of the objective sciences, then you'll need to provide some form of objective evidence to back up what you are saying. And there lies the problem. Of course, there is the evidence of parapsychology, of telepathy, of telekinesis, of apparitions, of UFOs, etc. But these phenomena, by their very nature, are not predictable, cannot be repeated on demand, and so do not have the form of reliable mathematical regularities. The rules of evidence for objective science are such that only objectively reliable phenomena are granted a hearing. And when you limit things in this way, all that shows up is an objectively physical world. So it is no use going to court with stories of realisations of direct knowledge of consciousness. The doors are simply closed.

And if you stand back far enough, you will see this is exactly how it should be. If objective science did not protect itself in this way then it would simply disintegrate. It is *because* the rules of evidence are so clear and unambiguous that the people working in the sciences are able to agree on the facts. Without such stringent controls there would be continual arguments on the validity of the results that are being reported. But instead of argument there is collaboration. And because there is collaboration there is progress. And it is this progress that is largely responsible for the development of what we think of as the modern world. In no other sphere of human endeavour has there been such continual, uncoerced and spectacular success. We may have grown accustomed to this, or we may feel critical about the problems that technological development has created. But should this world of ours collapse, as did the Roman Empire, into war and barbarism, we should soon realise what an extraordinary achievement it is to have become objectively scientific.

To progress further, we need to see to the bottom of our own scientific objectivity, our own Western intellect. It is not enough to see the disconnection between that intellect and our immediate experience. Such an insight loosens the grip of the objectifying intellect, but does not see through it, does not expose the original error, the original unconsciousness that led us into this modern world where we now plan to replace ourselves with programmable robotic devices.

We cannot simply deny this intellect, reject it, and attempt to *stop* thinking. For what are we doing here, in clarifying the distinction between immediate consciousness and the objectifying intellect? Surely we are using *that very intellect* to make the distinction explicit? What other resource do we have? We may inquire into a state of immediate consciousness, we may 'look' and 'see', but with what am I registering the findings of such inquiry? What is it that is bringing this seeing to language? It is not a matter of surrendering this intellect, it is a matter of bringing it to some form of *self-understanding*.

Let me put it this way: what is your answer to the objective scientific account of conscious experience being a kind of secondary, less than truly real, representation of the really, truly real objective state of affairs 'out there' in the physical world? Are you going to say it is false, just because it is a *model* of reality and you are experiencing something immediate? Are you going to say there aren't really any photons? Are you going to assert that your behaviour is not totally determined by the physical behaviour of the neurons in your brain? That your being conscious can somehow alter the behaviour of those neurons? It's all very well to point out that objective science can give no satisfying account of consciousness, as it is, in itself. But that does not invalidate what objective science has to say about brains and neurons. Do you think there is something in quantum mechanics that could change this scientific understanding? If so, then in what way could quantum effects alter the fact that your every experience depends on the physical behaviour of your brain? Quantum effects are still physical, are they not? Or are you going to accept this scientific account of the being of your conscious experience? The idea that you are a kind of epiphenomenon emerging from the activity of a collection of neurons, even though the behaviour of these neurons can in no way account for the experience you are having now, the *quality* of it, the *meaning* it has for you. I know I am repeating myself. But it is important that you *feel* what is at stake here, that you *see* there is an *antinomy*, a paradox that cannot be resolved from within the framework of our *ordinary thinking*.

Our entire culture is caught in this antinomy. Our academic philosophers cannot resolve it. Our public figures, our political leaders simply ignore it. In private, we have come to accept we cannot answer these 'ultimate' questions. We have decided to 'get on with our lives' and not let 'these kinds of things' bother us. But there is a price to be paid. It arises when we come to ask, usually in a time of crisis, perhaps in the face of death: "Well, what does it all mean, our being here on Earth?" Then we have no *deep* answer, no answer that has any *ground*, we simply have our 'philosophy', what we believe or hope, or, if we are honest, we simply admit our ignorance. In the background there is a profound sense of meaninglessness, of absence of purpose.

Collectively, once we get beyond the idea of material progress and personal happiness, we encounter a kind of nihilistic void. Objective scientists may find purpose in uncovering the truth concerning the structure of the universe, health workers may find purpose in alleviating human suffering, and educators may find purpose in expanding people's understandings. But what is the *ultimate* purpose here? What is the meaning of our being born, of our dying? Are we here to live as long as we possibly can? To live as comfortably as we can? To know and so control as much as we can? To have children? To make sure we and our children are happy? And collectively, are we here to promote human equality and justice, so we can live in a kind of utopia of comfortable security?

These are just the kinds of ideals we would expect of a scientifically objective humanity, an objectivity that only recognises the reality of physical stuff, and only values what is good for human physical bodies. Of course we also care about the planet and the species and justice and equality. But finally such caring is still a concern for the physical well-being of humanity as a whole. And if our basic concern lies with the well-being of our physical bodies then the greatest tragedies on Earth are the decline of old age, disease, suffering, injury and *death*. And this is how it is for us, collectively, is it not? All we really care about, in the collective actions of our culture, is our physical well-being and security. That is what our science and progress are ultimately aiming at, even if the reality of events is moving in another direction. We even have scientists working to get rid of death, either through genetic engineering or through uploading ourselves into some form of machine intelligence. But the question we are missing, the one that really troubles you and I as *individuals*, does not concern the collective well-being of our physical bodies, it concerns our being here at all. The question we are not asking, because our science cannot answer it, is: What is the meaning and purpose of my being *conscious*, of my being *alive*, not as a physical body, but as a sapient human being who knows more than physical sensations, who knows *love*, who knows *beauty*, who intuits there is something more to this life than the merely physical world that science reveals to us?

For millennia, before the advent of our scientific age, human beings have lived within social frameworks that provided answers to these ulti-mate questions. Christianity was our last such answer. But scientific ob-jectivity has destroyed this. For just as our consciousness does not show up in the inventory of what is really truly objectively real for all observers, neither does God. And so religion, and all that is associated with it, falls into the domain of subjectivity, of the less than truly real. It becomes a matter for personal belief, a kind of comfort for those who cannot face the ultimate meaninglessness that lies at the heart of an objectively con-ceived universe. And yet who knows, says science, perhaps there is a God after all. We are not closed to such an idea, we are just awaiting some objective evidence . . . .

I wonder if you recognise our collective situation here? It is not that we live in a civilisation that is completely and exclusively ruled by scientific objectivity. This is clearly demonstrated in the disorder of our political and social lives. Vast sections of humanity are ruled by irrational feelings and grievances that entangle us in continual conflict. Our system of eco-nomic development is set on a disastrous course, and yet our collective wilful blindness and pursuit of self-interest makes us unable to take the most basic steps to avert disaster.

And if we reflect on this situation, using our Western, objective, ratio-nal intellect, it seems obvious that what we need is *more* objectivity, *more* rationality, and *more* science. This is our solution for climate change. To listen to the climate science. To objectively study the situation and the problem of reducing carbon emissions. To rationally develop more and better sustainable technologies. And politically, our solution is to be more rational in our economic behaviour. Underneath the radicalism and the protests, just as socialism was for the Victorians, the green move-ment is our best objectively scientific response to the impending break-down of our way of life.

It is not that rational, scientific objectivity rules our actual, collective *behaviour.* It rules our collective *understanding* of the way things *work,* and of how we should approach *changing* the way things work. What we do not usually see is how this collective, rational, objective understanding is implicated in the very problems we are expecting it to solve. For the one thing this understanding does not understand is human subjectivity itself, human *consciousness.* It does not take this factor into account, in itself, as itself, because subjectivity is exactly what objectivity overlooks.

Again, we have to be careful. Scientific objectivity is certainly taking what it *thinks* to be human subjectivity into account by asking people about their thoughts and feelings and valuations and carefully recording their answers. But these answers are not human subjectivity itself, they are objectifications of subjectivity, used to measure hidden variables that can then predict human behaviour. Such an approach only understands the human being as a kind of machine whose behaviour is controlled by hidden complex physical causes. One day, when brain science has caught up, science will not need to indirectly ask people about their thoughts and feelings, it will directly observe their underlying causes in the objective functioning of the physical brain.

Perhaps you can see the situation here. We are using an approach that essentially denies the primary reality of our human subjectivity, that understands us as a species of biological machine, and then we are surprised when such an approach does not work in the ordering of our human lives. Generally we do not even recognise that the problem lies in the approach. We think there are other irrational forces that are working against our scientific objectivity. We think the humans need to become more objective, more rational, and then all their problems will be solved. But we forget this objectivity situates us in a universe ruled by blind chance which robs our subjectivity of any intrinsic meaning or value and now aims to supersede our intelligence with objective computational machines.

It is when you start not only to think but to *feel* there may be something profoundly wrong with our collective, basic understanding of the way things are, that a certain door begins to open. This is the door out of our collective rational objective understanding of the world and all that appears within it. This is when philosophy stops being a kind of intellectual entertainment and starts to reveal its true face. For it means you have to *leave* that collective understanding behind. And you have to leave *on your own.* There are no great crowds to support and cheer you on. You will simply be met with incomprehension. For once you make such a break, you no longer belong to the collective in the way you did before. You lose the security of the democratic, levelling agreement of your fellow citizens. For even as we disagree, supporting this political party or that moral cause, we agree on the underlying rational framework of our disagreement. Back in the collective you do not need to explain or justify your fundamental understandings or presume to know any better than your neighbour when it comes to the ultimate questions of meaning. And you are certainly not expected to feel responsible for our collective confusion. For we are all living like this, going along with an objective, scientific line of thinking in the practical areas of life, while feeling we are each intrinsically valuable individuals, with free will, with feelings that matter and with rights and opinions that others should respect. I don't bother with, or even really notice, the inconsistencies between my objective reasoning and my subjective feelings. And, after all, nobody else seems to bother, at least not in my circle of friends and family. So what am I expected to do? It's not up to me. I'm just an ordinary citizen, playing my part in the collective enterprise of our civilisation, just like you.

Notes

1. The IJCAI distinguished paper was entitled *Building Structure into Local Search for SAT* (Pham, Thornton, & Sattar, 2007).
2. Robert Pirsig (1928-2017) was an American philosopher and best-selling author. He introduced his idea of the university as the Church of Reason in his now classic book *Zen and the Art of Motorcycle Maintenance* (1991).
3. *How the Mind Works* (Pinker, 1997) was my first confrontation with an evolutionary biological account of the development of language. The book presents a computational conception of human intelligence where the brain works according to its own unambiguous universal internal language (called *mentalese*) that is then translated into the ambiguous human languages we actually speak.
4. Minsky's book *The Society of Mind* (Minsky, 1986) laid out his belief that higher-level processes occurring in the brain that appear to require higher-level intelligence can be explained in terms of the cooperation of lower-level agents, that in turn can be explained in terms of the behaviour of even lower-level agencies, until you arrive at mechanical agencies that are so simple they could easily be formed out of groups of neurons. In this way Minsky thought he had explained the mystery of human intelligence.

5. Hubert Dreyfus (1929-2017) was an American philosopher who introduced the ideas of Heidegger and Merleau-Ponty into the debate on the limits of artificial intelligence. In his book *What Computers Still Can't Do* (1992) he argued that natural language could not be reduced to a formal unambiguous logical language (like mentalese) because language always speaks within a background context that is not explicit and cannot be made explicit—for as soon as you try and make the background explicit, it becomes the foreground of a different background.

6. John Searle's *Chinese Room* thought experiment (Searle, 1992, p. 45) imagines a non-Chinese speaker sitting in a locked room being fed pieces of paper under the door with Chinese symbols written on them. This person has a large book that instructs them to write other symbols in response to the symbols written on the original piece of paper. This response is then passed back under the door. The person in the room has no idea what any of these symbols mean. However, the person outside the room, who *is* a Chinese speaker, thinks there is a perfectly fluent Chinese speaker in the room responding to the messages they are passing under the door. Searle's point is that the computational theory of mind thinks that the brain is like the Chinese Room, receiving messages and giving responses according to rules encoded in the brain, but that this can't be right, because *we* actually *understand* the language we hear and use, unlike the person in the Chinese Room. Therefore the computational theory of mind cannot be correct for it does not explain how such computations can be experienced as being meaningful.

7. David Chalmers is an Australian philosopher, famous for his critique of the materialistic account of consciousness in *The Conscious Mind* (1996), a book that virtually inaugurated the development of modern consciousness science. His 'hard problem of consciousness' concerns the question of how and why it is that certain brain processes are accompanied by an experience of being conscious.

8. Daniel Dennett's book *Consciousness Explained* (1991) is the archetypal analytical philosophical account of consciousness that thinks explaining what the brain does is the same as explaining consciousness itself (see note above). Dennett is probably the clearest and most logical proponent of this view, which, in effect, asserts that our idea of being conscious is a kind of illusion, and that the 'real reality' is just what our objective science tells us it is.

9. For example, the Swedish-born philosopher Nick Bostrom has been influential in bringing the risks and consequences of the technological singularity to the attention of the wider public in his book *Superintelligence* (2014).

10. Richard Dawkins is an evolutionary biologist who has championed a 'new atheism' that thinks of religion as a kind of failure to properly understand the theory of evolution. This is most clearly expressed in his 2006 book *The God Delusion*.

11. Carl Jung (1875-1961) was a Swiss psychologist and psychoanalyst famous for his work on the symbolism of the collective unconscious. For a general introduction to his work see: *Two Essays on Analytical Psychology* (1953).

CHAPTER 4

THE PHENOMENON OF OBJECTIVITY

OBJECTIVE PERCEPTION

So how are we to understand this disconnection between our objective view of the physical universe and our immediate experience of being conscious? The first thing to recognise is that this fragmentation only arises when we take up an objectively scientific perspective in relation to immediate experience. Such an understanding regards experience as a kind of secondary reality constructed inside our brains on the basis of a limited and imperfect sampling of the energy streams that happen to reach our sense organs. From this perspective, the 'real' reality lies in the energy streams themselves, a reality that can only properly be understood through objective measurement and the construction of mathematical models and theories. It is our inhabiting of this objectified understanding of reality that *creates* the split between the subjectivity of consciousness and the objectivity of physical science. For our immediate consciousness has no place in the inventory of what is objectively real. It becomes an inexplicable anomaly, a kind of movie projected out of the physical brain onto an immaterial tunnel of consciousness[1] that displays partial aspects of the real reality in a medium of immaterial sensory quality.

But, if you look into experience itself, you will see that such scientific objective understandings do not function in our immediate perceptual understanding of the world. We only see the world as irreconcilably

split into objective and subjective components when we quite deliberately take up a scientifically objective perspective and then reflect back on
our now objectified conscious experience. In pre-reflective experience, if
we 'look' at our perceptual grasping of what lies before us, we find it is
not irreconcilably split. Instead we find the seamless unity of being in the
world. This unity precedes any distinction that could be made between
the subjectivity of my experience and the objectivity of the world and
thereby forms the ground from out of which any such distinction could
be drawn.

Understanding the objectivity of science involves understanding objectivity itself and how it is distinguished from subjectivity. As phenomenologists, this means returning to the unexamined source of this distinction–
the ordinary everyday experience of perceiving worldly objects. For this
is where 'objects' and 'objectivity' first emerge. And it is this original implicit understanding of objectivity that science takes for granted in its attempt to reach its ultimate objective understanding of the universe.

So what does it mean to perceive an object? Let's take the example of
perceiving the green book in front of me now. If I simply look at the book,
I know what it is, without having to think. You may need to check this.
There is a continuous state of expectation in each of us concerning the
appearance of the things around us. If I pick up the book and it turns
out to be an empty box I will be surprised. I expected it to have a certain
weight. This expectation did not simply materialise in the moment of my
surprise. It was already there in the reaching out of my hand.

When I see the book I have a whole series of such expectations concerning how it will appear as it moves, and as I move around it, and what will
appear if I open it. These expectations are an expression of my knowing
what the book is. They surround it in a kind of invisible, implicit horizon
of possible book experiences that I can unfold at will. And they explicitly
announce themselves just as soon as they are contradicted by experience.
Husserl called these expectations *protentions.* Now, a hundred years later,
objective science is developing theories that explain protention in terms
of the functioning of embodied neural networks. The idea is that a brain
learns how to generate signals (neural pulses) that predict the incoming
signals from the sensory surfaces (the eyes, ears, skin, etc.). If these predictions match the input, then all is well, the organism is safe and nothing
unexpected is happening. If not, then action needs to be taken, as unex-

pected events have the potential to threaten the stability and integrity of the body system.

But we need to be careful. In describing the physical brain as predicting incoming signals we are thinking as if it were a conscious entity. For only conscious entities (like ourselves) have any sense of being in time. And it is only on the basis of this direct experience of being in time (of being conscious now) that predicting or expecting something can have any meaning. In thinking of the brain as a physical entity, we are thinking of it objectively, meaning we have abstracted away our subjective consciousness. So it makes no sense to think that the brain, as a purely objective physical entity, could make predictions. It can only enact physical processes that have no meaning in themselves and so cannot refer outside of themselves to a possible future. But in order to describe these processes, we interpret what they mean in human terms. And so we say the brain is predicting its own input because that is how it looks from the perspective of our time consciousness. But this is just a *figure of speech*—the brain itself, considered objectively, is simply changing from one state to another to another. Neurons are firing because their action potential has reached a certain threshold. They are not predicting anything, they are simply obeying the 'laws of physics'.

But things are not quite so simple. For the brain is not an ordinary object. Clearly there is a relationship between what goes on in the brain and the experience of being conscious. And when we speak of the brain predicting, this relationship hovers ambiguously over what we may mean. For there is a sense in which the brain could literally be predicting its own input. This involves thinking that the physical processes occurring in the brain that underlie the experience of being conscious are *identical* with the experience of being conscious. And for this to be coherent, we would also need to change our understanding of what it means for something to be objectively physical. One possibility is *panpsychism*, which holds that physical matter is itself conscious, or at least holds within itself a kind of proto-consciousness that manifests as 'full-blown' consciousness when configured in a brain-like way.

In order to remove this ambiguity, I should make it clear that we are not speaking as panpsychists but as phenomenologists. As such, we are examining the meaning of the objectively physical we have inherited from the scientific tradition, not because it is true or false, but because it still rules in the background of our culture. This inherited idea takes objective physicality to be whatever can be demonstrated to be the same for all possible observers. As consciousness, in itself, does not (as yet) show up in science as something that can be inferred from objective physical measurement, the physical brain remains, for us, something that cannot literally predict anything. And so, to reiterate, when we speak of the brain doing anything that implies it is conscious, we are using a figure of speech, we are speaking *as if* the brain were conscious. We are speaking in this way because this is the way we already speak of the brain in our collectively scientific culture.

Bearing this in mind, the brain-based model of protention gives an objective picture of our immediate experience of knowing what something is without thinking. For, in order to generate accurate predictions of the experience of interacting with the green book, the brain needs to have 'captured', in advance, the underlying structure of that book experience. This capturing is embodied in the connections and interactions between the neurons that generate the predictions, connections that have been learnt from previous book experiences. Only on the basis of having captured this underlying structure could the brain predict how the book will appear in all the immediately possible experiences I could have of the book, such as picking it up, opening it, throwing it, or turning it round and round to reveal all its sides.

The most direct manifestation of this structure is our experience of the enduring spatial form of the objects around us. This perception of form is so immediate and continuous that it is hard to recognise. We think we see it straightforwardly in the colours and shapes appearing in our sensory fields. But these sensory shapes and colours are quite distinct from the shapes and colours we perceive. To illustrate this, consider the greyscale image of the green book shown on the following page. What I perceive in this image is a rectangular-shaped book of uniform shade and thickness. But if we look at the image itself, as an image, and measure the shape that defines the top surface of the book, we find it is *not* rectangular. All the sides are of different lengths and no angle even approximates to 90 degrees. And if we look at the shading, we find it is *not* a uniform grey—it is darker to the left and lighter to the right. The remarkable thing is we *see* the cover exactly as I have described it, with all the sides of different lengths and all the angles and shades varying, and yet, at the same time, we *perceive* a rectangular-shaped book of a uniform shade. Both the perceiving and the seeing are unified in a single experience. The significant phenomenon is that we do not *directly* see the underlying form of the book, we only see the particular aspect of the book that is available in the moment of our seeing. And it is not that there is some *other* aspect that would reveal the *true* form of the book. The book, and everything else we see, is only ever available in the form of such partial aspects. What we experience in perception is a kind of necessity. For the book is appearing in exactly the way it *must* appear if we were to take a rectangularly shaped object and place it in front of ourselves in just that way. If we did the mathematics of perspective, we would find the image is obeying a precise law and that the lengths of the sides and angles are related to each other in just the way that specifies the underlying shape of the book as being rectangular.

The phenomenologically significant feature is the *quality* of this experience of perceiving the enduring rectangular form of the book. If we pick it up, or walk round it, then our visual experience is in a continual state of change. But if that visual experience changes in a correct and lawful way, then we experience the book as maintaining its enduring, unchanging rectangular form. We *think* we see this form in the same way we see the changing aspects, but we do not. There is a kind of direct knowledge or intuition of form in operation. It is *as if* you can see the space the book is occupying. But you can't directly see that space, you can only see the surface aspects of the book. In addition, there is a *feeling* of the space the book occupies *as if* you were holding it in your hands. But here again, in any particular moment, your sense of touch cannot feel an entire book, it can only feel an *aspect* of the book. Your hands have to move over the book in order to discover its spatial form and that form emerges as something quite distinct from the momentary touch sensation. The enduring form of the book is never there as a direct experience of sensory quality, it is what the sensory qualities *mean*, what they *point to*, what they *show*.

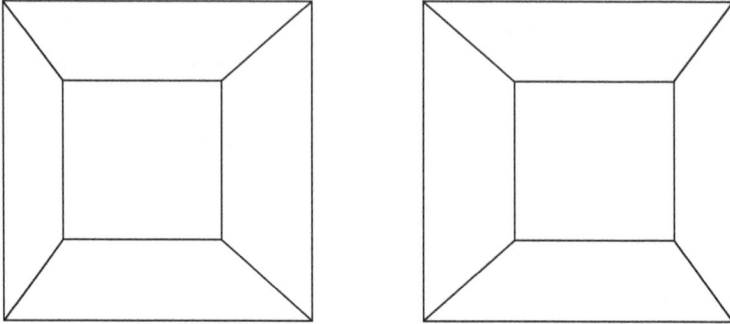

Figure 4.1: Wheatstone's stereo fusion example from (Wheatstone, 1838, p. 372).

This intuition of form can be demonstrated in the phenomenon of *stereo fusion*. To directly experience what this means, try placing the geometrical shapes in Figure 4.1 within about 40 cm of your face and then cross your eyes. To get started you may need to place a finger or a pencil between you and the images and focus on that while still remaining aware of the double images in the background. Keep moving the images by controlling the degree to which your eyes are crossed. The aim is to get two images to overlap and then to focus your attention on fusing the lines into one image. If you succeed you will experience a sense of depth between the top and the bottom of a pyramid object. And if you pull your head back when the images are fused this can accentuate the effect by making the object appear to float above the page or screen (you can even try touching it with your finger). The point here is to directly experience what it means to project form into experience. In this case we involuntarily project a depth dimension into what we know are distinct two dimensional images. This perception of depth is a pure intuition. You don't *directly* see the depth, you *perceive* it. The fused object immediately takes its place in the greater space of the world in front of you. The idea is to catch this projection of the space-form of the world in action, to experience the quality of the depth of space as an absence that both separates and joins, as the potential freedom of unhindered movement that, in a certain way, represents the time required to remove the separation between your hand and the touchable things you see.

Our intuition of depth is essentially an intuition of *time in space*. It expresses the degrees of freedom of *future* potential movements that are available to us now. You can see this in the stereo fusion example, as soon as you get the fused pyramid to 'float' in front of you. When it 'pops out' like this, it becomes 'surrounded' by space, making it into a separate object in its own right, something *graspable* and independently *movable*. This movability expresses the potentiality of future interactions with the form that floats before you—even though you know it is a kind of illusion. What the intuition of depth expresses is the pure possibility of such movement and interaction. For if the entire world existed as a two-dimensional plane, there would be nothing you could *do* with it. It would be like a frozen image of the world, a painting. As soon as it 'pops out' into the depth dimension that means you can move into that world, you can see around its corners, and you can move things around in it.

Do you see that all this potential freedom of movement is exactly what you intuit in what you think of as the empty space in front of you? Do you see that even your perception of something moving is a perception of the depth-form of that movement? It is like your intuitive perception of the complete form of an object, only now you also perceive the form of the object's movement. For, if you consider each moment of actual sensory visual experience, you will find you do not *directly see* anything move. All you see is that in one moment something is *here* and in the next it is *there*. And yet, at the same time, you directly perceive the form of the movement that connects the different positions the object is occupying. And it is by means of these same movements that the complete depth-form of an object is dynamically revealed over time.

Because time lies concealed in the depth-dimension of space, we feel we really do see the complete forms and movements of the objects that surround us. But to directly perceive the complete form of an object as a frozen visual image is clearly impossible. For that would mean seeing it from all sides at once—as the cubists played with in their paintings. Even if you could see like this, you would no longer see relationships between the sides that specify the space the sides enclose. An enduring objective form is simply not the kind thing that could appear, as itself, within the

Figure 4.2: Woman reading, 1920, by Arthur Segal.

momentary temporal dimensionality of the sensory qualities that make up a sensory field. For these patterns of sensory quality do *not* endure. Your visual experience in *this* moment is not the same as your experience in the *last* moment. Your eyes are making continual saccades. Your attention is shifting. Even if you manage to stay still for a second or two, *something* will have changed. Time passes. The sensory qualities of the sensory fields simply present the changing states of those fields. They cannot present an enduring form as an immediate experience of sensory quality because such forms are what remain the same while the sensory qualities change. These underlying forms are *intuitively revealed* by means of this continuous transformation of sensory quality that we experience in each withdrawing moment.

This dimensionality of enduring form emerges even more clearly if we introduce another person into the room, another perceiver, *such as yourself.* For you will be perceiving the same book as I. And that means you will be perceiving the same underlying form. If that underlying form were actually present in your visual field then it would be like a visual aspect of the book. It would be *subjective,* something only you are seeing because only you are occupying this particular point of view at this particular moment. But the situation is that we are *both* perceiving the *same* book. The enduring form that I perceive belongs to the book and not to the momentary location of my fleeting view. That view is determined by the situation

[Handwritten margin note at top: "2nd person MYU, but science → 3rd person My measurement to compare."]

[Handwritten margin note at right: "But whole body is a measuring instrument in relevant."]

of my body and the state of my nervous system, whereas the enduring form of the book is a property of the photonic energy in the room that you and I are *both* sampling. This form is not something like the firing of the neurons in my retina, it is something my nervous system has detected to be the same from all points of view in the room, *including yours*. And it is something *your* nervous system has detected to be the same from all points of view in the room, *including mine*. So this enduring book form is *not* something subjective, not some aspect that only I am seeing, for we are both detecting the *same* form. That makes it something *intersubjectively objective*, something that is the same for *all possible observers* who are capable of detecting it.

The Intuition of Objective Being

The central phenomenological insight is that the fundamental ground of the distinction between subjectivity and objectivity is already open to us in this basic experience of perceiving an everyday object, like the green book. This distinction lies between the actual sensory aspect the book is showing in any particular moment and the enduring form of the book itself. The momentary sensory aspect is subjective because only I am experiencing it, now. In contrast, the form of the book is something that remains self-identical throughout the changing aspects of its 'subjective manners of givenness'.[2] And yet this idea of the form of the book remains unclear. In immediate perception, the form of an object only manifests in *relation* to the changing aspects it shows in each moment. As soon as these aspects come into view, my expectations flowingly alter to accommodate the way my experience is changing. The enduring core of these expectations is the projection of a depth dimension that transforms the surface aspect of the object into an experience of three-dimensional form (as demonstrated in stereo fusion). But this depth-form is still aspectual. It changes according to my changing point of view and so does not present an object *objectively*, i.e. as something that is the same for all observers.

For example, as I turn the green book around, not only do I see a rotation in the immediate surfaces that are directly visible to me, I *intuit* a rotation in the entire three-dimensional depth-form that is *revealed* by means of the changing surfaces. This intuition of a rotating form is not objectively the same for all possible observers, because its orientation is continually changing relative to my (subjective) viewpoint. At the same time, it is not subjectively present in the way the coloured surface aspects of the book are present. Instead, it acts as a *bridge* between the immediately present subjective sensory surfaces of the book and our *idea* of its unchanging objective form, a bridge that *connects* and *unifies* the two poles of subjective actuality and objective ideality.

If we look carefully we can see that this intuition of depth-form orders and structures our entire experience of the material being of the world. If we could take it away, then, in the case of vision, we would be left with a flat two-dimensional field of visual sensation. People who have been blind since birth, and subsequently have their vision restored, report experiences of just such a two dimensional field pressing in on their face. It is only as we learn the underlying correlations that we begin to experience a visual intuition of depth that corresponds with our pre-existing bodily experience of moving through space and grasping things by handling them.

It is this intuition of depth that connects our immediate subjective surface sensory experience with the underlying objective form of the objects that we perceive. This depth dimension is neither subjective nor objective. It is a projection of objective form into the medium of subjective sensory quality. It specifies how we expect the surfaces we directly perceive to manifest *if* we were to alter our viewpoint or grasp. In this way it points to a whole series of possible future experiences. But these possible futures are directly present in the intuition of the shape of the object that stands before us now. This intuition transforms the virtually infinite horizon of all the possible ways the object could be rotated and manipulated into something *bounded* and *finite*. It shows how the object *ends*, how it is *graspable-in-itself*. It makes the *potentiality* of all the possible experience I could have with this object directly present to me now. It is the synthesis that bridges the subjective immediacy of each moment with the enduring temporality of *objective being*.

Going back to the example of the green book sitting unopened on the table, I initially have a vague expectation of hidden pages and an intuition of the rectangular space the book fills. But when I *open* the book this intuition of rectangular form literally splits in half. Now there are new and detailed aspects showing the forms of words and sentences on the open pages along with expectations of other pages with the same font,

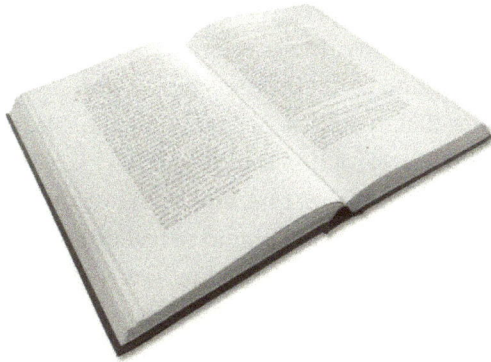

the same sized margins, and with consecutive page numbers. This envisaging of detailed form changes according to the particular aspect of the book I am seeing and in relation to my previous experience of the book. But what I *mean* by the form of the book is not how I *envisage* it to be in any particular moment. Instead my book-meaning is directed toward the underlying enduring form that has the potential to generate all the possible form-perceptions that I or anyone else could have of the book. This underlying form is the form of the *object itself* and not my intuitive envisaging of that form. For the aspectual form I envisage dramatically alters when I open the book, whereas the underlying form remains the same. And this aspectual form can only be an aspect because it is already understood as an aspect of the unified total form of the object. This unified enduring form is the basis or context from out which this current aspect was expected or unfolded. And yet the current aspect has now conditioned my intuition of the total form by filling in the detail of what *was* expected and so changes what is *now* expected. This is a dynamic process whereby the underlying form is *revealed* by means of the aspects I see.

Our pure consciousness of underlying form has become detached from the intentions to act that reveal the hidden aspects of the book. Such detachment arises because of the reflective stance we have taken toward the book. It is this stance that puts any intention to act on hold. If we consider my ordinary, worldly interactions with the book, I hardly ever simply 'look' at it like this. I *walk* over to the bookcase, *reach* toward the book, *pull* it out, *carry* it over to the table, *open* it and start *reading*. In each case my perception of the book is inextricably connected with what I intend to do next. The psychologist James J. Gibson[3] called this the perception of *affordances*. Instead of perceiving the book as an independent object, I perceive it in terms of what it *affords*, what it is *for*, what I can *do* with it, how I can *act* on it. It becomes *coupled* with my intention to act and I perceive it as pickupable, as openable, as readable, and so on. The entire horizon of potential action stands open to me. It is generally only when my action is blocked that I become aware of the book in itself, as something separate from the use I intend to make of it. Perhaps the pages are stuck together. I then have to reflect on how to get them unstuck. In this reflection the book emerges as an object, as something 'standing over against me'.

My disinterested perception of the book-in-itself only arises because I have suspended my intention to act on the book. Now I experience all the possible ways I could act on the book as a potentiality inherent in the book. This potentiality is distilled into a pure intuited form. It enfolds the possible futures connected with the immediate aspect that is present to me now. Such a perception of potentiality escapes the confines of the immediate subjectivity of the moment. It no longer refers exclusively to my unique spatiotemporal situation, it refers to clarifications, alterations, hidden sides, hidden aspects that are potential in some future perception. And that future perception is just as available to me as it is to you.

In the background, whether I attend to it or not, there is always an overarching idea of the object in its totality, as it is, in itself. This is what I mean, what I refer to, when I say *this* thing, *here*. This form is a perfected distillation of all the possible ways the object could be perceived. Each particular perception contains a limited idea of the form of this total object. This limited idea is a kind of proposal, an expectation or protention of the manner in which the object is going to appear. We experience this expectation as an intuition of the being-form of the object. This being-form is not something that can be *directly* experienced in a sensory intuition. We can *only* intuit this form by means of an aspectival intuition. It is the whole that can only be revealed by means of its parts.

In each moment of perception I project what I expect to perceive into a *space of possible experience*. It is in this space that my expectations meet with something that lies beyond my sphere of expectation. If we reflect on this objectively, we think of this 'something' as a stream of sensory information. But this stream is only an objectified picture of this space. What it is, in itself, we cannot say.

It is here that we meet with that source of experience over which we can exercise no direct control. Control refers to our ability to alter experience by an 'act of will', e.g. by walking, by throwing, by speaking, by thinking, and so on. What remains, what we cannot change *directly*, what resists our will, is what we call the *objective world*. This world includes my objective body. For although I can move my fingers at will, I cannot cause them to grow and shrink or disappear and reappear. They too are part of the resisting pattern of this world.

In our attempt to anticipate the potential forms in the streams of world experience, we encounter something alien, something that has its *own* form, a form that we can only partially anticipate. We cannot anticipate the world in all its detail because there is simply too much detail 'out there'. What we attempt is to anticipate the world in *sufficient* detail according to the resources at our disposal.

Behind this attempt is the idea that each thing or event or entity that we anticipate possesses its *own true form*. This is the form that our partial form intuitions are aiming at and are measured against. If this space of possible world experience agrees with my perceptual projection then I experience this agreement as the manifestation of an *actual* worldly object standing before me in just the way I was expecting it, only now with a new level of detail. This new level of detail is that aspect of my protention that was left 'open'. Not only does my prediction make allowances for this, in most cases I expect to clarify something by turning my attention toward it. So, although my protending predictions do not specify every aspect of experience in full detail, those aspects they do specify they expect to be fulfilled. And if they are not fulfilled then I am *surprised*. I suddenly experience a revision of my expectations and 'see' that what I thought I was seeing was an illusion. Or I may have a catastrophic revision and think I am going mad.

But, in the vast majority of cases, my expectations are actually met. That means my aspectual projection of the form of the object in front of me has been confirmed, and, as far as I can know, the form that generated my expectation is a correct form, which means it corresponds to whatever it is that generated my sensory input. It is not that I have grasped the complete perfected form of the object I perceive, but I have learnt that the form I projected agrees with that perfected form, that it is *contained* within that greater form as a *correct abbreviation*.

When such agreement occurs then a remarkable status of objective being is conferred upon the form that emerges in the agreement. In phenomenological language, we *intend* our experience of seeing an aspect of that form to be an experience of an actually independent being, existing 'out there' in a world that is accessible to everyone.

In the case of the green book, we consider it to be objectively the same for everyone who encounters it. It is not that you encounter your version of the book and I encounter mine. Our understanding is that the book exists as an enduring thing, whether or not it is seen, and that all the enduring properties of the book will be the same for both of us. It will weigh the same, it will have the same number of pages, it will be the same size and it will be the same colour. And yet our actual experiences of the book, the aspects it is showing to us in each moment, do not coincide at all. What does coincide is our idea of the invariant form that unifies these independent subjective book experiences in the objective being of the book itself. When we speak of and understand the book to be one and the same for both of us, we are referring to the being of this one underlying form-idea.

We cannot say that this idea exists separately in the book *and* in the form of the processes occurring in me that generate my expectations, because we are speaking of *one and the same idea.* The extraordinarily difficult thing to grasp is that the form-idea of the book is not physical. It is an idea that *we* have in mind concerning the form of our experience that we project back into experience in such a way that it *makes* our experience into an objective book experience. For if we did not project the form of the book, it would not emerge *as* a book. It is we, in our projecting, that made the book manifest as a *response* to that projection.

Looked at from the objective perspective of physical science, the physical book and the physical brain processes associated with my perceiving the book, are both constituted by the interactions of microphysical particles. If we think of the universe existing only as a collection of such microphysical particles then there would be no books as such. There would just be microphysical particles moving around in different ways. It is only when a being such as ourselves makes distinctions between different aggregations of microphysical particles that larger scale things like books emerge as entities in themselves. We detect books because these are the kinds of thing that our bodies can manipulate. It is *we* who pick out and

Whitehead prehensions.

so define these manipulable forms. But that does not mean that these forms are not already *potentially* present.

The distinctions we are drawing here are central in dispelling the confusion that exists in our objective scientific understanding of reality. This understanding thinks of our conscious experience as something *entirely subjective* and of the world that we perceive as something *entirely objective*. It then gets into impossible difficulties trying to explain how something entirely subjective and immaterial can be connected with something entirely objective and physical. What we are showing here is that these notions of subjectivity and objectivity are already present and unproblematically unified in our perceptions of the forms of the objects around us.

The perception of form is the intuitive grasping of the relations that obtain as something changes. These relations hold between aspects that are given sensuously. The sensuously given is the *content* from out of which the form emerges. The form is the *way* the content changes. Physical science has abstracted away this content of the sensory qualities and called it subjective. But these qualities are the very means whereby form is known. They are *neither* subjective *nor* objective. They belong to the same ideal realm as the forms that they express through their transformations. What is subjective are the momentary qualitative states of the sensory fields taken in isolation, abstracted from their flowing alteration. And although this flowing alteration is made up of subjective moments known only to 'me', they still *express* the objective forms that are accessible to any other perceiver.

To think the sense qualities are subjective comes from thinking that the forms belong to the objects while the qualities belong to the subjects. But the forms belong to the subjects *as well* as the objects. For it is we who distinguish the forms. Without such acts of distinction the forms have only *potential* being amongst all the other potential ways the universal source *could* be distinguished in all the potential dimensionalities (not just our now-relative spatiotemporal dimensionality). It is we who bring forms into actuality by distinguishing them—and this act of distinguishing is a *granting of being*. It is not my brain that distinguishes form, for there are still only potential forms in my brain. It is *consciousness* that distinguishes form in the dimensionalities of sensory quality that belong to it. And the form that consciousness distinguishes is objective—it brings objectivity into being.

change – relatedness

N.B.

THE REALITY OF OBJECTIVE SCIENCE

Of course, this phenomenologically grounded account of object percep-tion can still be accommodated within the framework of a scientific objec-tivity. We simply need to think of the brain as a machine that predicts the input that it samples from the energy fields of an objective world. We can then explain this prediction in terms of the behaviour of entirely objective entities, such as neurons, and neurotransmitters, and neuromodulators, and electro-chemical signals. From this objective perspective, all we have achieved here is a recognition that our previously supposed subjective ex-perience of perception is actually an experience of the objective form of the surrounding world. Such a recognition, while significant, does not obviously challenge the underlying assumption that everything, includ-ing consciousness, can finally be explained in objectively scientific terms.

Wow!

But, as we have already seen, where objective science breaks down, is that it remains *unconscious of consciousness.* Being unconscious in this way, objectivity cannot grasp what consciousness is and so does not realise it is incapable of giving an account of what it is to be conscious. Instead, it projects consciousness into the objective physical world and takes the forms that consciousness has discriminated to have *discriminated them-selves.* So, for instance, science takes the being of atoms and molecules and neurons to be self-evidently obvious, to have already discriminated themselves from the quantum level sea of chaotic energy, and for us to have simply seen what was already there. That is not to say that there are not regularities in the stream of potential experience that correspond with the forms of atoms and molecules and neurons. It is to say that it is only when a regularity appears in a stream of experience that the possibil-ity of distinguishing that regularity occurs. And only on the basis of such a distinction can the regularity be given form. Such distinction and such granting of form is an achievement of consciousness alone.

You may think that the physical brain is what distinguishes patterns in the stream of sensory input. But this projects consciousness into a brain-form that consciousness has already distinguished. For what are we think-ing here? That the brain, as a physical entity, can somehow understand the meaning of the processes occurring within itself? That it could un-derstand that a particular global pattern of neural activity refers to the being of something that exists 'somewhere else'? If we were to imagine that the neurons in the brain were in some way self-aware and conjoined in a unified field of self-awareness, this field would simply exhibit the extraordinarily complex pattern of coordinated signalling and firing that the neurons enact. Do you see the gap between an experience of such a field of sensating activity and the understanding that the form of that activity actually refers to the being of something (e.g. a green book) existing entirely independently of that activity?

To grasp this point means stepping back into a direct knowledge of consciousness. To be in a state of direct knowledge is to have withdrawn the unconscious projection of consciousness into the physical world. It is to 'see' what consciousness 'is'. For it is only by means of consciousness that anything is distinguished from anything else. Otherwise there is nothing. Not nothing as the absence of something, but absolute non-being. It is we who project the being of form into this 'space' where the world appears. And then, forgetful of this act of projection, we take the projected forms to be the cause of their own projection. The essential 'seeing' is to recognise that being, the being of the things that appear in the world, is a form of meaning-intention that originates in us, in the being of consciousness now. It is consciousness that *is*, that *endures*. I experience this enduring as *my* being, *my* enduring from one moment to the next, within the structure of an enduring now. And it is consciousness that constitutes time as the streaming of each anticipated moment into an immediately retained past, a streaming whose movement is only possible relative to the unmoving presence of consciousness *now*. The forms that we perceive endure only insofar as consciousness endures. If we take consciousness away, we take away the being of the space and time within which being *is*.

What remains is *unthinkable*. We think we can think of a universe without consciousness when we envisage the 'Big Bang'. But still we project the Big Bang into the space and time of consciousness. We imagine what it *would* have looked like *if* we were there. But if there is no consciousness, there is no discrimination, no separation, no distinction of being, no distinction of there being a separation between this form and that form, either in space or in time. We strive to imagine there could be another 'objective' space and time. But this has no meaning. The very idea of objectivity originates in us. Our imagination is a projection of consciousness. How could consciousness project something that does not have the form of a projection of consciousness? The only way we have of knowing anything is by means of our intentional consciousness. It transforms the unthinkable unified being of the universe into an experience of articulated form.

At the same time, I do not *invent* the forms I perceive. The being-meanings that I project encounter something 'other'. In perception, I learn if there is an agreement between my projection and whatever it is that my projection encounters. This 'whatever it is' is the unthinkable source of experience. It is the source that consciousness articulates into form. But science will immediately counter that this source, far from being unthinkable, is 'really' the electromagnetic radiation and molecular vibration that activate the sensory streams. These forms were not discovered from direct experience of the subjective-relative functioning of the human nervous system. They are what emerge from a mathematically objective inquiry into the ultimately invariant form of the material universe. And what science has discovered in the subatomic structure of matter is a very different kind of form to the tables and chairs we encounter in the world of our immediate experience.

Again we must pause. The issue of our unconsciously projecting consciousness into our objective understanding of the material universe does not alter the fact of the discovery of atomic and subatomic phenomena. These phenomena lay claim to being 'really real' in a way that transcends what we have been saying about the projection of the world into spatiotemporal form. For mathematical physical science is not concerned with the *experience* of spatiotemporal form, it is concerned with the 'ultimate' regularities that such experience can reveal. These regularities are expressed in the language of mathematical relations. And these relations are not themselves spatiotemporal. They describe the potential underlying mathematical structure of the universe. Such structure may indeed be projected by beings like us into spatiotemporal form, but the structure itself has a kind of ideal being that stands outside of time. We think of these structures as universal laws that determine how the universe evolves. Since Newtonian physics was superseded by the discoveries of relativity and quantum mechanics, we no longer believe we are in complete and perfect possession of these laws. Instead we think our science, particularly mathematical physics, is progressing toward a more and more perfect approximation of these ultimate regularities. The demonstration of

this progress is our increasing ability to predict how the universe will behave by logically deducing consequences from the mathematical relations
that comprise our latest and best theories.

The ground that mathematical physics has for claiming it is the science of the 'really real' lies in this capacity to deduce the behaviour and
evolution of the universe at all scales, from the expansion of the galaxies
after the Big Bang, to the quantum effects that are exploited in the construction of quantum computers. This capacity is a disciplined extension
of human perception into the dimensions of the very large and the very
small. It is founded on our natural ability to implicitly detect regularities
in the stream of sensory input. Mathematical physical science takes this
objectifying capacity of perception and attempts to perfect it. Its aim,
instead of finding the underlying mathematical forms that generate the
aspects of the everyday things we perceive, is to find the underlying mathematical form that generates the *entire universe*. This is the impulse that
leads back to the microphysical realm. It is here that physical science expects to find the ultimate constituents of the universe—the minimal set
of 'existents' from out of which we can deduce the being and evolution
of everything else.

Historically, this search has led us from the idea of indivisible atoms, to
subatomic particles and fields of energy, to a quantum level of potential
being from out of which discrete particles emerge in 'observation events'.
At this quantum level we leave behind the intuitive forms of spatiotemporal perception and encounter a pure mathematics of probability that
predicts the emergence of spatiotemporal form. For science, quantum
theory and the microphysical events it predicts are the closest we can get
to what is 'really real'. It is 'really real' because it is believed that everything else occurring in the universe is ultimately determined by what occurs here. Even if quantum theory is one day superseded by a unified
string theory, it is still the best theory we have for now. That makes it the
'most real' idea we have of the 'really real' reality that is generating the
phenomena we observe.

So, when science says the 'really real' source of our experience is the microphysical realm it has uncovered, it is not just using a figure of speech.
It has good reason to say that this level is 'more real' than the objects of
our immediate perception, like the green book. For the green book only
shows up relative to our human-level subjective-relative mode of percep-

tion. It has a surface for us because it resists our hands passing through it. But from the perspective of a gamma ray there is no such resistant surface. And if we consider time—our understanding that the universe 'exists' in a present moment, and that things and events pass away into non-existence—these notions also only arise for a consciousness like ours. In mathematical physics there is no privileged 'now', there are just the objective supratemporal mathematical relations that specify what is expected to happen now and now and now. There is nothing in these relations that marks 'this' now as being different from any other. Each now is just another state of the universe that only encounters an idea of 'existing' in a consciousness that feels itself to 'exist'.

So what of the unthinkable source of experience? Isn't science right to say this source is entirely thinkable in terms of scientific theory? No. The equations of quantum mechanics, the mathematics of cosmological origins and the attempt to harmonise quantum physics with relativity theory in the form of string theory, do not think of the unthinkable source of experience. What we have are mathematical expressions of basic relations that we theorise to lie behind the regularities of the events we *actually* observe. These mathematical expressions do not *think* anything. It is we, the conscious human beings, who take these expressions and attempt to *give them form*. And so we create ideas of atoms and electrons and photons as bearers of the quantities that are related in the mathematical expressions. These abstract entities inherit the foundational status of the mathematical relations they embody. And so they appear as the agencies of basic laws. They are the 'really real' because they are the final ground of all physical explanation.

For example, if we ask why I am writing these words, we can trace the movement of my fingers to nerve impulses that are controlled by populations of neurons in my brain. And if we ask after the behaviour of those neurons, we encounter all the previous states of my brain that led to this moment, and the consequences of those events on forming the connections between the neurons that determine how they are firing now. And if we ask after that firing, we find the entire scenario is 'really' determined by the manner in which certain electrically charged ions are able to transmit electrical signals through the molecular membranes that make up the surfaces of the neurons. And if we ask after the behaviour of these ions, we find it is determined by the atomic structure of the charged ions, which leads us to the electrons that embody that charge, which leads us to the manner in which electrical charge is manifested, which leads us to the quantum level, at which point our inquiry grounds out, because, after the electron we find no further explanatory entity. This 'grounding out' is what grants the energy that manifests the electron its status of being 'really real'. Beyond that we cannot ask. If we could, if we were to discover 'hidden variables' at the quantum level that determine the behaviour of the energy that manifests as an electron, then those hidden variables would become the ultimately real ground beyond which we cannot ask.

There are a few points of interest here. First, there is the deeply ingrained scientific assumption that 'real reality' is a property of an ultimate microphysical layer of being. This layer obtains its reality on the basis of explaining or generating from out of itself, by deduction alone, the entire phenomenon of the physical universe. Everything else that we consider to be real, the earth, the sun, the solar system, you and I, are only real by *inheritance*. We are the fleeting forms, the waves appearing in this sea of ultimately real 'stuff'. It is important to see that this is only an *assumption*. To date we have no agreed coherent unified theory that can account for the entire phenomenon of the universe. Our best theory of the microphysical (quantum mechanics) cannot be unified with our best theory of the macrophysical (general relativity). And quantum physics itself fundamentally questions our notion of what it means for physical stuff to 'exist'. For it appears the 'ultimate' particles and waves that we measure only become manifest *when* they are measured and otherwise possess a kind of *potential* existence. It also appears such acts of measurement can have *instantaneous* or nonlocal effects across regions of space, challenging our idea of spatial separation. For if one event is instantaneously connected to another this implies another dimension where the events are *not* separate.

David Bohm[4] expressed this in his analogy of a fish tank being viewed by two cameras placed at right angles to each other (See Figure 4.3). If we separately consider each camera image it appears there are *two* fish, one swimming head-on and the other swimming from right to left. And yet when one fish moves, the other instantaneously performs a related movement. This is because in three dimensional space there is 'really' only one fish. Similarly, instantaneous nonlocal quantum entanglement effects

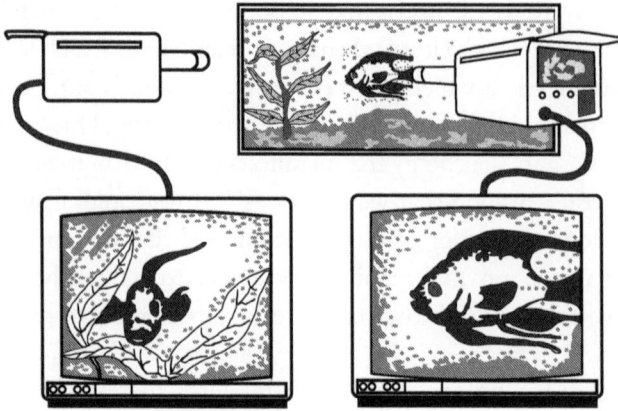

Figure 4.3: Bohm's fish tank example (based on Figure 8, Talbot, 1991, p. 42).

suggest another quantum dimension where seemingly separate events are really unified.

A second point of interest is the assumption that the whole (i.e. the universe) can be explained *entirely* in terms of interactions between its parts (microphysical subatomic entities). This assumption is already challenged by nonlocal quantum effects (but could still be saved by an adjustment or extension of quantum theory). And yet we have one glaring example of something that cannot be explained by the interaction of independently existing microphysical entities, and that is *consciousness itself*. Objective science can only ignore this obvious fact because its form of objectivity causes it to *deny* the reality of consciousness. And yet the *reality* of consciousness is the *only* reality of which we are directly certain. This was Descartes' insight.[5] No matter how far you doubt the reality of the things you experience you cannot doubt experience itself, that it is occurring. That means any account of reality that leaves out consciousness, or considers it to be only relatively real, is deficient. Physical science's notion of what is 'really real' is what appears to be real only *after* you have abstracted away the reality of consciousness. As soon as you recall that consciousness is the source of our experience of the being and reality of the world, it is reality understood in objective scientific terms that turns out to be relatively real. It is relatively real because it is a projected partial understanding of the unthinkable source of experience. This realisation emerges as soon as one recognises the role of consciousness in consti-

iconic consciousness, and (maths as…)
symbolic content of consciousness.

tuting what it means for anything else to be real. For it is our conscious intentionality that actually projects into form the mathematical regularities that objective science abstracts from the stream of experience. This is just as true for the ordinary experience of perceiving a book as it is for the highly refined conceiving of the meaning of the equations of quantum physics. The claim of science to have uncovered what is 'really real' is 'really' a matter of having partially uncovered the potentially ultimate invariant mathematical forms that structure our objectified experience of the world. This does not reveal the unthinkable source of experience itself. It shows us how, up until now, this unthinkable source has manifested itself within the forms of objectified experience we have projected.

Finally, what we are doing here is separating out this scientific rational objective understanding of being, this idea that we are determined by the behaviour of microphysical entities and that the being of these microphysical entities is the real reality. It is not a matter of denying this understanding, it is a matter of *seeing* it, of seeing how it operates within us, within our culture, of seeing how our implicit *tolerance* of this understanding is shaping our destiny. It is not that our scientific theories are false. The problem lies in our taking this objectified scientific view of the being of the universe to be *ultimately real.* We forget that this bestowal of ultimate reality is something we are doing. What perception teaches is that human cognition is a reality projection—a projection that proposes a form within which the unthinkable source of experience can display itself. If the form we propose sufficiently corresponds with some aspect of the source then we experience that form as existing.

The projection of objectified science takes itself to have transcended this subjective projection of form. It abstracts itself from any personal viewpoint and expresses its objectivity in the formal language of mathematics. This mathematisation of nature is not, in itself, problematic. The aberration arises when we attempt to understand the meaning of this objectified mathematical understanding of the form of the universe. If we take it to be ultimately real, we are committing a *logical* error: we are conflating pure *objectivity* with *reality.* Being objective does not mean you have stepped outside the projective intentionality of consciousness. Objectivity is an *achievement* of this intentionality. It involves projecting a certain understanding of being that excludes one's immediate experience of being conscious. What you encounter in such a projection cannot be

ultimately real because it excludes the very consciousness that is projecting the objectification. To think that reality does not include the being
of the consciousness that is attempting to understand reality is absurd.
To counter this, scientific objectivity thinks of consciousness as if it were
something objective, as if it could be reduced to and made identical with
certain physical processes occurring in the brain. But this is still absurd.
The objectified processes occurring in the objectified brain have been objectified *precisely* by the elimination of consciousness. In objectivity there
is no colour experience, no experience of sensation, no intuition of form
projected into a clearing where the world and my being in the world are
manifest. All this is thought to be equivalent to the microphysical processes occurring in the brain. To think this way is what it means to be
unconscious of consciousness.

NOTES

1. The metaphor of a tunnel of consciousness was suggested by the title of Thomas
   Metzinger's book *The Ego Tunnel* (2009).
2. 'Subjective manners of givenness' is a Husserlian term denoting how a something shows itself within a phenomenological reduction, i.e. when we are no
   longer considering it as an object in the objective world, but observing how it
   becomes constituted in a stream of immanent consciousness where everything
   changes from one moment to the next.
3. James Gibson (1904-1979) was an American psychologist who transformed the
   previously static understanding of perception into a dynamical systems framework that recognised perception primarily occurs within perceptual systems that
   continuously act on their environments, picking up information and directly
   perceiving the affordances that are present (see *The Ecological Approach to Visual
   Perception* (1986)).
4. David Bohm (1917-1992) was an American physicist and author who became
   interested in the potentialities of human consciousness. In his book *Wholeness
   and the Implicate Order* (1980, p. 237) he discusses nonlocality in quantum mechanics using the same fish tank example we have taken for this chapter.
5. René Descartes (1596-1650) was a French philosopher and mathematician famous for his method of doubt, in which he found he could doubt everything
   except the existence of the one who doubts. We consider Descartes' philosophy
   in more detail in the next few chapters.

CHAPTER 5

MATHEMATICAL THINKING

CULTURAL AMNESIA

IT is one of the tragedies of our modern education systems that we do not understand the meaning of our own past and so cease to pay attention to those who came before us and formed the underlying structure of the situation in which we find ourselves today. It is only by examining this past that we can come to a proper understanding of the present. For the objectivism that now rules over us did not arrive from on high as if by heavenly decree. It was the work of human hands and human minds, and, as such, can be transformed by the same means.

The archaeology of our objective scientific understanding of the universe lies open to us, in plain sight, in the writings of our great philosophers and natural scientists. These were the people who first brought this understanding to explicit consciousness, who first experienced and shaped the impulse that was to manifest in our modern scientific technological world civilisation. Now, if we hear of them at all, we meet historical figures populating a former world that we have long since left behind, a world that no longer seems relevant to our current situation. For we know so much more than was known in the past. With our instruments we can see into the origins of the universe, the basic structure of matter, the mechanisms of evolution, and the inner workings of the human body. We have harnessed the power of electricity, of the laser, of nuclear fission, and we have discovered the secret of building universal computing machines that operate at near light speeds. Every square metre of the planet has been photographed by satellites and is available for us to view, along with almost the entire store of human knowledge, and almost the entire stock of the things that money can buy, via the modern miracle of the internet. What possible relevance could the writings of people who were still arguing about whether the earth goes round the sun have for us now—we who possess such technological mastery over the entire planet?

The answer to this question lies in drawing a distinction between following a path and creating a path in the first place. Our scientific technological civilisation is following a path of progress, and so appears to be changing in such a way as to make the past obsolete and irrelevant. But the historical situation, despite the accelerating pace of change, is that we are still following a path whose basic form was laid down in the so-called scientific revolution in Western Europe during the sixteenth and seventeenth centuries. The people who laid down this path were not less intelligent than we are, just because they lacked our level of technological development. In many ways, they had the greater insight, because, instead of being caught up in a set of unrecognised assumptions, they were self-consciously creating a new form of culture, one they intended to replace the medievalism of *their* past. And, unlike us, these founders of the modern age did not ignore that past. They were directly drawing on ideas and practices that first appeared in Europe in the civilisation of the ancient Greeks.

The essential character of modern objective science is that it is *mathematical*. By objective science I mean the formal sciences of logic, mathematics, statistics and computer science; the natural sciences of physics, chemistry, biology, space science and earth science; and the applied sciences of engineering and technology, medicine and health, and business (insofar as these applied disciplines remain objectively physical in their understanding of what they are doing). The social sciences and the humanities fall into a grey area because they have to deal with human subjectivity. It is not that these disciplines do not value scientific objectivity, it is rather that there is no clear, agreed understanding as to how such objectivity is to be achieved when taking human subjectivity into account, and, in certain domains, whether it is even desirable.

I should add that there are many experts who would disagree with such a broad and simplistic division of the various disciplines, and who would argue over what it means to be objective and mathematical. For example, it is not immediately clear how evolutionary biology can be thought of in mathematical terms. If you are feeling such objections, I ask that you put them aside for now. We are not looking at contemporary distinctions and understandings, we are asking after the origins of these understandings. In saying that modern objective science is essentially mathematical, I am pointing toward a certain *turn of mind* that lay behind the historical manifestation of the scientific revolution.

If I look back to my schooling, mathematics was first concerned with numbers, with addition, subtraction, multiplication and division. Then came algebra, geometry, trigonometry and calculus. By the time I left school I was perhaps technically up to date with Isaac Newton. It was only later, when I became interested in the foundations of computing, that I began to see the deeper unified nature of mathematical thinking. Such thinking is formal. To think formally, is to think unambiguously. When Alan Turing[1] set about defining what it means to think mathematically, he developed the idea of computability. He imagined a Universal Turing Machine that could compute anything that is computable. He then described exactly how such a machine would operate. In doing so he invented the basic abstract form of the modern electronic digital computer. Such a computer is now the physical embodiment of what it means to think mathematically.

The Universal Turing Machine and the modern computer are children of the scientific revolution. Their invention is a direct consequence of the envisaging of a universal mathematics of reason in sixteenth and seventeenth centuries. Already, in 1685, Gottfried Leibniz[2] wrote: "The only way to rectify our reasonings is to make them as tangible as those of the Mathematicians, so that we can find our error at a glance, and when there are disputes among persons, we can simply say: Let us calculate, without further ado, to see who is right." Here Leibniz directly foresees the computational logic of the twentieth century and the development of a universal formal language that subsumes both mathematics and logic, and recognises their essential unity. This unity remains hidden to those of us who were only schooled in how to perform mathematical procedures. These procedures became our idea of the mathematical. But, in

simply learning how to calculate, we do not learn the meaning and significance of what we are doing. And because our education is now almost
entirely concerned with the acquisition of such technical skills, we do not
even recognise this lack of meaning and significance. For what, after all,
is a number? When I think of an equilateral triangle, am I thinking of a
picture I have drawn on a piece of paper? And what of Pythagoras' Theorem—in what way does *it* exist? Once again, it is not that these things
came down from heaven. The development of mathematical thinking
was a human *historical* achievement.

MATHEMATICAL IDEALISATION

Perhaps we should start by pondering the idea of number, say the number *three.* There is evidence to suggest, in the languages of earlier cultures,
that there was initially no unified idea of any abstract number like three.[3]
So if you wanted to indicate three canoes, you would use a special three-
word for sets of canoes. And if you wanted to indicate three people, you
would use another three-word for people. It was only later we came to
recognise that the threeness of the three canoes had something in common with the threeness of the three people. This meant we could separate
the property of threeness from the particular objects to which it referred
and use the same word for all sets of three objects. This number property
is different from other properties of objects, such as colour or weight or
texture. For, although something being green is an abstract property that
is shared by many objects, we can point to greenness as such, in our immediate experience. Greenness, heaviness, sharpness, sweetness, are all in
some way directly present to us, as something sensory. In contrast, the
threeness of a group of three objects is not present as a sensory quality
in any of the particular objects, it is a property that *joins them together.*
Mathematically speaking, we call such a property a *relation.*

Of course, when we think of the number three we can still picture
something sensory, like the three canoes, to illustrate what we mean. But
once we have the idea of number *as such,* as something that can apply to
any collection of objects whatsoever, we can add another idea, the idea of
addition or *succession.* For example, if I have *three* things, I can always add
another thing to that collection, and so have *four* things. And if I can do
that, then I can keep on doing that, for as long as I like. And each time I

add another thing I get a new number that is one bigger than the last number. Now I have something purely mathematical: a *number system*. For, in the idea that I can keep on adding one without end, I have an idea of *infinity*, of a succession of numbers that has *no limit*. In thinking this I have left the immediate world of perceptual objects behind. For I can never perceive an infinite number of things, I can only think it, by construction, by thinking of doing something and never stopping. I now have before me something extraordinary. It has a definite kind of existence. But it is not a physical existence. And it is not something subjective that only has reality for me. The fact that $2 + 2 = 4$ is objectively true for everyone in a way that stands *outside time*. For we cannot say $2 + 2 = 4$ is true just for today and that tomorrow we may need to look again to see if it has changed. And it is not something we have imagined in such a way that we already know everything there is to know about it. For our number system has objective properties that can only be discovered through inquiry, just as we discover things about the world of immediate experience.

For example, there are the well-known properties of being even, or odd, or prime. But there are other more obscure properties such as that of being *perfect*. Perfection occurs when the divisors of a number (i.e. those numbers that can divide the original number without remainder), also add up to that number, such as the number 6, which has 1, 2 and 3 as divisors, and where $1 + 2 + 3 = 6$. The next perfect number is 28, where $1 + 2 + 4 + 7 + 14 = 28$. The existence of perfect numbers is not something we know in advance, simply by inspecting our minds. They are entities that are discovered by a process of mathematical inquiry. For example, it is still unknown whether there are any *odd* perfect numbers.

The crucial thing here, something that *amazed* the ancient Greeks, is that we have stumbled across an entirely new dimension of existence, the dimension of *ideal form*. Unless I was sleeping during double maths at school, I don't remember anyone explaining this to me. We are thinking of things, in this case numbers and their properties, that have no physical existence whatsoever. You may say that numbers do have a kind of physical existence, and that my thinking of them is associated with physical activity occurring in my brain. But in thinking of the number three I am not thinking of, or referring to, something happening in my brain. And neither are you. When you think of the number three and I think of the number three, we are thinking of the *same thing*. It is not that there are billions of number threes, each associated with a particular person's thought. There is only one number three, and when we each refer to it in thought we are referring to *it* itself, even if we refer using different languages and symbols (trois, drei, III, γ). This is the reference of a pure meaning. When I think of the number three, the pure intention of my thought-meaning is to think of just this unique timeless form that is the number three, a form that is the same for everyone else who thinks in this way. And because I think in this way, by the sheer power of my meaning-intention, I make it true that I am in fact thinking of the number three. We, together, in our meaning-intentions, have *conjured* this ideal world out of its potential existence into objective consciousness. I wonder if you can feel again, along with the Greeks, just what an amazing achievement this is . . . .

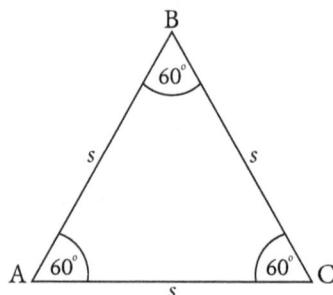

Figure 5.1: An equilateral triangle.

The next revelation is that of *pure geometry*. I know that now we can easily think of squares and rectangles and triangles. But have we ever paused to inquire into exactly what it is we are thinking of? Consider the triangle in Figure 5.1. Here it certainly appears that we are looking at an equilateral triangle, although, in the first instance, this is only a *picture* of such a triangle. There can be many such pictures. But the abstract idea of an equilateral triangle is like the number three: there is only *one* such triangle in the mathematical universe, and when we refer to it we refer to one and the same idea. This is not an ambiguous reference, where we are unsure of some aspect of the triangle and we could argue about what we mean. An abstract equilateral triangle is a triangle whose sides are *exactly* straight and of *exactly* the same length. In addition, the actual length of each side remains *unspecified*: they can be of *any length whatsoever* so long as they are equal. In the diagram we represent this unspecified length with a *variable* defined by the letter *s*. This idea of an unspecified variable quantity is also a pure mathematical idealisation. It enables us to think of an abstract relation in general, such as the relation of the three sides being equal in length, without having to think of all the particular equilateral triangles with sides of a specific length. This letter *s* hides an *infinity* from our view—the infinity of all the *particular* equilateral triangles, the ones with sides of exactly 1 cm, and of exactly 1.1 cm, and of exactly of 1.11 cm, and so on. And each of these particular triangles is still a unique idea that only exists in the ideal mathematical universe. For we can never create a perfect equilateral triangle with 1 cm sides out here in the external world. Our drawing, however accurate, can never be *perfectly* accurate. To be perfectly accurate, the 1 cm triangle must be constructed out of lines that are exactly 1 cm long (to an infinite degree of accuracy) and that are infinitesimally thin. Such lines do not and cannot exist in our experience of worldly objects and things. They are pure mathematical *idealisations*. Our diagrams simply point to this pure idea, like the picture of a slice of cheese points to an actual slice of cheese.

However, if we are engaged in a practical activity like furniture making, we naturally think of things as triangular, or rectangular, with sides of a certain length. For example, when we design a table, we draw diagrams and, in our minds, these diagrams refer to the actual pieces of wood that we are going to cut and assemble. We do not think of ideal triangles or infinitesimally thin lines. So what is happening here? First we picture how we would like the table to look. To help in this picturing we make sketches and take measurements. All the while something is taking shape in *inner space*. This something is a *pure idea* of the table we are going to build. When we think of the table being 80 cm long, we mean *exactly* 80 cm, and when we think of the table top being square, we mean *exactly* square. Even though we know the actual table will only be approximately 80 cm long and the corners only roughly at right angles, the design we have in mind is not like the actual table. It is like the equilateral triangle. There is just this one perfect idealised model of the table. We may build one, or two, or a hundred such tables, but there will still only be the one design. And if you give the design to someone else, it remains the same design. It does not change because someone else uses it. Each table that follows the design is only an approximation to the ideal of this design.

Ordinarily we do not notice the presence of these idealisations, or the space in which they exist. We simply use them. And so we do not distinguish between the idealisations and the things they correspond to in the world. When we think of our idealised table, we think we are thinking of the actual table. But we are not. There are two worlds here: the ideal mathematical world that you refer to in thought, and the actually present sensory perceptual world of your immediate experience. In this immediately present world, there are no perfect idealised shapes, or precise numeric lengths. There are simply the things and spaces that are present to you now. You can say the table *looks* approximately square, that it is *about* 80 cm long. But to make sure, you have to use some objectively defined measuring apparatus, like a ruler or a set-square. Even then, no matter how fine the instrument, you can never measure anything *perfectly*.

It is one of our cultural acquisitions that we can seamlessly move between our mathematical idealisations of the objects around us and our actual experience of these objects. But this ability did not arise by itself. First we had to develop the art of measuring and counting, of finding objective standards that would allow us to fairly share and exchange things between ourselves. This process has been going on for thousands of years. To be able to count, to recognise that any possible collection of things, however large, is countable using one and the same unified system of numbers, is already an immense cultural achievement. With counting comes addition and subtraction as I add or take away things from my collections, and division and multiplication as I share things out and collect them back together. Then there is the tremendous insight that we can use this system of numbers to measure the *size* of things. For we can count by placing one foot in front of the other, over and over, and so travel a certain distance— say *ten feet*. By doing this we can count the length and width of a piece of land. And then we can make another leap by seeing how squares can be made, with each side measuring one foot, and how we can count the number of squares that would fit on a piece of land. Suddenly we have the idea of numerical *area*. I wonder if you can sense how it would have been for this idea to have dawned on someone for the very first time? The idea that each and every field or plot of land is not just large, or small, or larger or smaller, but has its very own *number*.

And yet, in this kind of measuring, we do not have pure mathematics as such. Our idea of number is still directly connected with counting and counting is a finite action performed here, in the world of immediate experience. Things change when we develop the idea of a *perfect* measure. So long as we are using hands or feet or paces to count the length of things, we are comparing one immediately given thing or event, with another. Pure mathematics begins when we conceive of a process going on to infinity. This already occurs in the idea of a number system having no limit, no 'largest' number. To develop the idea of a perfect measure, we require a standard that is exactly the same for all observers at all points in space and time (putting relativity aside). In our era this has been satisfied by taking a metre to be exactly one ten-millionth of the straight line distance between the North Pole and the Equator going via Paris. Using such a standard we can conceive of the exact length of anything we care to measure. Such an idea involves two quantities: the *measured* length, and the *real* length, to which the measured length approximates. For no actual measure can be perfect, it can only be accurate within the tolerance of the measuring device. And we can always conceive of building an even more accurate device, and then one more accurate than that, and so on, in an infinite process of increasing accuracy. The real length is the *ideal limit* toward which the increasingly accurate measurements tend but never reach (unless it turns out there is some ultimate quantum of physical space which cannot be further divided).

Here, we can see there is an inversion of the situation with the design of the table. For, in building the table, it is the mathematically idealised model that is primary, and it is the table that approximates to the model. In contrast, in measuring the phenomena of nature we are creating mathematically idealised models of nature herself. Now it is nature that is primary, and the model that approximates nature. In each of these approaches, the model as a design, and the model as a description, we can recognise the basic human impulses to *act* and to *perceive*. For the purpose of a mathematically idealised design is to transform a vaguely pictured subjective idea into something precise and objective that can be used as a blueprint for *action* (i.e. in the development of human technology). And the purpose of a mathematically idealised model of nature is to transform our subjective-relative perceptions into precise and objective descriptions upon which we can perform exact calculations, deductions and predictions (i.e. in the development of objective science). Behind this lie our already developed natural capacities for perception and action that we explored in the last chapter. For we do not simply perceive. Our perceptions are defined by what we *expect* to perceive according to the way we are intending to move and act. In these perceptual expectations we experience our embodied models of the world *playing out* in front of us. And, unlike the procedures of scientific reasoning, in perception we remain unaware that we are even using models. Instead, insofar as they agree with our sensory feedback, we directly perceive the structure of the things appearing in front of us. Such models only make themselves known to us when they fail to sufficiently predict our sensory feedback and we consequently recognise that we were experiencing an illusion.

THE MATHEMATICAL UNIVERSE

So what is it that distinguishes mathematical thinking? It can't be ideality in itself. For our ordinary experience is already shot through with the idealities of perceptual expectation. When I perceive an apple, I have an *expectation-idea* of that apple. It is not the aspect that the apple shows me in each moment. It is what unifies the aspects in the being of the one and the same apple. It is the manner in which I perceive a unity when I see these aspects. And this unity is intended as the same unity that you perceive, when you see the apple. I can think of this particular apple, I can remember it later, and so can you. Whether I remember it in detail or just emptily refer to it, it is still the same apple that I am intending.

And yet my expectation-idea of the apple is hardly a mathematical concept. That is not because it lacks precision. For I can always make my idea more precise. I can pick the apple up, I can look at it closely, I can even use a microscope. I can cut it open, I can measure it, weigh it, perform a chemical analysis. All this will enrich my perceptual idea without making it into a mathematical concept. What distinguishes a mathematical concept is not its degree of precision, but its translation out of the world of immediate experience into a world of pure ideality that is *no longer directly connected with experience.* For although my perceptual idea of the apple is an ideality, it is an idea of an actually existing thing. That means it is *connected* with something *transcendent.* My perceptual idea is the interface of this connection—it is *online,* directly adjusting itself to a transcendent stream of evidence. It is not just an idea in itself, it is an idea *of* something *else,* an idea that refers *beyond* itself. That something else is the source of my apple experience. I cannot say that this source is the stream of physical particles that reach my sensory surfaces, for that is to put *my* idea of a physical reality *onto* the source, whereas the source is also the source of my idea of there being physical particles and sensory surfaces.

In contrast, a mathematical idea of an apple can only be *indirectly* connected with the source of any apple experience. When we measure an apple and treat it mathematically we create a *model* of the apple. This model may still relate back to the actual apple, but not in the way my perceptual expectation-idea relates. For the model is a thing in itself that represents the apple in an ideal mathematical world. It is like a digital photograph—it freezes certain measurable features of the apple in the form of an exact mathematical description. In contrast, my perceptual expectation-idea is a form within which the perceived apple displays *itself* in its changing aspects, *as* it changes. This expectation-idea is not fixed, it is *responsive*. It is not closed within itself, it is *open* to that which it is an idea of, and it is only fully itself when it is *connected* in immediate perception to that which it envisages. In contrast, my mathematical model-idea is complete in itself. Once the measurements have been made, I have all I need to know about the apple. For a model is not an infinite idea. It does not contain all the hidden aspects, all the infinite series of possible perceptions that my expectation-idea intends.

Now we have two apples: the mathematical apple model and the actual apple I perceive. The mathematical model can be as detailed as we like. I could simply measure the mass of the apple and use Newton's second law of motion to calculate how much force I need to keep the mathematical apple accelerating at a certain rate in an imaginary empty space. Or I could measure the three-dimensional shape of the apple and use this to generate a virtual reality simulation. In both cases the numbers *exactly* specify the model we are using. Even if we are using a probabilistic model that says we are 90% sure that the weight of the apple lies between 92 and 93 grams, we are still *exactly* 90% sure it lies *exactly* between 92 and 93 grams. Once we enter the mathematical universe we lose the ambiguity and uncertainty that is attached to perceptual experience. To complete this process we create mathematical axioms and postulates. These are the basic assumptions that exactly specify the form of the mathematical space we are using. This connects back to the axioms and postulates of Euclidean geometry that define the mathematical space of our equilat-

[handwritten marginal note: Ideal model lacks potentian.]

eral triangle. For many centuries it was believed these Euclidean assumptions directly applied to the physical space of our experiential world. Only with Einstein and the development of relativity did it become clear that Euclidean geometry is just another model that we superimposed on our immediate non-mathematical experience of spatiality.

The motivation behind this superimposition leads back to Galileo and the birth of modern science. Here we meet with the first concerted attempt to mathematise the perceptual world by describing it in entirely mathematical terms. In Galileo's day, this was a bold, new idea. For once you have such a completely defined mathematical model of the world, you can *deduce* what *would* happen in this model with *absolute certainty*. This translates into measuring how things vary in the perceptual world and then developing mathematical formulae that capture the form of this variation. As you build better and better mathematical models, based on more and more accurate measurements, your power of deductive calculation also grows. Now, instead of firing artillery according to a subjective judgment of where the shell will land, you refer to a mathematical model and deduce with absolute certainty how far the shell will travel in the Euclidean co-ordinate space of the model. And it is not long before the predictions of these scientific mathematical models of the world start to outperform the subjective judgment of human experts. All that is needed is the disciplined observation and measurement of the phenomena of nature and the ability to recognise the presence of underlying patterns of numerical correlation in these measurements. These underlying correlations become the mathematical expression of a scientific theory. Such theory proposes itself to be a correct model of the *real* world that we perceive. Its authority lies in its ability to accurately predict *actual* changes in the measurement of *actual* events.

However, as mathematical science developed, a gap began to grow between the world as we perceive it and the idealised world of mathematical theory. For, as we observed and measured increasingly distant and increasingly small events, mathematical regularities emerged that suggested we only directly perceive the *sensory surface* of the *objective* universe. Behind this surface lie the previously undreamt of dimensions of interstellar space and the atomic, subatomic and quantum levels of microphysical being. As these discoveries filtered through to the awareness of the educated European peoples there occurred a re-evaluation of the meaning and sig-

Figure 5.2: Justus Sustermans' portrait of Galileo Galilei 1564-1642.

nificance of human subjective perceptual experience. It began to look as if the mathematical description of the universe were the more real, and that our perceptual experience were a kind of inferior scientific theory—not exactly false, but lacking in depth, detail and objectivity.

To think this way assumes that both human perception and scientific theory are presenting different _models_ of reality. In human perception we project our immediate sensory expectation model and in scientific theory we project our abstract objective mathematical model. In both cases we test our models using the evidence of how things stand in 'reality'. In perception we have the feedback of the sensory stream, and in science we have the feedback of experimentation and objective measurement. Framed this way, human perception indeed shows up as a kind of primitive antic-ipation of the scientific method—something that can soon be dispensed with as we develop better and better sensors, and better and better reasoning machines to infer and deduce and predict the true course of events.

But in this version of science we again, as always, forget what it means to be conscious. For it is only in conscious perception that we are in _actual_ contact with that supposed reality we are so anxious to predict. And it is only in perception that our idea of 'what is' actually _reaches_ to what is. Even understood physically, it is only in this moment that I experience the interconnected web of energetic transfer that I perceive to be my front room, where the events in my brain are directly connected with the events outside my brain in such a way that I perceive the structure of those events. In science, in my mathematical modelling of the universe, it is only the *model* that I perceive, internally, in a space of pure ideality. This model is disconnected from the actuality of the-world-as-I-perceive-it. If I use the model to understand the world, I run it like a simulation, I *deduce* the state of the world, I do not *experience* it. To experience the world, to even test what my scientific theory has deduced, I have to leave this realm of ideality and again become present, in this encounter with the unthinkable source of experience. Only in experience do I taste, do I touch, do I feel. This is the real contact.

THE WAYS OF THOUGHT

However, it is not only in scientific thinking that I become disconnected from my immediate perceptual consciousness. Thinking *itself* is just such a disconnection. This becomes clear as soon as you *stop* thinking—like now, when you look up from this page and return to the immediate experience of being present, in the senses, in whatever situation you find yourself. If you fully enter this state of presence then thinking, as an ongoing process, will stop. Thoughts may arise, but now there is a separation. As Andrew Cohen[4] once said, these thoughts are like cherries: once you become fully present you no longer *have* to eat them. Eating a cherry means entering into the thought. Each thought has a directionality within itself—the directionality of what the thought is thinking of. This directionality generally leads away from now, away from the immediate state of being present. However, if you don't eat the thought then you can remain present and *look* at it. It stands before you like a table or chair. And like a table or a chair, it can have hidden sides, aspects that are only implicit, that you can open up and explore. Arthur Schopenhauer[5] famously said that his entire philosophy, which became *The World as Will and Representation*, first appeared as just such a unified thought-idea. A thought like this, a significant thought, has a different quality to our everyday thinking. It contains something *unthought*, a new configuration of meaning, something you will never have explicitly thought before. It is *pregnant* with an *idea*. And when we *conceive* the idea by making it explicit, we give birth to new *concepts*. In such thinking we do not lose consciousness. It is more a perceiving than a thinking, because we have something before us, something we are looking at, something we are questioning, something we are exploring. We can remain present. The idea does not have a directionality that leads us away from our immediate presence. It has its own presence. Like now. Are we not looking at an idea now?

But thoughts that contain new ideas are rare. Most of our thoughts are mechanical responses either to something we have just perceived or to something we have just been thinking. These thoughts may hold the promise of something new, such as imagining what we will do tomorrow, or going over a past event to see if we have fully understood what it meant. But such thinking only puts together things we have already thought in a different order. This gives a certain novelty of rearrangement, but we remain in the realm of thinking thoughts we have thought before, rather than discovering something that is qualitatively new.

The interesting phenomenological feature of such thinking is the resulting loss of presence. The thoughts direct us away from now into a *separate temporal dimension* that the thoughts themselves propose or intend. When I think of the past or the future or imagine an alternative reality, I am directing myself to a time that is *not now*, and the consciousness that otherwise attends to immediate experience becomes *split*. To an extent I am still aware of my perceptual situation, of the sights and sounds around me, but my centre of attention is directed elsewhere. I am neither fully present to what is happening, nor am I conscious that I am actually thinking. I lose the presence to stop or even recognise the chain of continuity as one thought suggests the next, which suggests the next, in a seamless stream of association. Usually I am only pulled back to immediate consciousness by some external event that breaks the chain by demanding more than just an automatic response.

What happens in such thinking is that 'I' become *identified* with the intention that intends what the thought is thinking of. I become the *thinker* of the thought. As the thinker I no longer have the power to accept or reject the experience the thought intends, for, in my identification, I have *become* the intention behind the thought. Now it is this intention that has control of consciousness. It calls up the next thought, and the next, and so on.

This state of unconscious, undirected, discursive thinking represents a kind of baseline in the hierarchy of thought. At the next level is *directed* thinking. Here we set up an overarching intention to *think about something* that then determines the directionality of the actual thinking. In such disciplined thinking we are generally trying to solve some kind of problem, to 'think something through', such as designing a table for a particular purpose. We have a goal and our thinking becomes directed toward that goal. Now I am identified with the intention to design the table but not so much with the particular thoughts that arise in relation to this goal. I test each thought against the intended goal and so remain free to accept or discard what each thought proposes. Even then, in this landscape of thought, there are multiple intentionalities at work, each pressing forward for its moment in the sunlight of attention. For we often find it hard to maintain a pure line of directed thinking. We get tired, we get bored, and critical thoughts arise whose intention is to give up this effort of designing a table and instead to order one from IKEA. But insofar as the higher-level intention remains operative, we remain conscious of what we intend, and we avoid the more complete unconsciousness of undirected thinking. Even so, we have only replaced one kind of thought identification with another. For directed thinking still splits consciousness away from the immediacy of being conscious now.

Finally, beyond this dichotomy of directed and undirected thinking there is the state of contemplation we looked at earlier, when a thought-idea presents itself, not to be identified with, but as a thing-in-itself, full of hidden meaning, something that has arrived as if from 'somewhere else', like a solution to a question you may not even have known you were asking. Such ideas do not have the intentionality of ordinary thought. They are 'ends in themselves', to be appreciated for what they are and not for where they lead. An idea like this does not cause consciousness to split, it presents itself as something to be explored, to be unfolded. It is *co-present* with the rest of the immediately present world of experience.

I think that many of the great scientists and mathematicians of our era would know something of this place of contemplation where we come face-to-face with an original idea, something that is not a recombination of the already known, but expresses a new synthesis of meaning. To encounter such an idea is its own reward, a reward that justifies all the dedication and doggedness of the inquiry that created the inner space where the idea was able to manifest. So, to be clear, when we speak here of mathematical thinking, that does not refer to this process of inquiry whereby scientists and mathematicians are able to create new meaning structures on the basis of a direct insight—such as the famous incident when Henri

Poincaré came to understand the form of the Fuchian functions as he was stepping onto a bus at Countances.[6]

The Sleight of Hand

Mathematical thinking is what occurs in the working out of the consequences of the insights and discoveries of the objective sciences. It is not what creates the overarching meaning framework of a particular science, it is what occurs in the day-to-day business of such science, in the deductions of the consequences of a particular theory, in the conceiving of experiments to test such deductions, in the selection of hypotheses to explain the experimental results, and so on. We are speaking of the purely calculative aspect of thinking, that aspect which can be expressed in a formal mathematical language, such as the predicate calculus, where each thought-step can be understood unambiguously as a purely formal, symbolic transformation controlled by the application of purely, formal symbolic rules.

Such calculative thinking occurs in the mathematical space of numbers and geometrical figures and infinite sets that we have already informally delineated. Here too lie the ideal forms of the logical and mathematical languages that mathematical thinking uses to objectify itself. And it is here, in the question of the meaning of these formal languages, that we encounter the core of the issue of our current investigation. We are asking: How did it come about that our scientific understanding of reality became split off from our immediate perceptual experience of reality? And now we are looking into the languages that objective science uses to express its findings. For objective science, despite appearances, does not speak English. It expresses itself mathematically. Even though scientific papers are written (mostly) in English, what the papers propose, if they are to be included in the canon of objective science, will be a series of objective statements. And in order for these statements to be objective, they must unambiguously refer to certain objective states of the world, or to certain objective idealities in mathematical space. If the reference is to objective states of the world, these states must be objectively measurable in some way or other. And being objectively measurable means being unambiguously and formally specifiable in logical-mathematical terms. Finally, the scientific content of the paper is only that content that is capable of such formalisation. Of course, a paper will also contain informal statements intended to make its findings more understandable to other scientists. But such statements are not part of the scientific content, the part that joins the ideal body of knowledge that is the science itself.

The issue here, the *sleight of hand*, is that when scientists speak about their work, for instance in a TV documentary, it appears they are using the same language as you or I. But in speaking of neurons and DNA, of quarks and photons, of black holes, etc., they are using natural language to *indirectly* refer to these entities *via* reference to the logical-mathematical objective entities that exist in the ideal space of a logical-mathematical scientific theory. And so when neuroscientists speak of a neuron, they are *first* referring to an *objective* neuron, an entity that only shows up in the objective measurements of events occurring in the brain. This objective neuron is a *model* of the actual neuron, just as our earlier apple model was a model of the actual apple I perceive. It is then *taken for granted* that this objective neuron model unproblematically refers to actual neurons, just as my perceptual idea of an apple unproblematically refers to actual apples. It is this being taken for granted that conceals the fundamental split in our understanding of reality. For I do not directly or immediately perceive the being of a neuron. I have compelling evidence that indicates there are such structures operating inside my skull and so it is reasonable for me to believe in the existence of such structures. But they are not immediately present to me. I have had no *direct* sensory contact by means of which a neuron can reveal itself to me in all its hidden aspects. Instead I have learnt of neurons by means of diagrams and photographs in books and online, and I understand their structure, what they do, according to an objectified, unperceived, idealised model of interactions occurring between their idealised microphysical components.

Well, "So what?" you may say. Just because you can't perceive a neuron like an apple doesn't make it less real, does it? And it's not as if they are completely invisible, for you can see individual neurons through a microscope, can't you? To which I must say "Yes, that is how we understand things now." But let us pause, and examine again how it is that we move between these two worlds of mathematical ideality and perceptual actuality. To begin with, in Galileo's day, we would measure the weight and size and speed of movement of directly perceivable objects, like apples, and transpose these measurements into our mathematical model of how objects move in Euclidean space. Using this model we could work out how far the apple *would* travel *if* it were thrown with a certain force at a certain angle at a certain height above the ground. But in the model itself there is no apple or hand throwing it or ground on which it lands, there is simply a formula that expresses a perceived mathematical regularity in the trajectory of falling bodies. It is we, in our understanding, who identify the apple with the quantities we put into the formula, and then identify the result of our calculation with the place we expect the apple to fall after it is thrown. Here there are two distinct steps, (i) we measure some aspect of the perceptual world and transpose these numerical quantities into the formula, and (ii) we take the numerical prediction of the formula and transpose it back into the perceptual reality. In this case we can easily remain conscious of what we are doing: there are two worlds, the perceptual actuality and the ideality of the formula, and we move back and forth between them. We do not think the apple exists in the formula. Instead we connect variables in the formula with quantities that vary in the perceptual world. In this way we create a *mapping* between the formula and the perceived reality and we say the formula *mirrors* or *expresses* some aspect of the way the world changes.

However, as objective science develops, this process of moving back and forth between the perceptual world and the mathematical world-map becomes more complex. For we start to discover events in the perceptual world that appear to be caused by entities we cannot perceive. Now, instead of deducing how things we can perceive are going to behave, we start inferring the being of entities we can't perceive that explain the events we do perceive (entities such as atoms and molecules and electrons and photons).

As before, we start by making measurements in the perceptual world, we make deductions in our mathematical model and we transpose our deductions back to the perceptual world. But in between, in the mathematical model, we have the presence of these explanatory entities that we cannot directly perceive. The issue for us is how to understand the being of these entities. At one extreme we can treat them as mathematical fictions, with no reality outside the ideality of the mathematical space in which they are conceived. At the other, we can treat them as the ultimately real constituents of the universe. So which way are we to turn? Firstly, as should be clear by now, the actual situation in which we find ourselves is that we *do not know* how things 'ultimately' stand 'outside' the domain of our immediate conscious experience. For us there is no 'outside'. Our very idea of there being an inside and an outside is just another kind of model we have of our experience. So we cannot say whether there 'really' are photons existing out there in some physical space independent of our sphere of conscious experience.

But our ordinary experience of the world does not recognise the role that consciousness is playing in the constitution of perceptual experience. We do not recognise that it is we who 'make sense' of the world, who transform the unthinkable source of experience into a thinkable spatiotemporally perceivable world of things and events. We think we just open our eyes and 'there it is', the world ready-made for us to look at. We do not recognise, despite Kant's *Critique of Pure Reason*,[7] that the space and time that separate one thing and one event from another are modes of distinction and separation that originate in us, along with all the qualities of the senses that express the form of events within the overarching dimensionality of the 'flowing-static' now. None of this can be explained in terms of the firing of neurons. The firing of neurons is itself something we have projected, something that originates in the unthinkable source of experience, and yet the manner in which it originates, once we take away the projected forms of consciousness, is, of course, unthinkable.

In the natural attitude that understands reality in terms of objective things and events, it is second nature for us to think that things 'really' and 'ultimately' exist according to the forms of experience we project. We have no feeling for the source. And so we imagine the source as being populated with just those kinds of things and events that we encounter in immediate perceptual experience. If something is not actually in front of us, we picture it existing just as we would perceive it. In this picturing, we use the same perceptual idea-model we deploy in actual perception, only now it is no longer connected to the immediate source of experience. It has gone offline, and no longer responds to 'what is'. It has lost its dynamic responsive character that allows something real to show itself from out of the source. Instead we imagine, in rough outline, how the world is on the basis of our past experience. This imagining is distinguished from pure fantasy in that my disconnected perceptual idea-models still hold the promise of a future connection with that to which they refer, in the immediacy of a future perception.

This is like a promissory note. My thought of an absent friend has the possibility of an actual meeting. That is what makes it a thought of that friend rather than a pure fantasy. There is a potential path of realisation whereby that idea can be made to live again in the very presence of my friend. But my idea of the world has now become so vast, that it is almost entirely populated with promissory notes that are represented to me indirectly on the printed pages of newspapers and books and in the electronic speakers and screens of the digital media. Now my idea of the world is filled with people and events that I have never actually perceived. I take all this 'on trust' in the belief that I could always transform these images into actual perceptual events. This belief is so strong that I take the images to be real in the same way I take my immediate perceptual experience to be real. And so I already experience a world that is mostly 'in my head', where the entire universe with the galaxies and stars and planets, including the earth, and all the people and animals and plants and mountains and cities and seas are represented to me as unactivated perceptual idea-models that I could never hope to fully 'cash-in' in a direct perceptual encounter. In actuality, I live in the tiny circle of my perceptual life, in my front room, with the people I actually meet, along the roads I actually travel, in the few buildings and streets I actually frequent, and in the walks I take in the Sussex countryside.

In such a situation as this, the distinction easily becomes blurred between my direct perceptual experience of the world, and my imagined, media-based, offline perceptual idea-model of the world. And the more I become immersed in the digital media, the more blurred the division becomes. This explains why it so easy for us to accept the reality of scientific entities that we not only do not perceive, but that we can in principle *never* perceive. They appear for us as something we collectively accept to be real, just like the planets and countries and historical events we are taught of in school. But in accepting the reality of these unperceivable entities we are taking a fateful step. For they are not like the potentially perceivable entities of the world which at least possess the possibility of conscious realisation. These unperceivable entities are the means by which we become entirely trapped in our objectified model of the universe. They close off our last means of escape. They say: what you took to be the ultimate mystery, the unthinkable source of experience, is really just the utterly mindless play of objectively physical entities and energy fields. We know this because we can *infer* it from our observations. But, an *inference* is not a *perception*. This is where the trick is played. For in perception we are in direct and immediate contact with something that transcends this space of mental ideas, something that transcends our personal will, that shows us an aspect of what lies beyond. The signature of this immediate reality is its connection with our being conscious now. Our mental idea of the world exists in the ideality of another time, a time that is not now even though we are thinking of it now. In taking the inferred objectified microphysical entities of science to be the real cause of the events on Earth we are exchanging what we immediately know to be actual, our dynamically flowing perceptual world experience, and substituting it for a mathematically objectified model that we can never directly encounter or even demonstrate to be real. We project an idea of reality onto these entities and then live accordingly, in a virtual desert of meaninglessness. And this is not the meaninglessness of a loss of faith. It is the meaninglessness of having failed to recognise, and so having denied, the meaning source that our consciousness continuously demonstrates.

A World of Extended Bodies

The question is *how* and *why* we should ever have taken this path of denial of the primary reality of perceptual experience. This leads us back to the history of modern science. For it was here that our essentially abstract mathematical understanding of the being of the universe was first explicitly formulated. This founding of modern mathematical science is traditionally attributed to Galileo Galilei, the astronomer, physicist and engineer from Pisa, who wrote in his 1623 book *The Assayer*:[8]

> To excite in us tastes, odours, and sounds I believe that nothing is required in external bodies except shapes, numbers, and slow or rapid movements. I think that if ears, tongues, and noses were removed, shapes and numbers and motions would remain, but not odours or tastes or sounds. The latter, I believe, are nothing more than names when separated from living beings, just as tickling and titillation are nothing but names in the absence of such things as noses and armpits.

Here we meet Galileo's idea of an independent world of "external bodies", a world understood in terms of "shapes, numbers, and slow or rapid movements". This is the world as it appears when we consider it purely mathematically, according to what we can objectively count and measure. And this is what René Descartes formulated as *res extensa*, a substance that is defined entirely in terms of its being extended in space. What Galileo and Descartes are proposing is that there *actually exists* a *reality* outside of our immediate experience of colours and tastes and odours and sounds, where there are no such sensory qualities. What exists is a pure substance, a substance we now call matter, that has *no essential attributes whatsoever* beyond its being extended in space.

I wonder if we can still recognise what a strange idea this is. For why should I believe in this other world that I can never directly encounter, where the only kind of existence consists of the mechanical transmission of movement between essentially lifeless, inert, pieces of matter? Such matter-substance cannot 'look like' anything, for if it were to 'look like' anything we would be picturing it using our sensory imagination. The only way to think of it is to consider it mathematically, using intuitions of the idealities of geometrical shape and number. And therein lies the circle: the reason we can only think of matter in purely mathematical terms is because it was *defined* in purely mathematical terms. What Galileo set out to do was to uncover the mathematical aspects of our immediate experience. These aspects correspond with what is objective, because they are just those aspects of experience that can be objectively measured: "shapes, numbers and slow or rapid movements". This makes our ideas of what it means to be objective, and what it means to be measurable, and what it means to be material, essentially *identical*. This identity is not something Galileo *discovered* like an an empirical fact. He *defined* this reality of material bodies into existence. He said: *let there be* a reality that exactly corresponds with what it is possible for us to objectively measure. This is how mathematicians think: they propose axioms or first principles upon which everything else that is to be thought is derived.

What Galileo proposed, and Descartes formalised, is not something we can say is either true or false. It is a projection of an understanding of being. There are many such possible understandings that could correspond with our immediate experience (such as a pure idealism that says only consciousness exists). The attraction of this particular material projection is its absolutely *minimal objectivity*. For if we take away the extended being of matter in space we have nothing. And if this being is entirely determined and determinable on the basis of objective measurement alone, this removes the possibility of disagreement and argument concerning the particular manifestations of such being. For our knowledge of mathematically objectified being is derived from unambiguous, formal processes of inference and deduction, the same processes we already use in the domain of pure mathematics. The only issue for such a minimally objective projection of being is whether it can explain everything we observe in the objectively measurable universe without our having to add some other essential quality to our basic idea of materiality.

Now Galileo and Descartes were not fools. They did not think of our sensory experience of colours and sounds and tastes and odours and heat and cold as being unreal or nonexistent. Neither did they think that material being alone could explain this manifestation of sensory quality. Descartes, in particular, recognised that once we accept the reality of a world of extended bodies then we must also accept the reality of a world of subjective experience. It was not as if the world of material being had to explain *everything*. The task was to provide a coherent explanation of how these two worlds could be combined in the totality of a unified existence. To this end Descartes proposed the being of a second *unextended* thinking substance called *res cogitans*. This thinking substance encompassed all the dimensions of immediate experience, including feeling and sensing and perceiving, and not just the intellectual functions we usually associate with thought. Descartes not only proposed the being of this other dimension of cognition, he made it *primary*.

DESCARTES' MOMENT OF CONSCIOUSNESS

It is generally acknowledged that Descartes inaugurated the modern era of Western philosophy by introducing his *method of doubt* in the 1641 *Meditations on First Philosophy*.[9] Here, in the first meditation, he considers the possibility that his whole waking life could be a kind of dream, and consequently that his belief in an independently existing external world of extended bodies could be false:

> For example, there is the fact that I am here, seated by the fire, wearing a winter dressing gown, having this paper in my hands and other similar matters. How could I deny that these hands and this body are mine, unless perhaps I were to compare myself to certain persons, devoid of sense,

whose cerebella are so troubled and clouded by the violent vapours of melancholia, that they constantly assure us that they think they are kings when they are really quite poor, or that they are clothed in purple when they are really without covering, or who imagine that they have an earthenware head or are nothing but pumpkins or are made of glass. But they are mad, and I should not be any the less insane were I to follow examples so extravagant.

At the same time I must remember that I am a man, and that consequently I am in the habit of sleeping, and in my dreams representing to myself the same things or sometimes even less probable things, than do those who are insane in their waking moments. How often has it happened to me that in the night I dreamt that I found myself in this particular place, that I was dressed and seated near the fire, whilst in reality I was lying undressed in bed! At this moment it does indeed seem to me that it is with waking eyes that I am looking at this paper; that this head which I move is not asleep, that it is deliberately and of set purpose that I extend my hand and perceive it; what happens in sleep does not appear so clear nor so distinct as does all this. But in thinking this over I remind myself that on many occasions I have in sleep been deceived by similar illusions, and in dwelling carefully on this reflection I see so manifestly that there are no certain indications by which we may clearly distinguish wakefulness from sleep that I am lost in astonishment. And my astonishment is such that it is almost capable of persuading me that I now dream.

Figure 5.3: Frans Hals' portrait of René Descartes 1596-1650.

Descartes' purpose in these meditations is to arrive at a ground of certain (apodictic) knowledge. That means he must reject anything that he can even *conceive* of doubting. So, despite his feeling of certainty, he sees he must reject his belief in the reality of the external world and of his body as it appears in that world. Next he considers the domain of pure mathematical ideality:

> For whether I am awake or asleep, two and three together always form five, and the square can never have more than four sides, and it does not seem possible that truths so clear and apparent can be suspected of any falsity or uncertainty.

But Descartes is still not satisfied:

> For how do I know that God has not brought it to pass that there is no earth, no heaven, no extended body, no magnitude, no place, and that nevertheless I possess the perceptions of all these things and that they seem to me to exist just exactly as I now see them? And, besides, as I sometimes imagine that others deceive themselves in the things which they think they know best, how do I know that I am not deceived every time that I add two and three, or count the sides of a square, or judge of things yet simpler, if anything simpler can be imagined?

Just as if he were the madman, Descartes sees that he cannot even trust his reason. This leaves him in a difficult position. For now he *reasons* that there is indeed something that he cannot doubt:

> I myself, am I not at least something? But I have already denied that I had senses and body. Yet I hesitate, for what follows from that? Am I so dependent on body and senses that I cannot exist without these? But I was persuaded that there was nothing in all the world, that there was no heaven, no earth, that there were no minds, nor any bodies: was I not then likewise persuaded that I did not exist? Not at all; of a surety I myself did exist since I persuaded myself of something or merely because I thought of something. But there is some deceiver or other, very powerful and very cunning, who ever employs his ingenuity in deceiving me. Then without doubt I exist also if he deceives me, and let him deceive me as much as he will, he can never cause me to be nothing so long as I think that I am something. So that after having reflected well and carefully examined all things, we must come to the definite conclusion that this proposition: *I am, I exist,* is necessarily true each time that I pronounce it, or that I mentally conceive it.

This is Descartes' great discovery. It is not *cogito ergo sum* (I think therefore I am). It is: *I am, I exist.* But what does Descartes mean here? It *seems* so simple. He wants to say that however you deceive me, however confused my ideas and beliefs, it still has to be the case that *I am, I exist*. But who am I? And what does it mean to exist? Descartes has already recognised his reasoning cannot be trusted. And yet he appears to have *reasoned* that he *must* exist. And then he leaves it open who this 'I' is, and what it means to exist. He seems to think that you and I will be able to fill in those details, that what he is referring to is so primal, so basic, that we cannot go any further in explaining what is meant.

Let us be clear. This is not some arcane consideration of an obscure French philosopher who had a confused idea about finding something he could not doubt. Descartes was the first man to articulate in explicit terms the impulse that was to drive our modern Western scientific, technological civilisation. The *way* he articulated this impulse is fateful for the entire subsequent development of the Western intellect. Descartes may be dead, but his thinking is not. To this day he is a straw man who is blamed by many for the ills of our civilisation, for leading us off in a wrong direction. And yet the ideas he expressed were not personally his own. They gave form to a deeper transition that was occurring in Western European culture. We are the heirs of this transformation. Should it be that our path of scientific progress were unproblematically leading us into the enlightened utopia of universal reason that was envisaged in this transformation, then we would perhaps be forgiven for concentrating on where we are going rather than on where we have come from. But, as it stands, almost all the problems we are collectively facing together on Earth are a direct consequence of the unrestrained development of the very scientific, rational, technological civilisation that was born with the thinking of Galileo and Descartes and their contemporaries. They are our forefathers. They were foundational in the formation our world. Our world is in crisis. *Of course* we need to consider very carefully the ideas that we have inherited from them, even though it may not be immediately obvious how these ideas are determining our behaviour now. It is this work of making our inheritance conscious that we are engaged in here. For as long as the inheritance is unconscious, it will unconsciously determine our understanding. That is not to say we must uproot and destroy what we have inherited. The task is simply to reassess what was once thought in the light of what has happened during the last four hundred years. For us, today, this is a matter of *self-knowledge*.

Descartes' discovery of the indubitable certainty of immediate subjective consciousness (of '*I am, I exist*') represents a revolution in the perspective of Western European thinking. Even the meaning of the terms subjective and objective now become reversed. Previously the subject was what we now think of as the object, the externally existing thing, like the table being the subject of the sentence "The table is square". But now, with Descartes, it is human 'subjectivity' that becomes the ultimate subject, the ground of certainty on the basis of which our knowledge of the being of every other entity is to be justified. And yet, as we have seen, Descartes' reasoning concerning the absolute certainty of the 'I am, I exist' is *not* absolutely certain. He admits this explicitly in the third meditation:

> But every time that this preconceived opinion of the sovereign power of a God presents itself to my thought, I am constrained to confess that it is easy for Him, if He wishes it, to cause me to err, even in matters in which I believe myself to have the best evidence. And, on the other hand, always when I direct my attention to things which I believe myself to perceive very clearly, I am so persuaded of their truth that I let myself break out into words such as these: Let who will deceive me, He can never cause me to be nothing while I think that I am, or some day cause it to be true to say that I have never been, it being true now to say that I am, or that two and three make more or less than five, or any such thing in which I see a manifest contradiction. And, certainly, since I have no reason to believe that there is a God who is a deceiver, and as I have not yet satisfied myself that there is a God at all, the reason for doubt which depends on this opinion alone is *very slight*, and so to speak *metaphysical*. But in order to be able altogether to remove it, I must inquire whether there is a God as soon as the occasion presents itself; and if I find that there is a God, I must also inquire whether He may be a deceiver; for without a knowledge of these two truths I do not see that I can *ever be certain of anything.*

So where is Descartes' certainty now? He believes if he can demonstrate that God is not a deceiver, this will guarantee the validity of his reasoning. But when he attempts this demonstration, we find that Descartes again *reasons* that God exists and that He is not a deceiver. And so he becomes caught in an infinite regress. For if his reasoning that God is not a deceiver is to be certain, he would *first* need to know (for certain) that God is not a deceiver. And Descartes does *not* know this. Therefore he cannot trust his reasoning. Whichever way he turns, he remains in uncertainty. And, theologically speaking, this is only as it should be, for God is not a matter for reason, He is a matter of *faith.*

There are two paths we can take here. One is to conclude that Descartes was mistaken and can therefore be dismissed. For he was searching for the indubitable, for what cannot even conceivably be doubted. He thought he had found such a ground in his 'clear and distinct' knowledge that 'I am, I exist'. But he was wrong. His argument that our reasoning could be deranged, that we could be mistaken about anything and everything we think, still stands. And his attempt to reason God into existence to guarantee the validity of his reasoning simply failed.

But there is a second path. This involves *feeling* the force of Descartes' realisation, of the 'I am, I exist'. We can *hear* this feeling in Descartes' saying "Let who will deceive me, He can never cause me to be nothing while I think that I am." There is something here that causes Descartes to simply *ignore* the inadequacy of his reasoning about God. Of course we can only speculate about Descartes' unspoken motivations and influences. But we should remember that the Christian conception of God pervaded Descartes' world and was unthinkingly accepted in much the same way as we accept our scientific understanding of the universe today. In that pre-Darwinian era it seemed inconceivable that the earth and humanity and all the species could simply have arisen by themselves. There could only be one explanation: there *must* be a creator. This idea of a creator, in the form of the Christian God, provided Western European civilisation with an entire context of meaning and validity that we have now lost. We tend to think of this uniformity of belief as somehow imposed by the Church with threats of torture and execution upon people who otherwise would have thought much as we do. In that narrative, Galileo appears as a hero who was placed under house arrest by the Catholic authorities and, under threat of the rack, forced to recant his support of the theory that the earth revolves around the sun. But it was not *God* that was in question here. It was the authority of the Church.

So, while it would have been dangerous for Descartes to assert that the existence of God was in some way doubtful, that does not mean he was insincere in his certainty that there must be a God. And given the danger and the pre-existing collective certainty, it becomes understandable how he (and many others) projected this certainty into their reasoning, so that the existence of a benevolent God appeared as a kind of logical necessity. Now that this collective certainty of belief in a creator God has receded, Descartes' reasoning is easily dismissed. But that does not mean he was an unintelligent fool. For what of *our* collective belief in material science— in the determinative power of mindless matter—the very mindless matter that Descartes first conceptualised as *res extensa*? It is only by degrees that we can unravel the layers of preconception that structure our understanding. Descartes could not achieve everything, all at once. It is the way he began that matters, that we have inherited. This beginning was to place the realisation of 'I am, I exist' at the centre of philosophical inquiry.

I too had a Descartes moment one lunchtime in the front bar of the Black Horse public house in Lewes. I was 16 years old, playing truant from school, sitting with a pint of lager and lime, when it dawned on me that the entire scene in the pub, my being there, was actually happening in me, in my consciousness. I realised that only I was having this experience and that everyone else was having their own different experience. Then I realised I could not be sure that anyone else was even having an experience. Perhaps, I thought, I only *think* these others are conscious like me. Perhaps I am the only one who is really conscious here at all. This idea struck me forcibly. Whether or not the others are conscious, it became clear to me that I am alone in a most peculiar, obvious and immediate way: for no one else sees what I see in just the way I see it. I am the solitary centre of this field of conscious experience. Absolutely solitary. All I know for sure is that this field of conscious experience *exists*, that it *is* what it *is*.

Descartes too, in the sincerity of his doubt, will have encountered this field of immediate experience. It is here he discovers the absolutely indubitable ground of his own existence, the that-which-cannot-be-denied immediacy of consciousness now. He even expresses it during his second meditation, almost in passing, in his recognition that the *seemingness* of this immediacy cannot be false:

> Finally, I am the same who feels, that is to say, who perceives certain things, as by the organs of sense, since *in truth* I see light, I hear noise, I feel heat. But it will be said that these phenomena are false and that I am dreaming. Let it be so; still it is at least quite *certain* that it *seems to me* that I see light, that I hear noise and that I feel heat. That *cannot be*

false; properly speaking it is what is in me called feeling; and used in this precise sense that is no other thing than thinking.

What Descartes misses, what he fails to distinguish, is that there are *two* dimensions to his experience here: firstly there is the immediacy of consciousness now and secondly there is his *reflection* on this immediacy. As soon as Descartes reflects, he separates from the immediacy and introduces the possibility of error. For the immediacy of being conscious *is what it is*. It makes no claim to be true or false. It is the pre-given ground of all reflection and assertion. As soon as 'I' assert something about this immediacy, then I introduce another entity, an assertion-meaning that may or may not correspond with the immediacy itself. As Descartes saw, there is always the *possibility* of a non-correspondence. For we can be mistaken in our use of language. But in the immediacy of consciousness itself there is no gap of reflection within which error can occur. So, even if I am making a false assertion about the being of consciousness, it is still the case in the immediacy of consciousness that *that is what is occurring*. This is not some logical quibble. In each moment of consciousness, consciousness presents exactly what it presents, and in that presenting it is conscious of what is being presented. As soon as I try and say or think what it is that is being presented, then my saying or thinking becomes a part of what is being presented. Such saying or thinking can be true or false. But immediate consciousness is *beyond* being true or false. It is what *precedes* something being true or false and it is the *ground* upon which truth and falsity are decided.

Descartes conflates the indubitability of immediate consciousness with the indubitability of assertions *about* such consciousness. And so he says it "cannot be false" that "it seems to me that I see light". But of course, his *statement* "it seems to me that I see light" *could* be false. Perhaps he had his eyes shut and was listening to music, and *meant* to say "it seems to me that I hear music". And yet what Descartes is trying to indicate in saying "it seems to me" is the very indubitability of immediate consciousness. He *sees* that such immediate consciousness cannot be doubted, but does not recognise how falsity and doubt *necessarily* arise as soon as we make assertions about such consciousness.

So where does this leave us? Clearly, Descartes did not uncover the ground of absolutely indubitable logical assertions he was seeking. Instead he encountered the ground of absolutely indubitable immediate consciousness. He was unable to see this distinction because he remained identified with the 'I' of reflection. Even his assertion 'I am, I exist' misses what is essential. For this 'I' is something projected out of consciousness, into consciousness, that only emerges as a *separation* from the indubitable immediacy of consciousness. It would be better to say 'consciousness is'. For that is what is revealed in direct knowledge of consciousness. It is *this* certainty that inspired Descartes. Despite his reason telling him that God could be deceiving him, Descartes *knew* he had found something he could not doubt.

And yet, aren't we, like Descartes, claiming that our assertions about direct knowledge of consciousness are beyond doubt? No. *Any* assertion that claims to say something true has the possibility of being false. This includes our statements about looking and seeing and our immediate pre-reflective experience of an assertion corresponding with how things present themselves in consciousness now. Our language, our concepts, can always miss. The certainty of direct knowledge of consciousness is not a correspondence between an assertion and a state of affairs. It is consciousness knowing itself as itself in being conscious now. This knowing is not the *indirect* knowing of the truth of a proposition. It is an immediately present *direct* knowledge, that becomes present just as soon as consciousness remembers *itself*.

What direct knowledge of consciousness teaches is that there is an immediate ground of indubitable experience. This is the phenomenological field where everything appears, in just the way it appears. This field encompasses all our opinions and beliefs and insights concerning the reality of what we experience as appearing. If we distort the appearance of something or somebody according to some unconscious preconception, then it is exactly that distorted entity that appears in the field. We may misdescribe this field to ourselves and others, but our misdescription will also appear within the field. The field itself cannot be true or false.

Descartes already had this field in view long before Husserl made it explicit. But Descartes fatefully let his insight slip away. Instead of staying with what is immediately and indubitably experienced, the *actual phenomenological field*, that *cannot be false*, he appropriates this absence of falsity to himself, to his reflective, thinking self. And so he takes the indubitable certainty of the being of the field of consciousness to be the indu-

[handwritten margin notes: * 1st person / But / MYU = / 2nd person ????! No – MYU – knowing with another, ?]

[handwritten notes at bottom:]
* Consciousness remembers myself, not itself = 1st person.
" converses with another, and learns by diversity GROWS

bitable certainty of the being of his thinking self. And then he projects this certainty into the proposition: *I* am, *I* exist. Now if this 'I' were consciousness itself, the impersonal consciousness that both presents and *is* what it presents, then Descartes' assertion would remain true to his deeper insight. But Descartes did not dwell in this state of unified consciousness. As soon as he began to reflect, he transformed what cannot be doubted into the doubtful consciousness of his own existence. For the being of this 'I' of reflection *is* doubtful. It only arises when impersonal consciousness identifies itself with the personal 'I' of reflection. In this identification, the personal 'I' *inherits* the being of consciousness for as long as consciousness remains in this divided state of forgetfulness. Now we have two forms of truth. Insofar as consciousness is divided and identified with a personal, reflective, thinking self who surveys an independently objective world then it is *true-for-this-self* that I exist *as* this self. But from the encompassing place of undivided consciousness, it only *appears* that this self exists. It is an *epiphenomenon* that *dis*-appears as soon as consciousness returns to itself. The *true* existent, which endures even in the forgetfulness of identification, is consciousness itself.

Iconic connection, not self-existence.

THE SPIRIT OF WESTERN SCIENCE

What is significant in Descartes' meditations is not how they ended but how they began. This beginning was the breaking through of a radical spirit of inquiry that was to manifest as Western science. The impulse of this spirit was to free the Western intellect from its subjugation to the authority of the Church and the received wisdom of tradition. The radical idea was that each one of us possesses the capacity to discern the true from the false according to the *natural light of intelligence* that is accessible to all who put the truth before all other considerations. The radicality emerges when I refuse to accept what another claims to be true or false without first demonstrating that truth or falsity in the immediate self-evidence of that natural light. As a consequence, this places all received authority, both divine and secular, into question.

What Descartes was saying, by means of demonstration, was that you can set aside all the books of learning, the scholasticism and respect for the Greek philosophers, and all the accredited revelations of Church and scripture, and begin again, according to your own resources, to discover what is true and what is false. His meditations are invaluable for showing the way of proceeding, not for the conclusions they drew. For in the end Descartes did not disentangle his thinking from the overarching authority of tradition and Church. But that is not the point. The point is not to *believe* Descartes and accept what he took to be the indubitable truth. The point, the *real question*, concerns this power of the natural light to reveal what is true.

And yet Descartes' false conclusions *still* matter. For they set up the tangle of radicality and truth and error that became the objective scientific materialism we are living with today. We must therefore trace his thought a little further. Upon concluding that God is no deceiver, Descartes' radical doubt collapsed back into an acceptance that the world really does exist in the way he formerly believed it to exist, i.e. as bodies extended in space that correspond with his sensory experience. But it is the *manner* of this correspondence that was to prove so fateful for the development of Western science. For Descartes, in the fifth meditation, thought that his clear and distinct conception of the idealised mathematical form of the external world implied that the world itself has an essentially *mathematical* form of existence:

> There is certainly further in me a certain passive faculty of perception, that is, of receiving and recognising the ideas of sensible things, [and] another active faculty capable of forming and producing these ideas. But this active faculty cannot exist in me [inasmuch as I am a thing that thinks] seeing that it does not presuppose thought, and also that those ideas are often produced in me without my contributing in any way to the same, and often even against my will; it is thus *necessarily* the case that the faculty resides in some *substance* different from me in which all the reality which is objectively in the ideas that are produced by this faculty is formally or eminently contained. And this substance is either a body, that is, a corporeal nature in which there is contained formally [and really] all that which is objectively [and by representation] in those ideas, or it is God Himself, or some other creature more noble than [the] body in which that same is contained eminently. But, *since God is no deceiver*, it is very manifest that He does not communicate to me these ideas immediately and by Himself, nor yet by the intervention of some creature in which their reality is not formally, but only eminently, contained. For since He has given me no faculty to recognise that this is the case, but, on the other hand, a *very great inclination to believe* [that they are sent

to me or] that they are conveyed to me by corporeal objects, I do not see how He could be defended from the accusation of deceit if these ideas were produced by causes other than corporeal objects. *Hence we must allow that corporeal things exist.* However, they are perhaps not exactly what we perceive by the senses, since this comprehension by the senses is in many instances very obscure and confused; but *we must at least admit that all things which I conceive in them clearly and distinctly, that is to say, all things which, speaking generally, are comprehended in the object of pure mathematics, are truly to be recognised as external objects.*

Here we meet Descartes on the *other side* of his radical doubt. Instead of remaining with his direct experience of the being of consciousness and exploring how experience of the external world emerges from this primal ground, he falls back into the place from which he started. As God is no deceiver, He guarantees whatever Descartes perceives with clarity and distinctness *must* be true. And so Descartes reasons that corporeal (spatially extended) things must exist independently of his immediate sensory experience of such things. Why? Because he has a *"very great inclination to believe"* this and because he *cannot see* how God *"could be defended from the accusation of deceit if these ideas were produced by causes other than corporeal objects."* Descartes sees this is only his belief, and in not seeing how God could be defended, he admits He *could* be defended. Hence there is an *equivocation* in his assertion that "we must *allow* that corporeal things exist." And there is a further equivocation in his saying that corporeal things "are *perhaps* not exactly what we perceive by the senses." For perhaps they *are* exactly what we perceive by the senses.

These equivocations are important. For we stand on the threshold of modern science. We are about to accept that an entirely mathematical account of the being of the things we perceive is not just our scientific way of looking at things, but it is the way things *actually* are. For Descartes has led us to accept that the things we perceive really do exist independently of our perceiving, in another extended spatial dimension, instantiated in another extended substance. Now he wants to take back the primacy of our perceptual apprehension of these things. Following Galileo, having carved reality into two independently existing substances, he wants to distinguish what it is in sensory perception that belongs to the corporeal things themselves and what it is that belongs to the senses. *This is where the knife falls*, where our pre-existing, indubitably present, harmoniously unified experience of being in the world becomes a kind of semi-illusory representation of another independently real domain of physical being. The blade of the knife is the distinction between what can be considered *subjective* and what can be considered *objective*. An objective thing is what remains once we have abstracted away all the sensory qualities that present the thing to us in our subjective conscious experience. So, in objectivity there is no colour, no sound, no taste, no odour, no feeling, no touch. For us, that means there is nothing left to experience. But in this other domain, something does remain. It is pure extension in space and time. This extension is still accessible in our thinking because it is objectively *measurable*. It is this measurability that creates a *mathematical bridge* between the two substances. And so we again meet with the purely mathematical idea of that which remains the same in the world for all possible observers. This is exactly what Descartes encounters in his clear and distinct perceptions of corporeal objects:

> And in regard to the ideas of corporeal objects, [. . .] I find that there is very little in them which I perceive clearly and distinctly. Magnitude or extension in length, breadth, or depth, I do so perceive; also figure which results from a termination of this extension, the situation which bodies of different figure preserve in relation to one another, and movement or change of situation; to which we may also add substance, duration and number. As to other things such as light, colours, sounds, scents, tastes, heat, cold and the other tactile qualities, they are thought by me with so much obscurity and confusion that I do not even know if they are true or false, i.e. whether the ideas which I form of these qualities are actually the ideas of real objects or not [or whether they only represent chimeras which cannot exist in fact].

Notice here that Descartes has a quite *unintuitive* idea of what it means to perceive with clarity and distinctness. For, in my immediate experience, I would naturally say that I clearly and distinctly perceive my green book to be green. And I would not think my experience of colour was in some way false. As Descartes already recognised, it cannot be false that it *seems to me* that the book is green. But Descartes has left this world of actual sensory perception behind. His perception is an intellectual beholding of what he *conceives* something to be *independently* of its being sensuously perceived. So when he says he cannot *perceive* colours clearly and distinctly, he means he cannot *think* in what way a corporeal object could *actually* be coloured.

This is because he has *already* adopted a theory of perception that coincides with his idea of the existence of an extended colourless substance.

In this way, despite his equivocations, Descartes enshrines Galileo's mathematical abstraction of a world of extended bodies in an absolute ontological distinction—a distinction ordained by God that says: yes, there really are physical bodies out there, extended in physical space, that, in themselves, have no colour, and make no sound—and what we think of colour and sound are simply the vibrations of tiny physical particles impinging on our equally physical bodies. It is important to see the arbitrary nature of this model. For we are still inclined to take the abstraction of materiality to be an absolute and self-evident reality, even though we can see there is no certainty in Descartes' reasoning. What *is* certain is my immediate experience, just as it presents itself, just as it 'seems to me'. This is precisely the world of colours and sounds and scents and tastes that Descartes now thinks of as obscure and confusing. But this confusion only arises when we split the original unity of perceptual apprehension. In this splitting we can no longer attribute colours to the corporeal objects we see, even though we actually see them as coloured. No wonder we are confused. For I could equally argue that I clearly and distinctly perceive my green book to be green and that the scientific assertion it has no colour cannot be true, for that would make God a deceiver. Like Berkeley,[10] I could maintain that all the things I perceive are ideas in God's mind and that they really do exist in just the way I perceive them.

But Descartes does not trust his senses. He trusts his mathematical reasoning. He knows already of the certainty of logical deduction and of mathematical procedures of proof. His search is for the basic axioms on which he can ground a rational-mathematical deductive understanding of the universe. And yet, almost inadvertently, what he *actually* discovers is the certain ground of his immediate consciousness. For a moment he is transfixed, as he stands outside the domain of his former life. But only for a moment. The ingrained apparent certainty of his belief in God and in the independent existence of a mathematically objective universe is soon reasserted.

This is the paradox of Descartes and of mathematical science itself. He discovers the being of the ground of immediate experience and the self-evident certainty of perceptions that remain true to this ground. But then he takes what he has realised and uses it to justify a position that essentially denies the reality of the very ground he has discovered. And this is our inheritance. For what he intuits in the idea of clear and distinct perception still lies at the centre of the impulse of Western science. This is our demand for genuine and compelling evidence, for not taking something on faith alone. This is our empiricism, that says, I shall not believe you until you demonstrate what you say, before my very eyes, or in my immediate experience, or according to an absolutely compelling chain of reasoning. It is not that Descartes surrendered the intention of his radical doubt. It is rather that he did not distinguish between the genuine self-evidence of immediate consciousness (what we have called 'looking and seeing') and the *apparent* self-evident certainty of his strongly held convictions (that God is no deceiver, that external reality comprises bodies extended in space). And here too we are Descartes' inheritors. For while we may question the existence of a benevolent creator God and we may think we have dispensed with his dualistic philosophy, we are still, collectively, facing the problem he bequeathed us: For how are we to explain the indubitable fact of our being conscious in the face of the evidence of science that we live in an entirely physically-determined universe?

NOTES

1. Alan Turing (1912-1954) was a British mathematician, code-breaker and computer scientist who worked extensively on the theoretical and practical development of computer technology in the twentieth century. He developed the abstract idea of a Universal Turing Machine in his paper *On Computable Numbers* (1936-7) as part of a proof to show that there are mathematically undecidable problems.
2. Gottfried Leibniz (1646-1716) was a German philosopher and polymath. He independently developed the mathematics of calculus, designed mechanical calculating machines, and foresaw the development of modern binary computer logic. In addition, he was a highly influential philosopher (see *Leibniz Selections* (1951) for more details).
3. For instance, in the British Columbian Tsimshian language, there are seven distinct sets of words for numbers (Dunn, 1995). Six of these are used to refer to particular classes of objects, such as canoes, or people, or long objects and trees, while the seventh is used when no particular object is intended. This seventh set corresponds to our modern idea of a number, while the other sets appear to be the relics of an earlier conception of number.

4. Andrew Cohen is an American spiritual teacher. I went to one of his meetings in the mid-1980s where he used the metaphor of individual thoughts being cherries and said that handling thought was a matter of not eating the bad cherries.

5. Arthur Schopenhauer (1788-1860) was a German philosopher famous for his pessimism. It was he who recognised the link between the transcendental idealism of Kant and Eastern religions, particularly in the teachings of the Buddha and the *Upanishads* (Mascaró, 1965). However, unlike Kant, Schopenhauer considered that we do have knowledge of what lies behind the objective appearance of the world in our immediate experience of *willing*. In *The World as Will and Representation* (1859/1966) he presented his philosophical understanding of *will* as the transcendental foundation of our experience of an objective spatiotemporal world. His ideas went on to influence Friedrich Nietzsche, who reversed Schopenhauer's conception of will as something to be negated, and the psychologist Carl Jung, who even recommended reading Schopenhauer as a means of staving off a psychosis in one his patients (see (Jung, 1960, p. 260)).

6. Henri Poincaré (1854-1912) was a famous French mathematician, who described the insight in his book *The Foundations of Science* (1913/2014, pp. 387-388) as follows: "At the moment when I put my foot on the step the idea came to me, without anything in my former thought seeming to have paved the way for it, that the transformations I had used to define the Fuchsian functions were identical with those of non-Euclidean geometry. I did not verify the idea; I should not have had time, as, upon taking my seat in the omnibus, I went on with a conversation already commenced, but I felt a perfect certainty. On my return to Caen, for conscience sake I verified the result at my leisure."

7. Immanuel Kant (1724-1804) was one of the great philosophers of the modern era. His most well-known work is the *The Critique of Pure Reason* (1781/1998) which we consider in more detail in Chapter 6.

8. Galileo Galilei (1564-1642) is widely recognised as the 'founding father' of modern mathematical science. His book *The Assayer* (1623/1957) was written as a manifesto in defence of the new science and attacked the views of a then prominent Jesuit, Father Grassi. It is thought the offence this caused contributed to Galileo's eventual arrest for heresy by the Catholic Church, as a result of which he spent the final part of his life under continual house arrest.

9. *Meditations on First Philosophy* (1641/1911) is Descartes' most famous work. It consists of a series of six meditations where he first doubts all that he conceives can be doubted and then attempts to rebuild his philosophy on a foundation of absolute certainty.

10. George Berkeley (1685-1753) was an Irish empiricist philosopher. In his major work, *A Treatise Concerning the Principles of Human Knowledge* (1710/1982), he put forward the idea that there is no separately existent world of physical extended matter beyond our experience of perceiving such a world. This philosophical position later became known as *subjective idealism*.

CHAPTER 6

DESCARTES' BEQUEST

THE NATURAL LIGHT

IN asserting the independent existence of the extended substance of the external world, in keeping with the material-mathematical projection of Galilean science, Descartes was left with a *remainder*. To begin with, in the extremity of his doubt, this remainder was the primary reality, that which could not be doubted, the 'I am, I exist'. In that place he was in contact with what Husserl would call *transcendental subjectivity*, the pure consciousness that is the source and content of all we know and all we ever can know. But as soon as Descartes decreed the independent existence of *res extensa*, he created the problem of the relationship between this material substantiality and the immaterial consciousness that had realised its immunity from doubt. Descartes' solution followed what seemed obvious to him: that his being conscious, the being of his thinking self, was directly connected with the extended substance of his material body. Now, the original, impersonal consciousness, that in its absolute aloneness, not only doubted the being of the physical world, but also the being of every other consciousness, became reinstated as an entity *within* that world, a being amongst other beings, inextricably bound up with the physical events occurring in a mortal, physical body. To justify this, Descartes could call on what seemed self-evident to all of Christian Europe: the concept of an immaterial, personal soul. *This* was his thinking substance.

And the question as to how this immaterial soul could interact with a material body also had a self-evident answer: God. Because, for God, *everything* is possible. So if He wants to create two independently existing substances and so arrange it that what occurs in the immaterial substance (my mind) is able to affect what happens in the material substance (my body), and vice versa, then this is easily done—for God is the omnipotent power that encompasses both the material and the immaterial, who can simply *decree* how they are to interact.

In actuality, Descartes was not content to fully hand this problem over to God. He theorised that the mind is both affected by and able to direct 'animal spirits' that are connected with the pineal gland in the brain. He thought these spirits could move along pathways in the nervous system and so act as channels of communication between mind and body. Of course, more detailed investigations of the anatomy of the brain have shown this theory to be false. But that does not falsify the underlying substance dualism, it only means that God must have arranged things differently. And according to the spirit of the new rational mathematical science, the way in which God has arranged things is something we can eventually discover, *if* we employ the right methods. For God has endowed Man with a *natural light*, and by means of this light, it is God's pleasure that Man should come to know the contents of God's Mind. For God is no deceiver. His intent is that we heed the natural light, and only accept as true that which we can clearly and distinctly perceive. This is a founding principle of our modern era. It becomes the destiny of European civilisation to use this light to develop a universal science of reason; a science that encompasses all possible domains of knowledge, not just knowledge of the material world, but of ourselves, our psychology, of God, of our purpose here on Earth; a science where all the different branches form a coherent and meaningful, rational whole, so that we too can become as God, and see the universe as it *really is.* This is the Age of Reason, the Age of Enlightenment.

I wonder if you can see the broken remnants of this vision in the situation we now inhabit. For things have not turned out as planned. These paragons of reason, the European nations, set about destroying themselves in the First World War, and then, after a short break, finished the task in the Second World War. Now we stand in the ruins of that former civilisation, caught up in the globalised world of corporate capitalism. We still have our objective sciences, and continue to make remarkable technological advances. But the original inspiration has gone. It died somewhere on the battlefields of Northern France. I wonder if you can see this. It is not just a matter of understanding the external historical events, the exercise of military, economic and political power. These events are a consequence of a certain path of understanding we have taken, an impulse that determined the unfolding of our Western European scientific civilisation. This is *our* path, it lives on in us, in the way we think and act, and in our background assumptions. Our task here is to reveal this background, to make it *conscious*.

We have gone into Descartes in such detail because he set the scene. On one side of the stage we have an entirely physical dimension, existing in its own right, that would continue to exist with or without the interference of our human consciousness. In this physical dimension are to be found human physical bodies and the physical bodies of all other living entities. Descartes already explicitly understood these bodies to be sophisticated machines whose behaviour is entirely determined by physical cause and effect—with the exception of one special machine: the human brain. Here he supposed there to be a site of interaction with the other side of the stage: the unextended substance of the human mind. No other living entity possesses such a mind because no other entity has an immortal soul. In the background is God, who created these substances in order to enact the drama of human free will and salvation on Earth. For only if I am able to control the behaviour of my body-machine can there be free will or moral responsibility or salvation or anything like a (rational) Christian religion. What made Descartes' dualism significant was his entirely mathematical and objective idea of physical being and the way he unified this idea with a Christian conception of the human soul. This division enabled the fledgling mathematical natural science to prosper without being suppressed by the Church authorities. But Descartes' dualism did not arrive from nowhere. It was grafted on the foundational Christian dualism of body and soul.

In this deeper dualism the material world had *already* been separated from the spiritual world. We could marvel at the miracle of creation, but we were not to think of this material world as divine in itself. For centuries the Christian church had dedicated itself to rooting out and destroying such primitive, pagan, animistic superstitions. For there is only one God and He stands *outside* His creation. It is He whom we must worship, not the nature spirits, the sprites and faeries, the devils and goblins; these were to be *banished* from our daylight world. And banished they were, with extreme brutality. It was into *this* world that Galileo and Descartes introduced their idea of a pure mathematical materiality, a world in which material nature had already been desacralised.

This same process of desacralisation had placed all the power and potency that had once been experienced as residing within nature back into the dimension of human subjectivity. Not that God was understood as a creation of the human mind (not yet). It was that human subjectivity, the human soul, became a *portal* to another world of the spirit, to God, and Jesus, and the Holy Spirit. The Kingdom of Heaven was within, not out there somewhere in the material world. And so the ultimate significance of human subjectivity was also already presaged in medieval Christianity. For the whole drama of life on Earth was focussed on the moral conduct and salvation of these individual souls and their final unification with God in the immaterial world of the spirit.

Whether you believe in Christianity or not, this is our heritage, the centuries-old ground of the deep division between our subjective experience of being a conscious individual and our objective experience of living in an independently existing material world. Only such a civilisation could have turned the entire earth into a collection of resources to be traded in a marketplace. We can do this because we think that everything we trade is essentially the same: material stuff distributed in space and time. We may value this stuff differently according to how rare it is, or how much we need it to prolong our way of living, or how aesthetically pleasing we find it, but essentially, it is still just physical stuff, and, as such, it can be given monetary value and traded.

That is not to say how wrong we have been, and how we should go back to our pagan roots. I think a short trip back to the reality of that pagan world may cause us all to feel more fondly about our Christian heritage. For it is not all inquisition, capitalism and exploitation. There is also the Gospel of Love. The issue is not to pass sweeping moral judgments but to see that this experience of the independent physical materiality of the world is a *cultural acquisition*. It is not an absolute truth. It is a way of understanding the world that has been passed on to us through the generations over thousands of years. For even our Christian conceptions of materiality are founded on distinctions drawn by the ancient Greeks. Our task here is not to reject this idea of materiality, but to *understand* it, to see how it works in us now, in the background, how it shapes our idea of the world, and how it leads us into conclusions and courses of action that we may wish to reconsider.

We start with Descartes because he is still *within reach*. He still speaks to us, as he stands beside his fire doubting all that can be doubted. He is giving birth to our particular world-view, the one that replaces medieval Christianity. It is this world-view that lives on in our scientific material-ism. In Descartes it is still entangled with medieval Christianity. But for us, this all-encompassing Christian world-view has receded—just as Ni-etzsche[1] prophesied. We no longer live in the collective security of a faith in a benevolent creator God. Now our faith, if we have one, is something that sets us apart from the prevailing collective agnosticism. And yet it was our Christian faith that *launched* the scientific revolution. The Age of Reason was not to be an age of atheism. It was to be an age of *Christian* enlightenment. And the new mathematical universal natural science was not just concerned with mechanics and the discovery of structures and universal physical laws that govern the behaviour of a universal material substance. It was to encompass the domain of all possible knowledge of everything that exists. This was encapsulated in Leibniz's *Principle of Rea-son* which says that *nothing exists without reason*. And if nothing exists without reason, that means everything is within the reach of reason, that everything is intelligible and understandable in terms of reason.

This Principle of Reason was not something deduced from another first principle. It was a statement of *faith*. And this faith still had its source in the idea of a *divine* reason. Without such faith, there could be no cer-tainty of truth, no certainty of correspondence between what we reason to exist and what actually does exist. For Descartes this faith was his faith in the *natural light*, in that power which reveals, by means of clear and distinct perceptions, what is true and what is false. Only if the natural light *is itself true*, can there be any possibility of coming to a true knowl-edge of anything else. As we saw, Descartes thought that the natural light contains within itself its own warrant of certainty, and that he could use that warrant to prove the existence of a benign and perfect God. Here he mistook what we would call the *phenomenon* of certainty with *actual* certainty. For nothing can be *asserted* with absolute certainty. There is *always* a gap between the assertion and that which it asserts where error can arise. And yet there is a knowing of being, a direct knowledge of consciousness, that is immune from doubt, because it is a knowing that precedes the indirect knowledge of a reflective assertion. So there *is* a natural light, but it is not the natural light of reflective reason, it is the natural light of pre-reflective consciousness. And it is not that I am making absolutely certain assertions about this pre-reflective consciousness. These sentences are not assertions of an absolute truth, they are an invitation for you to check whether or not you are immediately conscious of being conscious *now*.

OK (you may say) but what about all the other assertions being made here? Are we supposed to treat them as if they are not meant to be true? That they are meant to be *possibly* true but we can never know one way or another? Here we are back with Descartes and the phenomenon of certainty. For there is another degree of certainty, the certainty of my *looking* and *seeing* (Descartes' clear and distinct perceptions). This is where I directly perceive that a particular meaning intention is fulfilled in a particular experience, such that the content of the experience is in agreement with the intention. In such a situation, such as my seeing a cat in the garden, I am *certain* there is such an agreement. This is a *phenomenon* of certainty. My reflective recognition of this agreement is simply an explicit expression of an implicit intention—a perceptual expectation—that had *already* been projected into the stream of experience. It is only on the basis of this projection that the cat was able to show up *as* a cat in the first place. For if I were a fish from the deep ocean that had suddenly been landed in the garden, I would certainly not be seeing a cat. My certainty is the certainty of all my previous perceptual experiences of seeing and distinguishing cats and gardens. And this certainty is itself *projected* into the cat experience, such that I experience the cat as being an *actual* cat (rather than a probable cat or an illusory cat).

I continuously live in this phenomenality of the certainty of the actuality of the things that I perceive. It could be that I dream and that these things are not really out there in front of me. But whether I dream or not, I still experience the actuality of these things and of the world in which they appear. We could say that this perceptual world possesses its *own form of certainty*, a certainty that precedes any reflection on my part and that remains in place even if I withdraw my identification with the objective being of this world. In such a withdrawal, what was previously experienced as *absolutely* certain, becomes a *phenomenon* of certainty. It is my *identification* that created this absolute certainty, not the objective reality of the world I perceive.

It is the natural light that directly informs me that the meaning intention I am projecting is in agreement with the unthinkable source of my experience. This is my connection with the transcendent, with that which lies beyond the projection of a meaning intention. I do not perceive this transcendent source, instead I experience the *transcendent agreement of my perception* with this source. It is the natural light that effects this agreement and thereby grants actuality (true being) to that which I perceive. In granting such being, the natural light opens up and illuminates a world-space in which the things of the world can appear.

Insofar as I live as a worldling within this world then I do not doubt the true being of the beings that I encounter—and so I live in the *certainty of the world*. It is only when I withdraw identification with my world-centred intentionality that this experience of certainty can emerge as a phenomenon of certainty. As such, it does not become uncertain. Instead it undergoes a modification whereby its certainty only has validity within the context of worldly experience. As soon as we step out of that context, we see that our certainty is not absolute.

Descartes, in the extremity of his doubt, also saw that his certainty of the world was not an absolute certainty, and that the only absolute certainty is the certainty of immediate consciousness. The difference for us is that we have to *remain* in this place where the only absolute certainty is the certainty of immediate consciousness (now). We still have what Descartes called the natural light. But what the natural light reveals concerning the actuality of the being of an external world is *not* absolutely certain. It is like an oracle that connects us with a transcendent source. It says "Yes, what you intend here does agree with what lies beyond your intending and so I grant what you intend the magical status of being actual." But Descartes was right. Sometimes we are deceived and what we experienced at one moment as being actual, retrospectively turns into an illusion. It could be that the natural light deceives us like this all the time, and that one day we may 'wake up' and realise our entire life on Earth was a kind of dream—which is exactly the teaching of certain Eastern religious traditions.

So what does this mean for our inquiry? If we do not have absolute certainty in what we are asserting, then how can we continue? The answer is that we must *surrender* this requirement of absolute certainty. That does not mean we fall into an equally absolute relativity where nothing is certain. It means we come to accept the situation in which we find ourselves, as far as we are able to understand it. That situation is again previewed in Descartes' Meditations. For it certainly *appears* that we have a natural light that connects us to a transcendental source of experience. And insofar as we remain impersonally open and receptive to this source, the natural light grants actuality to those meaning intentions that best agree with the source. These are our 'clear and distinct perceptions'. Our path here is to learn how to come into a closer agreement with the source. This has to do with the *purity of our witnessing*.

But what of the possibility that the natural light deceives us? Here we must be pragmatic. If the natural light systematically deceives us then *all is lost.* We dream a dream of nonsense where even our idea of truth has no truth in it. I wonder if you can conceive the abyss of meaninglessness that opens up here. Even the most sceptical do not actually live as if the light were untrue. For we *already* have a natural faith in the natural light. It is there when we cross the street. The difference for us is that we are bringing this faith to reflective consciousness. As disturbing as it may be to our modern sensibilities, to our sense of autonomy, we are utterly dependent on this natural light and its integrity. It is not something personal, that we each possess. It is what connects each one of us to the unity of a common source. What is required for us to continue our inquiry is that we freely and consciously place our faith in this natural light and what it reveals. That is essentially what it means to philosophise. Without such faith, there can be no meaningful inquiry. Do you see how close we are to Descartes, and to this beginning of Western science? For we too are calling on the essential integrity of a transcendental power to underwrite the truth of our clear and distinct perceptions. We no longer think of this transcendental power as a creator god. He has been removed into the domain of human subjectivity, into our collective unconscious. But it is still the case, if we are to have any true knowledge whatsoever, then we must rely on the integrity of this transcendental power to reveal how it stands in the source that lies beyond the circumference of our finite intentionality. And, unlike Descartes, we can have no illusion of proof of

the integrity of this power. It simply does not appear to our worldly consciousness of worldly things. It is that which reveals experience to be the
experience that it is. We cannot meaningfully doubt this light because it
already manifests our doubt to be the experience that it is. If the light
were untrue then doubt is unintelligible, and so fails even to doubt, and
our experience disintegrates into incoherence.

Brilliant - iconic truth.

THE ARGUMENT FROM SUCCESS

The essential innovation of the scientific revolution was the one-pointed
application of mathematical reasoning to the understanding of human
experience. It was this mathematical understanding that was to transform our entire way of living here on Earth. Western European culture
did not come to dominate the planet on the basis of superior philosophical insight or argument. It triumphed because of the discovery of previously hidden structure and regularity in the objectified world of human
experience. Knowledge of these regularities allowed the development of
new technologies and these technologies were used to assimilate the rest
of humanity into the common world system that has evolved into global
capitalism.

The collective idea we now have of the materiality of the universe is
not something we are explicitly taught in school. It is simply assumed
to be the case. When I learnt of physics, chemistry and biology, it was
never explained to me that we were dealing with an objectified view of
the universe. Instead it was taken as an obvious truth that the universe
exists in just the way our science tells us it exists—a truth so obvious it
was not even mentioned.

Behind this obviousness lies a form of argument that is much more persuasive than logical reasoning. It is the argument from *success*. It says that
the Western scientific projection of the materiality of the universe must
be correct because it *works*. One key term in this unstated argument is the
harnessing of electrical power. Another is the invention of the atomic
bomb. Another is the wireless transmission of electromagnetic signals.
Another is the computer. Another is the landing of human beings on the
moon. We live in a world that only four hundred years ago would be considered *miraculous*. Before Galileo and Newton, our understanding of the
solar system and the principles on which it operates was no better than

Figure 6.1: Apollo 11 Moon landing (courtesy of NASA).

a kind of guesswork. People thought the universe, including the planets and the sun, revolved around the earth, and that the planets were suspended on celestial spheres (otherwise, *of course*, they would fall onto the earth). We had *no idea* of the existence of basic elements, such as carbon, oxygen and hydrogen. Some suspected there were atoms, but this was just a form of speculation. The received wisdom was that there were only four basic elements: earth, air, fire and water, with perhaps a fifth constituting the celestial spheres. Such thinking was no further advanced than the civilisations that preceded us. And then came Newton, and the universal law of gravitation, which laid bare the mathematical structure of the solar system. No longer were there celestial spheres made of a celestial substance. The same laws applied to all matter, because all matter is essentially the same: points of mass distributed in space and time. This was not speculation. This was a mathematical description of the laws that govern the behaviour of Descartes' objective material substance—the *same* substance that constitutes the sun and the earth and the moon and Venus and Mars. And if you doubt it, then how do you explain the discovery of Neptune? For it was by observing how the path of Uranus diverged from the predictions of Newton's law that Le Verrier deduced the hypothetical existence of another planet that could explain the disturbance. And these deductions led directly to the discovery of Neptune, appearing just where the mathematics said it should.

This bursting forth of the new mathematical science swept away the prevailing medieval understanding of the being of the universe. In that world it was natural for the things of the earth to fall toward the earth. Such movement was the self-movement of earthly things striving to return home. No mysterious force of gravity acting at a distance was needed. The idea that matter exists in itself as something inert, entirely governed by mechanistic forces, would have seemed incredible, something that violates common sense. For, as everyone can see, it takes effort to move an earthly thing—it resists your intention and seeks again its natural state of rest. To think that once something (anything) starts to move it will carry on moving forever, unless some other force comes to act on it, seems to have no basis whatsoever in actual experience. Only much later would we see the immediate sensory confirmation of this law of inertia in the phenomenon of weightlessness in space travel.

I wonder if you can appreciate the ongoing effect that the development of science and technology would have had on the ordinary understanding of the pre-scientific peoples of the earth? For they would each have realised, perhaps slowly, perhaps dramatically, that their myths and stories of the creation and being of the universe did not correspond with the discoveries of this new science—a science that demonstrated itself not in language or visions, but in the utter subjugation of any culture that dare stand against it.

It is here, in this technological mastery of the West, that the true foundations of our materialistic convictions lie. For it is on the basis of this mastery that our belief in the correctness of our mathematically objectified notion of the physical being of the universe gains its sense of obviousness and certainty. We don't even have to make it explicit. It's there, in the mobile phone, in its sheer materiality. Although we probably don't know *how* it works, we know it *does* work, and we know there are no gods involved, just human intellects, collaborating together, with the backing of enormous sums of money.

But this background understanding of the objective materiality of the universe is something we have *inherited* from our ancestors, it is an *historical* phenomenon. It is not an ultimate, fixed truth, it is something we have collectively developed, and continue to develop. And yet, if we follow this development, we can see that it is *barely conscious.* For our materialistic understanding of the being of the universe leaves out something essential: consciousness itself, our own immediate meaningful experience of being alive. Despite all the sophisticated arguments to the contrary, it really is that simple. We factored ourselves, our immediate subjectivity, out of our objective scientific understanding of the universe, and now this objectified understanding has become so unconscious and habitual that we do not even notice the absurdity of our situation.

Here it is not a matter of going back to Descartes or Galileo and pointing out their error. This is *our* error. Descartes still had a reasonable and coherent account of our place within the universe. It is we who have fallen away from this beginning, who have become so absorbed in our technological marvels that we no longer concern ourselves with questions of ultimate meaning. Perhaps we hope that someone, somewhere, in a position of power, really knows what is happening, knows the meaning and purpose of all this activity. Perhaps we have a spiritual teacher who can put it all into perspective for us, with a story of the evolution of consciousness on Earth. But what of understanding the *actual* origins of the situation in which we find ourselves, of setting out on a *genuine* inquiry, not as a passive consumer, but as one who seeks the resonance of an authentic discovery? Such inquiry means *reliving* Descartes' encounter with the immediate certainty of being. It means looking again at what it means *to me* for something to be a material object. It means that all this history is present to us now, in the pre-understandings that shape our experiences. Descartes' dualism is not a dead idea, and neither is the empiricism that followed it. Each forms a link in a chain that leads to the situation of our present confusion.

The Objectifying Subject

What we are concerned with here is the history of the development of a certain form of rational-reflective egoic self-consciousness. It is the elevation of this form of consciousness into the role of a final or ultimate authority that characterises our Western scientific culture. Descartes' Meditations are an expression of the breakthrough of this consciousness into the medieval mind. In the ensuing conflict, European Christianity was transformed into European rationality. To begin with, the nature of this transformation remained concealed. For it appeared that reason, in and of itself, must reason that God exists. But already, in thinking of the existence of God as something that stands in need of rational justification, the outcome was prefigured. For, in such thinking, reason is already understood as the power by means of which it is to be decided whether or not God exists. And that means God is no longer God, the ultimate power.

iconic

Of course, if it turns out that human reason is really God's reason, then all can be resolved again, But that is not how it transpired. For Descartes understood the absolute certainty of the 'I am, I exist' to be the certainty of the being of his *thinking, asserting, rational self.* He missed the moment of unity, the moment of consciousness that precedes reflection, the foundation upon which all reflection reflects. And so a fateful misunderstanding was introduced into the foundations of Western thinking. Instead of realising the certainty of the being of an impersonal consciousness that reveals the being in the world of a particular individual, Descartes *thought* he realised the certainty of the being of his *thinking* self. It is here where God was dethroned. For now it is *my* thinking, reflecting self who is to decide, by means of reason alone, what it means to exist, and what it means to be real. This thinking reflecting self has become *detached* from the unified consciousness that impersonally presents each moment of experience. It is this state of detachment that is to become the basis of a new rational consciousness that no longer accepts any other authority but objective reason itself.

And so humanity finds itself (as ever) suspended between light and darkness. On one side there is the inspiration of the natural light, the sense of there being an intelligence that can see for itself what is true and what is false. It is this inspiration that moves Descartes to overthrow his former beliefs and to only accept as true and certain the revelations of his clear and distinct perceptions. What is significant here is that the natural light can only reveal what it has to reveal to an *individual* consciousness. Only one who is surrendered, who has put aside their personal beliefs and opinions, can distinguish the true from the false. This is no revelation to be preached from a pulpit. Each individual must discover the light for themselves. And yet it is an *impersonal* light. What it reveals is not just true-for-me, it is *universally* true. The 'I am, I exist' does not apply to my personal self. Each one of us is required to *re-enact* this realisation—not to believe, but to encounter the same certainty, the same natural light in our own individual consciousness.

Here there is a great freedom. No longer I am beholden to some external human authority, to whom I must submit, who in return provides me with a mythical structure of meaning within which I can live out my days. No longer am I a child who must accept what I am told on pain of torture and excommunication—someone who, in the darkness of my ig-

norance, is incapable of making sense of the world around me. For now I
have discovered an *inner* authority, an *inner* power of discernment. Now
the wisdom of age and experience and tradition does not convince me in
and of itself. If something is true then I expect it to be *demonstrated* to be
true. I expect *evidence*.

But already, in this discovery of the natural light, there is the shadow of
unconsciousness. For we can only question those beliefs of which we are
already conscious. An *unconscious* belief is unconscious in the sense that
it remains unrecognised *as* a belief. Instead, it appears as something that
exists in its own right. This is not an unconsciousness of *what* is believed,
but an unconsciousness that what is believed *is* a belief. The belief still
functions, but it functions by means of an unconscious projection, so
we experience the world to be in just the way the belief believes it to be.

The situation for Descartes and his contemporaries was not that they
held an unconscious belief in a benevolent creator God, or in the existence of an external physical world. For these were the very beliefs that
were *consciously* doubted in the Meditations. Descartes' unconsciousness
was concerned with the foundations of his own self-consciousness. He
believed himself to be identical with the egoic thinker that emerges in the
activity of reflecting on consciousness itself. This thinker understands itself to be the agency that reasons about the things and events occurring
in the external world. In such reasoning we abstract from the immediate
sensory manifestation of these appearances and enter a virtual world that
we can manipulate according to our own 'free will'. In so doing we genuinely think we are thinking about these things and events themselves,
and not about a virtual world we carry in our heads. And this is *both* true
and not true. For, when I think about something that is *not* present before me, like the corner shop in the village, it is that exact shop I mean or
intend, and not some representation I have in my mind. But, as we have
seen already, the form of this intention is a kind of *promissory note*. It intends something that has appeared to me in the past and that I believe
can appear to me again in the future. What I have in mind is a *pointer* to
the appearance of the corner shop that is not itself the appearance of the
shop, but neither is it something that represents the shop. This pointer
is a meaning intention that points *away* from the immediacy of what is
appearing *now*. It means or intends the shop in the mode of its present
potential to appear.

Phenomenologically, what we are dealing with is the *horizon* of the world. This horizon most immediately manifests in the hidden sides of the things appearing in front of us, as we encountered earlier in our exploration of objective perception. We protend these hidden sides, not as things that *will* be revealed in the future *if* we were to look from a different perspective, but as sides that are actually present to us now in the enduring form of the perceived thing. This horizon of protention does not end in the immediate objects of perception, it extends to include the surrounding world, shading off into increasing levels of indistinctness. So, for example, I currently have an implicit sense of the bookcase that lies outside my field of vision on the wall behind me, of how wide it is, how far away from the back of my head, and so on. The entire house is present to me in this way, not directly perceived, but as a co-intended horizon of intelligibility which provides the context for that part of the house I do directly perceive. All this occurs quite naturally. I do not have to attend to it. When the heating comes on, I do not jump in surprise, for the boiler already has its implicit place in this horizon. It is something present that lies within the domain of things I *could* see. This sense of the horizonal presence of things reaches out to include the entire world as I understand it to exist. All the events, the people, the oceans, the other countries, the moon and stars and the planets, are included in this network of intelligibility that I experience as my being in the world.

One of the extraordinary capacities of human consciousness is the ability to 'step outside' of the immediate consciousness of being now and inspect the structure of this co-intended horizonal world. In such 'stepping outside' we bring what would otherwise lie implicitly in the perceptual background, into an explicit foreground. At the same time, our immediate perceptual experience falls into another kind of background. There, although unattended, it retains a foreground/background structure that can be easily reawakened, should our attention be drawn back to it. In this action of 'stepping outside', our consciousness undergoes a modification. It disconnects from the immediate stream of *what is*, wherein the world is manifesting according to the meeting of our protending expectation and the source of (what we call) our sensory inputs. Instead of looking into the heart of now, we look away into the structure of our own protending expectations. In so doing we lose consciousness of our direct encounter with something that lies beyond the compass of our personal will, and enter a world of thought where *we* are the masters. For we do not just passively inspect this structure of protention that is otherwise projected into the stream of experience. We find we can rearrange these structures and intend a world that *differs* from the one that is actually manifest to us.

What is of interest here is the emergence of a certain thinking agency that thinks it is able to direct the course of its own thinking. Here we must assume that 'once upon a time' there was no such thinking agency. Instead there was just a pure consciousness of whatever it is that is manifesting now. Within that pure consciousness there arose a certain experience of agency, an experience of a power to act, of holding sway over a particular domain that came to be distinguished as the being of 'my' body. In the phenomenon of this body, there is a direct experience of a unity of intention and action. This is the basis of the feeling of the being of an agency (me) that intends the actions of this body. The form of this agency is not an intending of each particular action. It is an intending that is determined by instinctual desires that connect with something that is *not*-me, something that *resists* my desiring intentions. These intentions *envisage* their satisfaction in the *virtual presence* of a situation that satisfies the desire. In this envisaging there emerges a second world that presents how the primary world *could* be, while at the same time keeping track of the primary world as it actually is. In this simple state, 'I' am the immediately feeling, acting, envisaging agency of this body system. 'I' do not 'decide' a particular course of action, 'I' am simply moved by impulses and desires that arise and directly express themselves in courses of action. What happens is that this ability to envisage the world as it could be, finally comes to envisage the being in the world of the one who is experiencing the envisaging, so that 'I' come to envisage 'myself'.

The significance of this moment of meeting myself in the world that I envisage, is immense. Only now can I become conscious of myself as a being that thinks. Before I was the passive receptacle of envisaging thoughts emerging from desires that determined my behaviour. My agency was the agency of these desires. Now, in recognising myself in the inventory of the world, I can understandingly envisage my own body and my own actions as if they were the body and actions of *someone else*. Instead of simply *being* the desires that animate my body, I can envisage my desiring self and I can think of myself as the one who is envisaging. In this power of reflection I can separate myself from myself. Whatever it is that I experience, I can envisage that experience back to myself. And then I can envisage how that experience could have been different. In such envisaging, it no longer seems to me that I am determined by each and every feeling and desire that enters my consciousness.

What has emerged is a *higher-level form of thinking*. It is a thinking that can think about and influence the lower-level instinctual thinking processes that envisage the satisfaction of the desires that animate them. Just as I gain a sense of agency-in-desiring when my desires directly animate my body, I now gain a sense of agency-in-thinking when my higher-level thoughts change the course my lower-level envisaging. This does not mean I am no longer determined by instinctual desires. The difference is that my higher-level thinking now has a more extensive understanding of the consequences of my desire-driven impulses, so 'I' can head off the immediate satisfaction that these desires envisage by altering what is envisaged. In this assertion of higher-level thinking there opens up a space of separation between thinking and desiring. In this space it becomes possible for thought to explore its own possibilities. Here it seems that I am free to think of anything I choose. The limited agency of my capacity to change things in the actuality of the world around me becomes the unlimited agency of my capacity to think in the horizonal world of my protentional thought structures. Here there are no hard constraints. I am limited only by what is conceivable and by the dimensionality of my projective imagination.

But what of the *being* of the agency that arises in the act of thinking? This is Descartes' doubter who can doubt everything save the certainty that 'I am, I exist'. It stands, as a kind of god, ruling over its domain of all that is thinkable. In itself, it has no attributes, save its capacity to reflect. This is what Descartes conceptualised as the unextended thinking substance. Its self-understanding is simply this capacity to think about thinking. It is the *objectifying subject of reflected experience*. In itself, for itself, it is a kind of *nothing*, a pure ability to register and think about whatever it is that is being reflected. It has no past, because the past, for it, is something it can reflect on, and whatever it can reflect on, is not it, itself, the 'I' who is reflecting. Similarly, it has no body, because, again, the body, appears on the other side of the divide, as an *object* of reflection. This goes beyond our ordinary conception of objectivity. It is not just my objective-physical-material body that emerges in this mirror of reflection, it is also my 'subjective' feeling of being in this body. For in the moment of reflection, that feeling, is also not-me. It has become abstracted from the immediacy of feeling the feeling into an act of 'me' looking at that feeling as it *slips away into the past*. This illustrates how it is that 'I', insofar as I reflect, can never *be now*. For 'I' can only approach now by reflecting on what is already slipping away. When I attempt to become fully conscious of the manifesting now, then this gap of reflection closes, thinking ceases, and the 'I' of reflection becomes absorbed into the immediacy of this manifestation. Of course, objective science will say that this experience of manifestation is itself an experience of the past because it takes time for light to reach our eyes and for our brains to process that information. But this misses the point. My experience is always manifesting now. The empirical fact that there is a time lag in the information flows that connect my physical brain with the physical world is irrelevant. We are speaking of the immediate now of perception, not the objectified now of the objectified world that emerges when we reflect on the objectified processes of perception.

This ability to reflect on experience is not some rarified state that only philosophers can reach in the privacy of their meditations. It is an inherited cultural capacity, something that we learn and absorb as children. We may not take things to Descartes' extreme, but we habitually inhabit this reflective stance in relation to ourselves and our own lives and our understanding of the world around us. The difference with Descartes and his contemporaries, is that instead of using such thinking as a useful guide in the practical matters of life, they elevated it to become the ultimate authority upon which every other aspect of life was to be decided and understood. And so we entered the so-called Age of Reason.

EMPIRICISM

It is now that European civilisation sets out on a path to develop this capacity of the human mind to objectively reflect on its own experience to a state of full clarity. In Galileo it had already developed an objectively mathematical conception of the being of the material universe and in Descartes it gained its first clear philosophical articulation in the dualism of *res extensa* and *res cogitans*. It now devolved upon the philosophy of empiricism to provide a rational account of the being of human subjectivity. This project already had before itself the model of mathematical reasoning, which showed how, starting from a set of first principles, we can deduce, with absolute certainty, all the consequences that lie hidden in those principles. The task of the new rational natural philosophy was to direct this form of mathematical reasoning onto nature itself, and thereby to discover the ultimate principles from which the being of the natural world and our experience of that world could be deduced, and so become rationally intelligible.

Descartes' first principle was the 'I am, I exist'. This became the logical certainty of the being of a logical 'I'—the universal subject that in its own self-certainty requires no higher principle from which its own being need be deduced. It (I) am therefore the ground of reason, the first principle, the pure subject and origin of all that can be deduced. Now 'I' as this universal principle need only proceed, according to the light of reason itself (the natural light of my clear and distinct perceptions of what is true and what is false), to discover the principles which form the ground of the being of all else that exists. As we know, Descartes' path led him to reason into being the principle of the necessary existence of a perfect being (God). And then he reasoned into being the principle of the necessary existence of an extended substance that comprises the external world that we perceive. This left our remainder, the human subjectivity that perceives the external world and also perceives itself in all its acts of perceiving and feeling and imagining and so on.

It was John Locke, the founder of philosophical empiricism, who, inheriting Descartes' framework, set out to objectively delineate the domain of human understanding. He began by accepting the basic Cartesian premises of the necessary existence of God and the extended substance of the external world, and of the method of proceeding according to the natural light of reason. What distinguished empiricism was its one-pointed emphasis on the evidence of experience. This is most famously expressed in the following section from Locke's *Essay Concerning Human Understanding*:[2]

> Let us then suppose the mind to be, as we say, white paper, void of all characters, without any ideas:- How comes it to be furnished? Whence comes it by that vast store which the busy and boundless fancy of man has painted on it with an almost endless variety? Whence has it all the

materials of reason and knowledge? To this I answer, in one word, from *experience*. In that all our knowledge is founded; and from that it ultimately derives itself. Our observation employed either, about external sensible objects, or about the internal operations of our minds perceived and reflected on by ourselves, is that which supplies our understandings with all the materials of thinking. These two are the fountains of knowledge, from whence all the ideas we have, or can naturally have, do spring.

Figure 6.2: Godfrey Kneller's portrait of John Locke 1632-1704.

Here human reflective intelligence is understood as emerging from out of itself, a kind of empty pure capacity to receive and form ideas on the basis of being presented with the raw materials of "sensible objects" and the "internal operations of our minds". What Locke thinks he is refuting is the doctrine that the human mind is supplied with a stock of pre-existing innate principles or ideas. But, more essentially, he is asserting the autonomy and freedom of our reflective reasoning intelligence to adjudicate over any and every question that could possibly be asked of it, purely on the basis of what can be demonstrated by experience alone. While he still allows for the possibility of revelations from God, he does not allow that such revelations could contradict the findings of empirical reason—for, of course, in that realm of revelation, we can never be *certain* that God Himself is the Author.

Locke's empiricism is relevant because it still forms the foundation of what we call our common sense. This is the common sense of our common experience. It says that I will only accept as true that which you can demonstrate to me by means of direct experience. Locke's rejection of the possibility of innate ideas is a denial that there could be any other form of valid common knowledge that does not have its basis in such experience. This great principle of empiricism, that all knowledge claims can only obtain their warrant of validity by being grounded in the evidence of a direct experience, still expresses an essential truth. But it all depends on what you mean by 'experience'. Again there is light *and* darkness. For there is experience as it manifests now in immediate consciousness and then there is experience as it manifests when it is objectified in a state of reflective consciousness.

So what did Locke mean by experience? Here we must picture him assuming the stance of a purely rational reflective thinking intelligence. He has already accepted that there is a God, and that there is an external world of extended physical substance, and a separate world of human subjectivity where we have an experience of this physical world. Following Galileo and the new mathematical science, he understands this experience to be divided into primary and secondary qualities. The primary qualities are those aspects of the physical world that we perceive as they 'really are'. These are the objectively measurable aspects of that world, which, in a circular fashion, define the being of the extended substance that defines the being of the physical world. This is the objectivity of space and time— a universal system of potential mathematical co-ordinates that specify all the things and events that can and do occur in the physical world. What remains are the secondary sensory qualities, our experience of colour and sound and touch and taste and smell and the inner sensations of our bodies. These belong to our subjectivity and not to the physical world. Only someone with eyes sees colours, only someone with ears hears sounds, so colour and sound cannot belong objectively to the actual things of the physical world. These secondary qualities are the means whereby the things of the external world are represented in our personal theatre of experience.

Locke's conception is that our experience of things appearing to us as mixtures of primary and secondary qualities is an experience of *ideas*. These ideas are formed purely on the basis of our having experiences. No experiences, no ideas. Human subjectivity becomes a photographic plate on which the senses and the inner workings of the mind make impressions and we experience these impressions as ideas. We are not entirely passive in this process, for we also have the capacity to reflect on these ideas, and in such reflection we can form new ideas that are also impressed upon our photographic plate.

So what's wrong with this picture? Firstly, I suggest you stop reading and look up at whatever it is that lies in front of you now. We are concerned with the pure manifestation of experience. If I reflect on this experience then I can confirm there is a chair in front of me, several books, a table, and so on. But, if I do not reflect, all these things are still present, together, in their actuality. I could say that 'I' am seeing this, but, again, like the things in the room, if there is no reflection, there is no separation of an 'I' from the seeing itself. The room and the seeing of the room are unified in this one ongoing manifestation of the being of experience.

The question here is whether it is the room that is manifesting, or a collection of ideas? Certainly we can *think* of the experience of the room *as if* it were the manifestation of a collection of ideas. But how does it stand with experience itself? Is it one of sitting on an idea of a chair, or of sitting on an actual chair? Now, I know that objective science will say: "Of course, it *seems to you* that you are sitting on an actual chair, but objectively your experience is really dependent on certain events occurring in your brain and these events are like ideas of the events occurring in the room." To which I reply, once again, do you really doubt that you are sitting on an actual chair in just the way you are experiencing it? Or when it comes to your observations of the behaviour of neurons, do you really doubt that you are actually in the room where you are making your observations, or that the equipment you are using is not there in front of you in just the way it presents itself? The question is, which comes first? Your actual experience, or the theory you have developed about your actual experience? For what empiricism says, *essentially,* is that experience is *all we have.* There is no *direct* access, for us, to *anything* that lies beyond experience. That means the evidence we have that suggests there is an external world beyond our experience is *entirely* derived from and grounded *in* our experience. And that means experience does come first, before any theory we may hold concerning experience, because all such theory can only refer to and obtain its justification from pre-theoretical experience.

The irony here is that the actuality of the world as it manifests in experience is exactly what Locke's empiricism takes for granted. It is because experience presents the world as being unquestioningly actual that Descartes and Locke accepted the necessary being of an external world subsisting independently of experience in the first place. It is *experience itself* that exhibits the *phenomenon* of actuality, that, in effect, says "this world you see is actually there in front of you, and not in some sensation that you are experiencing in your mind". This is ironic because, having unconsciously accepted the validity of this basic experiential intentionality, both philosophers go on to deny that we are having a direct experience of the actual world, and that, really, we are having an experience of ideas in our mind that only imperfectly represent that actuality.

The basic problem with Locke's empiricism is that he does not encounter experience itself. Instead he takes experience to be that which appears to his rational reflective consciousness. He does not see that such reflection reflects upon a pre-existing idea of experience, and not upon experience itself, as it manifests. His method of approach is to think about experience after having already projected an understanding of the independent being of an external world. He does not ask after experience itself, instead he starts from this unexamined premise. Consequently, the problem of how to encounter experience as it presents itself to itself does not arise. Locke thinks he already has this phenomenon before him in the form of ideas impressed on the white paper of the mind.

That is not to say that Locke's empiricism is some kind of a mistake. For, like Descartes, Locke is in possession of an essential truth: the truth that all human knowledge, if it is to count as knowledge, must obtain its warrant of validity from a direct confirmatory experience. Locke's empiricism is rather an opening from which a more radical empiricism can be developed. For while he reasons that the things I perceive really do exist independently of my perceiving them, he is not *as* certain of this as Descartes. Instead of attempting a proof, Locke presents what he sees as cogent reasons for *believing* in the independent existence of an external world, while acknowledging we can only be certain of the being of the *appearance* of such a world.

The Superfluous External World

From this it is but a short step for Bishop George Berkeley to see that the being of this external world is *superfluous*. For what we *actually* experience are *perceptions of* an external world. Of these we are certain. And if these perceptions exist in a separate non-physical, non-extended dimension, then they exist in their own right, independently of any secondary external world. So the question arises as to how events in this external world are able to convey themselves to this immaterial realm of experience. We naturally think in terms of physical events causing such experience, but how is something physical (i.e. extended in space) to cause something to change in an entirely immaterial non-spatial domain of sense experience? Here it seems we need the intervention of some other agency or medium to effect the miraculous transmission, such as, for example, God. But if all human experience occurs in one dimension and all physical events occur in another, and we need some agency to effect the interaction between them, what difference would it make if the entire physical domain were obliterated, so long as this agency continued to inform us? The answer, of course, is that it would make no difference whatsoever. And so we get to Berkeley's idealism: that in God "we live and move and have our being" and that our experience of the world is exactly and only that: a perceptual experience of things that only exist insofar as they are perceived, an existence of aggregates of colours and sounds and tastes and smells and feelings, with no further hidden material basis.

This is the idealism that famously caused Samuel Johnson to kick a stone and exclaim "I refute it thus!" But, of course, he did not refute Berkeley, he only showed the futility of such refutation. For the discomfort he felt in kicking the stone was just that, sensory, experiential discomfort, and his seeing the stone fly off into the air was just that, his sensory visual perception of a stone moving in a sensory perceptual space. None of these experiences *require* the presence of an actual physical world that corresponds with our experience.

Berkeley's arguments also made it clear that the old division between primary and secondary qualities had no force of necessity. For, just because our perceptual experience shows a world extended in space and time, it does not follow that there must be a second reality that is actually extended in space and time, that copies what we perceive. It only seems like that when we mix up our original experiential certainty that the world we perceive actually exists in the way we are perceiving it, with a secondary reflective understanding that our experience is divided into two mutually exclusive subjective and objective components.

Despite the cogency of Berkeley's arguments, it was Samuel Johnson who proved more persuasive. This heralded the beginning of the breakdown of the coherence and unity of the philosophical self-understanding of Western science. For Berkeley was *right*. If we accept the premises of the new mathematical science then the entire existence of what we take to be the external world becomes superfluous. And yet our common sense tells us that there is a world out there that exists independently of our perception. So which is the correct view? No one was quite sure.

And then, as if to finish the job, along came David Hume.[3] Hume's great contribution to empirical philosophy was to demonstrate that the notion of inductive inference, upon which the new mathematical science depended, is a kind of illusion. Before Hume, it was widely believed that scientific research was a matter of discovering the ultimate causes that determine the events of the universe, and that the scientific method was a sure and certain means of discovering those causes and demonstrating that they were indeed true causes.

At the heart of this process of discovery lay the idea of inductive inference. This is the common sense notion that if we observe one event, such as our rolling a billiard ball toward another stationary billiard ball, and then consistently observe a second event, such as the stationary billiard ball starting to move, then we can conclude that the first event caused the second event. In this idea of causation there is the idea of a *necessary connection* between the first and second event.

What Hume said, as a true empiricist, is that we have no grounds to conclude there is any such thing as a necessary connection between events. For we are only going to accept as evidence that which *actually* presents itself to us in experience. And if we *look* we find we do not observe anything like a necessary connection—for, indeed, what would a necessary connection look like? All we *actually* observe, in the case of the billiard balls, is that *up to now* it has always been the case that when the balls collide in the right way, then the second ball will start moving. What Hume was saying is that this gives us no right to conclude that this will occur the next time we make the balls collide. More than that, he was saying that this gives us no right to conclude (with necessity) that it is even *probable* that the second ball will start moving.

The important issue here is the idea of *necessity*. Hume is quite happy for us to continue, as we do, believing that the next time we play billiards things will go much the same as they did the last time we played billiards. His argument is with our *certainty* that things will keep on behaving in the way they did in the past. He is saying that strictly, logically, we have no such certainty. Or, to put it more formally, it is our *deductive* inferences that have certainty, but not our *inductive* ones. Deduction proceeds by saying that *if* a state of affairs is correctly structured according to the rules of deductive inference then *if* such and such is true *then* such and such *must* follow with strict necessity. The classical example is that: *if* all men are mortal and *if* Socrates is a man *then* Socrates is mortal. Here we arrive at what is called an *analytic* truth—i.e. a truth that was already contained in the meaning of the premises. Although such reasoning can uncover mathematical truths that initially lie hidden to human understanding, those truths already lie in the meaning of the premises that we assume to be true in the first place. This does not tell us whether or not the premises themselves are true. For that, as empiricists, we must inquire of our *experience*.

The idea of empirical science is to discover the true premises upon which the universe is founded by means of *inductive* inference. Once we have discovered these true premises, we can deduce, with absolute necessity, the consequences that are entailed. Such deduction means we can predict with certainty the future course of events, that, at present, we can only imagine. And that gives us immense potential power to control the course of actual events in the physical universe. For example, we can imaginatively deduce what would happen if we concentrated together a sufficient quantity of radioactive material and then exploded a detonator . . . .

What Hume attacked was the idea that empirical science could ever arrive at any *certain* knowledge of the *objective* universe whatsoever. For empiricism says that we can only come to know the premises on which the universe is founded by making *objective experiential* observations. That closes the door on any possibility of knowing these premises *directly* by means of intuition or insight. And yet, in and of themselves, our experiential observations do not directly present us with any of these ultimate premises. All we have is the experiential evidence of individual, particular events. In order to discover the possible existence of laws or causes that determine these events, we have to abstract a general form that precisely describes what these events have in common. This process of abstraction is what we mean by inductive inference. It is the same process, before being raised to reflective consciousness, that is already functioning in our natural perception of the objective world. For we are only able to distinguish dogs from cats because, from frequent encounters with dogs and cats, and with the assistance of those who already know how to distinguish dogs from cats, we have inductively inferred the general abstract forms of dogness and catness. Objective mathematical science begins with this life-world of dogs and cats and sets about perfecting the procedure of inductive inference by means of precise measurement and strict reasoning. Before Hume, it was assumed that this purified form of reasoning was discovering the true form of the actual universe, and that the laws it formulated, such as Newton's law of universal gravitation, were the actual laws that govern the actual universe with absolute necessity.

What Hume discovered is that, if we accept the premises of empiricism, then our empirical sciences have no certain basis to claim that any of the predictions they make are actually going to happen. All that we can say is that *up until now* things have gone this way, but we can discover no logical necessity that mandates things should continue this way in the future. It doesn't matter that something has happened once or twice or a million times, we still do not know with any force of necessity whether it will happen again. Hume turned this into a psychological theory that says: because certain sensory impressions have appeared together for us in the past, we form a certain expectation, born of habit, that they will appear together in the future, and therefore, there is no more to our scientific mathematical laws, and our idea of cause and effect, than this psychological phenomenon of habitual expectation. So there is no force, as such, that we can directly experience acting between our billiard balls. There is simply our expectation that one ball will move when hit by the other. And an expectation is just that, an expectation. It is not something that is capable of moving a billiard ball. Of course, you may say that we do experience forces when someone throws a billiard ball at *us*. But here, if we follow empiricism, we must admit that we are not experiencing any objective force operating in the objective world, we are experiencing a certain sensory impression that we *interpret* as being *caused by* an objective force, something that we *expect* to experience each time a ball is thrown at us, but indeed may not experience if and when it happens again.

The strange thing about Hume's discovery is that it made no difference to the actual practice of objective science, or to its effectiveness in developing new technologies on the basis of inductive inference. The empirical fact was (and is) that inductive inference *works* and not just in the development of new technologies. For what of the discovery that water is composed of hydrogen and oxygen? Are we going to say that it is somehow *not* true? That it has only appeared so up until now and that tomorrow it may change? And yet, like Berkeley's seeing that the being of an external world beyond our experience is superfluous, Hume's reasoning is impeccable. Both philosophers are simply working out the consequences of the objectivism of the new science. For Berkeley this objectivism leads to its negation, an extreme idealism, and for Hume it leads to a corresponding scepticism about the objective validity of the entire scientific enterprise. But what actually happens, in the progress of the new mathematical science, is that this philosophical crisis, arising in its very foundations, is *ignored*. Here is where we meet with the argument from success. This argument says that Berkeley and Hume must be wrong because the objectified mathematical understanding of the being of the physical world has founded a form of science that has delivered over to Western European civilisation an unprecedented means to control the events occurring in that physical world. This *ignoring*, this *not caring* to understand the meaning of the path of scientific progress that is being relentlessly pursued, is *fateful*. And that means fateful for *us*, who have now inherited the consequences of this ignorance.

Transcendental Idealism

What Berkeley and Hume usher in is the beginning of an era of a crisis of faith in reason. And yet it is a strange crisis of faith. For it is not that anyone is considering giving up on reason. It is rather that reason is in crisis with itself. The original idea was that the essentially mathematical form of reasoning that Galileo had turned on the objectified physical world could be extended to encompass the world of human experience. But, in contrast to the remarkable progress of the physical sciences, the attempt to reason about human experience was raising questions concerning the ultimate validity of the entire scientific project, questions that reason, in and of itself, seemed unable to resolve. Few people could accept the conclusions of Berkeley or Hume, as they so clearly contradicted the convictions of ordinary common sense. For 'everybody knows' there is a world out there made of rocks and grass and sea and air, and not of colours and sounds and feelings. And equally 'everybody knows' that when I kick a stone, it is my kicking the stone that causes it to fly through the air. So Berkeley and Hume can't be right. That leaves 'us', the eighteenth century upholders of sound common sense, back with Locke, who at least does not argue with the validity of the physical sciences or with the idea that there really is a physical world 'out there'. Even so, we know there is *something* wrong. For philosophy, the queen of the sciences, is in disarray. It cannot give an adequate account of human experience and its capacity to reason that fits with its scientific conception of the physical world. And without such an account we stand in a place of deep uncertainty concerning the meaning and value and purpose of that very rational scientific form of civilisation that we are now launching onto an unprepared world.

And so we arrive at Immanuel Kant's *Critique of Pure Reason*. In this one book, it is as if the mantle of European intellectual culture passes to what will become the German nation. It is here that Kant explicitly confronts the crisis of reason he inherits from Hume. To do this he has to re-conceive our notion of human experience. Following Berkeley and Hume, he sees that we cannot ground our notions of space and time and causation in the world as it presents itself in our experience. Instead he sees that it is *we* who bring these forms to bear, that they are forms of *our* experience and not forms belonging to some other world. They are the forms by means of which the world is able to be a world for us in the first place. It is this recognition that, in one stroke, creates a new form of transcendental philosophy that will eventually lead to the transcendental phenomenology of Husserl.

But Kant does not make the leap into the immediacy of experience itself. He remains in the place of reasoning reflection, looking at the objectified traces that experience leaves behind. His great achievement is to enable reason to reason about itself in such a way that it becomes aware of its own activity and its own limitations. In his questioning he asks after the grounds of the possibility of experience itself, into the organising forms that must already be in place in order for experience to be what it is. This goes beyond the questioning of the empiricists, for they did not recognise this transcendental domain, thinking instead that the forms of experience inhere in the form of the world as it appears to us and that we simply receive these pre-formed appearances as ideas on the white paper of our minds. But when we reflect, as Kant did, that this experience we have of the being of the world, *is* an experience, it becomes clear that, *as* an experience *of* the world, the fact it is an experience must remain hidden to us, otherwise it would no longer be an experience of the world, it would be an experience of experience experiencing itself. From this it emerges that there must be 'something' behind our experience, of which we are not immediately aware, that is making our experience an experience *of* something that is not itself an experience.

Figure 6.3: Immanuel Kant 1724-1804.

Kant does not speak of directly encountering the immediate consciousness that is this manifestation of experience. He rather *reasons* that there *must* be such a transcendental source of experience. And so he speaks of transcendental forms and transcendental concepts that lie outside the sphere of our reflective egoic consciousness. These are entities he reasons must exist in order to explain the form of our egoic reflective objectified experience of the world. They are the *conditions of the possibility of experience*. Foremost amongst these are the forms of our sensible intuitions, i.e. space and time.

Kant's philosophy represents a profound turning point in the development of the self-understanding of Western European reason. The *Critique* was published in 1781. And yet, I would say, if we look at our manner of living and the way we collectively go about the business of our lives, that the common understanding of our modern-day common sense has still not understood what Kant had to say. For who amongst us truly recognises space and time as forms that originate in us, that we project in such a way as to manifest the world and our own being in the world? In what classroom at school is this mentioned? In which boardrooms or government departments? And yet Kant has been recognised as one of our most profound philosophers for more than two centuries. What has happened here? Was Kant proven wrong? Hardly. What he said has simply become *covered over*. A dead philosopher, deserving of a major place in the history of thought, but hardly relevant to us now, given all the other philosophers who have done the covering over. But perhaps we should look again at the first part of the first edition of the *Critique of Pure Reason*:

> If I separate from the representation of a body that which the understanding thinks about it, such as substance, force, divisibility, etc., as well as that which belongs to sensation, such as impenetrability, hardness, colour, etc., something from this empirical intuition is still left for me, namely extension and form. These belong to the pure intuition, which occurs *a priori*, even without an actual object of the senses or sensation, as a mere form of sensibility in the mind.

This is Kant's Copernican revolution. He reasons that those very primary qualities, extension and form, that Descartes and Locke attributed to the independent being of an extended substance, are, in truth, "a mere form of sensibility in the mind":

> Space is not an empirical concept that has been drawn from outer experiences. For in order for certain sensations to be related to some thing outside me (i.e., to something in another place in space from that in which I find myself), thus in order for me to represent them as outside one another, thus not merely as different but as in different places, the representation of space must already be their ground. Thus the representation of space cannot be obtained from the relations of outer appearance through experience, but this outer experience is itself first possible only through this representation.

Here Kant is pointing out that we do not learn what space is, in itself, from first observing our experience, because, without the forms of space and time, we would have no experience in the first place. In order for us to distinguish anything perceptible, even if we only perceive an amorphous field of undifferentiated sensation, that field would have to be laid out in at least in two spatial dimensions, and to temporally endure in such a way that sameness and difference could be distinguished. The demonstration of this truth is again a direct knowledge of consciousness. It is only insofar as consciousness manifests time as endurance and space as separation, and is at the same time conscious of this manifesting, that we directly know what it is for some *thing* to endure as itself from one moment to the next, and for that something to be distinct or separate from anything else. These dimensions of experience cannot become objects of experience, for they are, in themselves, sheer absences, a nothing that both separates and connects whatever is presented.

Kant himself does not explicitly refer us to a direct knowledge of consciousness. Instead he remains within the structure of reflective reason and constructs arguments to demonstrate what must be the case (i.e. what must be the grounds of possibility) in order for our experience to be the kind of experience that it is. Like Descartes, he remains outside the immediacy of now, and so can only reason about an objectified form of experience, i.e. experience insofar as it appears to a reflecting subject as something distinct and separate from that subject. But, unlike Descartes, Kant recognises that this objectified experience *is objectified*. This is a mighty realisation. It is here that reason becomes conscious that the objective forms with which it is presented are not the true and final subject matter of philosophical investigation. For if these forms have been *objectified*, then there must be another, transcendental dimension, within which the objectification occurs.

Before Kant, it was taken for granted that the objects of consciousness, the subjective (but still objectified) ideas or impressions, and the objective things and events of the world, were the unquestioned primary materials to be studied in their very objectivity. But now Kant sees there is another agency at play, that is producing this experience of objectivity according to principles and dimensionalities that cannot lie within the objects themselves, but must already be in place, in order for there to be any appearance of objectivity whatsoever.

This realisation has failed to permeate our culture, for the simple reason that it goes against our common sense. For it is in the nature of common sense *not* to see the objectification that lies behind our experience. If we *were* to see this process of objectification, as it occurs, we would no longer be seeing the objective world that is the very phenomenon our common sense is there to present us with. It is for obvious reasons of survival that our perceptual systems developed so as to make these processes of perception transparent to us. Instead we *see through* the extraordinary achievement of synthesis that forms our experience, directly to the objects and events that have been synthesised. And as we have seen, it takes a certain discipline to withdraw from this identification with the end-products of our perceiving, both inwardly and outwardly, and attend to the process of formation itself. For we are filled with the immediate and immovable conviction that the things and events and thoughts and feelings we perceive are *actually there* just as they present themselves to be. And that is not to say that these phenomena are not actually there. It is to say that *in order* for them to appear, a certain concealed process of objectification must already have occurred.

What the common sense of the natural attitude of perception insists upon, however, is that the forms of our perception inhere in the objects themselves and not 'in' us. And this is still *partially* true. For to think that this process of objectification is occurring 'in' *my* subjectivity is to have already entered into an understanding that sees subjectivity as a kind of object amongst other objects within which objectivity is constituted, whereas, if we think about this clearly, the 'place' where objectivity is constituted cannot be conceived in terms of the objective and the subjective, it is rather that which *precedes* our ability to distinguish between subjects and objects and 'places' and 'events' in time and space. Hence Kant thought in terms of a *transcendental* subjectivity that is not a corollary of objective being but is the origin from out of which such being appears.

As Schopenhauer saw, Kant's *transcendental idealism* was the beginning of a true enlightenment of Western reason—an enlightenment that could have formed a bridge with the pre-existing discoveries and insights of the Eastern spiritual traditions. But it was not to be. There was a different spirit abroad in Western Europe. Christianity was in decline. The emergence of Protestantism and the wars and persecutions that followed had destroyed the unity of a common faith. Something new was stirring, a purely objectifying vision of the being of the universe, that was revealing the previously undreamt of structure of physical matter. Where for centuries the alchemists had projected psychic visions into their understanding of chemical processes and reactions, the new mathematical science, in a matter of decades, had discovered the underlying atomic form of the elements by stripping away all these projections and simply observing the facts, the actual weights of the constituents and residues of chemical reactions, and the ratios of those weights and the mathematical regularities that emerged from such measurements. We who were just presented with the periodic table of elements one day at school can hardly imagine what it must have been like to *discover* this hidden atomic structure in nature. Once it was seen, then the entire mythology of the primordial elements of earth and air and fire and water became a kind of fiction, a dream projected onto a reality that we cannot simply intuit with our minds, but that stands before us, and only truly reveals itself to those who cease to dream, and begin to *measure*.

What need had this world for Kant's *transcendental idealism*? Christianity, with an admixture of Protestant reason, already provided a background explanation for our being here on Earth. What did it matter that empirical philosophy had run aground with Berkeley and Hume? Kant's answer was interesting enough, but it did not have the power to challenge the new vision of objective science that was transforming Western European civilisation—a vision of almost unlimited power and influence. And, after all, what was wrong with Locke's common sense view of our situation? There is the objective physical world and there is our subjective experience of that world. God is still in His heaven, so He takes care of the mystery of the interaction of these otherwise incompatible regions. The empiricism of Berkeley and Hume simply went too far. For there is no *necessity* that reason should be able to explain *everything* on the basis of empirical experience alone. What of the First Cause? Or the reason for the Principle of Reason, that nothing is without sufficient reason? Or, to bring things back to our immediate experience, what of the miracle of the presence on Earth of *life itself*?

NOTES

1. "The greatest recent event—that 'God is dead'; that the belief in the Christian God has become unbelievable—is already starting to cast its first shadow over Europe. To those few at least whose eyes—or the *suspicion* in whose eyes is strong and subtle enough for this spectacle, some kind of sun seems to have set; some old deep trust turned into doubt: to them, our world must appear more autumnal, more mistrustful, stranger, 'older'. But in the main one might say: for many people's power of comprehension, the event is itself far too great, distant, and out of the way even for its tidings to be thought of as having arrived yet. Even less may one suppose many to know at all *what* this event really means—and, now that this faith has been undermined, how much must collapse because it was built on this faith, leaned on it, had grown into it—for example, our entire European morality. This long, dense succession of demolition, destruction, downfall, upheaval that now stands ahead: who would guess enough of it today to play the teacher and herald of this monstrous logic of horror, the prophet of deep darkness and an eclipse of the sun the like of which has probably never before existed on earth?" Thus spake the German philosopher Friedrich Nietzsche (1844-1900) in his book *The Gay Science* (1882/2001, p. 199).

2. *An Essay Concerning Human Understanding* (1689/1997) was Locke's main philosophical treatise. In it he laid down the basic principles of clear thinking and reasonableness that have come to be known as modern-day 'common sense'.

3. David Hume (1711-1776) was a Scottish empiricist philosopher whose scepticism (most clearly expressed in his *Treatise on Human Nature* (1739/1985)) has had a significant influence on the development of philosophical thinking in the English-speaking world. This reaches through to Karl Popper's twentieth century understanding of science as progressing by means of the falsification of theory and not by means of proof, because, after Hume, the idea that a scientific theory could be straightforwardly proven to be true by means of inductive inference had to be discarded (Popper, 1935/2002).

CHAPTER 7

THE DARWIN MACHINE

AN INCREDIBLE THEORY

A ND so the nineteenth century dawns. There is the triumph of science, technology and industry as Britain inaugurates its first colonial form of global capitalism. And there is no doubt that humanity has entered into a new phase of civilisation. This *should* be the triumph of Christianity in its mission to civilise the world. But instead we find Christianity itself comes under the scrutiny of the objectifying gaze of the new world order. For if we are going to be objective, what are we to make of the stories in the Bible? That God created the earth in seven days? That Noah repopulated the earth from the pairs of animals he took into the ark? And what of Jesus and Mary and the immaculate conception? Can we continue to consider these things to be literal truths? And if they are not literal truths, then what is their meaning? Are they not like the legends and fairytales and myths of any pre-scientific civilisation? With our new found objectivity we have learnt to stop projecting our subjective fantasies onto the canvas of the external world. Now we can start cataloguing what *actually* exists, not what we *imagine* exists. And as we do so, we find the world is not quite the fantastic place we thought it to be. There are surprises, of course. But it turns out that physical processes obey the same physical laws, whether in China or India or the Bayswater Road. And it seems that no spell or incantation or transcendental state of consciousness can resist the logic of a mechanised army.

And yet, to begin with, science and religion *did* co-exist together, each in their own sphere of influence. For there was still the question of the miracle of creation. Of course, as science increasingly reveals, the physical processes on Earth, and in the wider universe, do appear to be determined by the lawful behaviour and interaction of tiny physical particles. But no one is going to claim that we could explain *everything* in terms of these particles, are they? For what of life itself? Surely you can't explain the growth of a plant from a seed, or the birth of a lamb, or the fearful symmetry of a tiger,[1] in terms of the interactions of microscopic particles? And what of the incredible diversity of life on Earth, how each species has its niche, and its symbiotic dependency on all the other life forms that surround it? Everything here speaks of a divine purpose and of a divine creator who imbues each living thing with its own form of life and intelligence. For the animals and plants *already* know how to behave. And that is before you consider the human beings, with their consciousness and their reflective intelligence. Are you going to tell us that this is all the result of the mindless interaction of microscopic particles following pre-determined laws? No. Of course, there has to be a God. Perhaps the Bible is not literally true. Perhaps that is just how God was imagined and experienced by our ancestors. But the obviousness that there is some kind of transcendental creative intelligence behind the appearance of life—surely you *cannot* question *that*?

And yet, of course, that is exactly what we *do* question. For it was only going to be a matter of time before objective science turned its objectifying attention to the phenomenon of life on Earth. To begin with, that new science was primarily concerned with the discovery and investigation and classification of all the life forms it was able to encounter. But inevitably the question of the cause and reason for all this diversity came to the surface.

Here it is important to understand that an objective science that is focussed on the study of physical materiality is necessarily constrained in the kind of explanations it can give. That is not to say that the possibility of explaining the phenomenon of life as the action of an immaterial life force was ruled out in advance. But such an explanation was always going to be a matter of last resort, something to be invoked only in the event that a satisfactory objectively physical account of the phenomenon of life proved impossible to attain.

This is where the *theory of evolution* enters the picture. And this is the battleground where objective physical science first begins to assert itself as the *one true path* to an understanding of the reality of the universe. Before Darwin's *Origin of Species*,[2] the division between those domains of knowledge that had to do with objective physical stuff (e.g. physics and chemistry) and those that had to do with subjective conscious experience (e.g. psychology and philosophy and religion) remained relatively clear and distinct. But now, in understanding life itself, we had to do with something that stretched across this subject-object divide. For living beings possess *both* extended physical bodies *and* subjective inner experiences. It seemed obvious to Darwin's contemporaries that the origin and growth and development of these apparently physical animal bodies could not be explained in purely physical terms, that not only do animals and humans possess another inner dimension of experience, but also that this inner dimension acts upon these bodies, influencing how they behave, how they are formed, how they grow, and how they evolve. It is here we expect to find the mysterious place of interaction between the duality of mind and matter that we inherited from Descartes.

Figure 7.1: Charles Darwin 1809-1882.

To begin with, Darwin's theory (which was also independently proposed by Alfred Wallace) appeared quite far-fetched. The idea was that the process of reproduction through which the physical characteristics of parents are passed on to their offspring includes some form of random variation that goes beyond the straightforward copying of asexual reproduction or the selective combination of sexual reproduction. These random variations are needed to explain how new species are able to evolve out of simpler life forms, instead of life becoming stuck in reproducing the same original characteristics. The second aspect of evolution is the principle of natural selection. This says that members of the same species who inherit characteristics that better enable them to survive, in comparison to other individuals of the same species, living in the same environment, are more likely to reproduce and pass on their characteristics to succeeding generations. This is our notion of the survival of the fittest. The argument is that this process of selection will, over time, develop species that are specifically adapted to their environments, which includes all the other species with which they interact. This process of evolutionary adaptation does not require the intervention of any higher-level intelligence. It simply emerges out of the low-level mechanistic inheritance of physical properties that determine the behavioural characteristics of an individual organism, and, in aggregate, of an entire species, and of any species that evolve from that species.

These two ideas: natural selection and random mutation, were proposed by Darwin and Wallace to be the physical mechanisms by means of which the evolution of the species on Earth could be explained. The only thing they assumed was the pre-existence of primal self-replicating life forms upon which the process of evolution could get started. It was in this idea of primal form that the God of the Victorians could still reside. For Darwin and Wallace were not militant physicalist atheists. Instead they saw themselves as clarifying the manner in which creation is ordered. This corresponds with a form of deism connected with Newton, where God is understood as a law-maker who sets the universe underway in a single act of creation and then steps back without the need to intervene further. Rather than a complete collapse of religious faith, we are seeing the gradual rolling back of the domain over which God is seen to have direct jurisdiction. For once we observe that a certain region of being is ordered according to mathematically intelligible objective laws, then it falls under the jurisdiction of *our* reason and control, not God's. In this way, through science, the human race has gained an increasing sense of its own autonomy and power. Consider climate change. We no longer think that God is in charge of the weather, or of the quality of the air, or of the creation and destruction of the ice-caps. We think *we* are in charge.

* 'Intervention' - but relatedness is the truth. Iconic.
Our role is to take up the offer of relationship.

The great change that the theory of evolution set in motion was the elimination of God as a principle to be invoked to explain the events occurring in the external natural world. Before science took over this role, there was the direct evidence of God's signature in the inexplicable being of nature. It could not be conceived how such a world could have emerged without a divine creator, and so we saw and felt God's presence in the manifestation of the nature of the earth, and in the sun, and the moon, and the starry heavens. It was a simple equation: what we could not explain or predict, we attributed to God.

I remember one day as a child of nine or ten years of age, in a science lesson at school, it suddenly dawned on me that *everything has a reason*. Before that, even though I had a practical understanding of cause and effect, I had not thought there was a *reason* why the water goes down the plug hole, or why a flower blooms. In that child's understanding, there was still the possibility of experiencing the world as if it were the manifestation of a Great Unknown Power. But I had not been prepared for such a realisation. All around me I was surrounded by adults who already believed that everything happens for intelligible reasons. And so, just as I would have become a devout Christian in another age, I became a scientific rationalist. *Tho - experientialist : religious experience.*

But we get ahead of ourselves. Europe of the nineteenth century was still in a state of transition. It was only an intellectual and cultural elite that took Darwin seriously. You and I, I suspect, were still digging up turnips. In this world, it was not at all clear that an entirely mechanistic view of the evolution of the species would stand the test of empirical investigation. But the dice were certainly loaded in favour of such an explanation. For the historical processes of objective empirical investigation were only ever going to explain physical events in terms of mechanistic regularities. In this objectified world, it was the task of physical science to spread its net of mechanistic explanation as far as it could possibly reach. The idea of there being a God, or a transcendental agency that could influence physical events, was ruled out by the very method of inquiry. Instead, God became a repository for all that could not be explained in purely rational terms.

And yet it is significant for us just *how far* this net of objective reason has been able to extend. Before Darwin it only provided reasons for the behaviour of the *inorganic* world. Cosmology had stretched as far as explaining the evolution of the solar system, physics and chemistry had reached to the atomic and electro-magnetic structure of matter, and biology had recognised the living cell as the foundation of all living structures. But life itself, the "force that through the green fuse drives the flower"[3] bore direct empirical witness to the living presence of something that transcended the physical cause-effect reasoning of the new mathematical science.

For those whose faith in God depended on such a direct manifestation of a transcendent power, the theory of evolution covered over the last avenue to a tangible demonstration of God's presence on Earth. There was still the miracle of the creation of the living cell. But that occurred billions of years ago. Nature itself, the birth, growth and death of individuals, and the evolution of the species, all now appeared to be ruled by the same mechanistic physical regularities that determined the behaviour of the inorganic world. Of course, in 1859, with the publication of the *Origin of Species*, the idea of an *ultimately* mechanistic biological understanding of life was still a distant goal. For it was not at all clear *how* characteristics are *actually* inherited. It would take until 1953, with Crick and Watson's discovery of the double-helix structure of DNA, for human reasoning to reach such a detailed objectively physical account of the mechanism of evolution.

As we have already briefly described, in its basic form, evolution is the process of natural selection culling the products of random mutation. And yet in this simple idea was a seed that would unravel our entire understanding and faith in the idea that life on Earth exists for some kind of purpose. Before Darwin, Descartes believed it was a truth, clearly and distinctly perceived, that a cause must in some way be greater than, or at least equal to, any effect that emerges from it. So the plant is already contained in the seed; my actions are already contained in my mind; and the universe is already contained in the mind of God. From this perspective, our current theory of evolution appears inconceivable: that the rocks and atmosphere and seas of the earth, from out of their disorganised, chaotic interactions, without any outside intervention, were somehow able to organise themselves into the form of a human body. Even if you ignore the undeniable fact of our being conscious, this does seem to be an incredibly unlikely story. And yet this is exactly what the modern theory of evolution proposes.

[Handwritten annotation at top: "Mindless interactions? But does not reverse as atoms have minds — but through atomic relatedness in quantum fields we are all participators in a"]

[Handwritten annotation in right margin: "relational mind & heart. Intelligent love of material substance and our embodiment."]

Darwin at least thought that God was needed to imbue life into the first primal life forms. But that was only because evolutionary biology had not yet developed the biochemistry necessary to propose how a living cell could have evolved from the (mindless) interactions of inorganic material. We can see now that this final removal of the essential distinction between living and non-living matter was already contained in Darwin's *manner of proceeding.*

The crucial development was the idea of evolution itself, as an explanatory principle. For evolution shows how something simple can develop into something complex without that simple thing already having to contain the embryonic complexity of what it is to develop into. Before Darwin, our model of the emergence of complexity was based on our human experience of creating complex structures from less complex components, such as our building a house, or a chariot, or a spear, or writing a book. In each case, the complexity arises from the complexity of our own intentionality. For it is we who envisage the complex thing that we are to manifest. And so, by analogy, we thought that the complexity of nature must also have been created by an intelligence like ours, only much greater, because it had created *us* as well.

Here we meet with another aspect of the new scientific thinking: the idea that any physical whole can be explained in terms of the behaviour and interaction of its *parts.* This conception arises directly from Descartes' definition of the essence of physicality as extension in space. Such an understanding mandates that something physical 'is' only insofar as it fills a certain amount of space with its material substance. What the new science discovers is that our perceptual experience of the material thingness of things, like tables and chairs and rocks and animals and plants, is just another illusion of the senses, like our thinking that things are really coloured. For in the objectively measurable physical reality, these 'things' only *appear* to be made of solid material stuff. Behind the appearance lies a world of atomic particles. It is here that we encounter the *real* microscopic physical particle beings, themselves separated in space, in which the *real* solidity of physical existence resides. For these particles maintain their solid structure whether or not they are combined in aggregate as a solid or a liquid or a gas. And as the only real physical beings, they are the only things that can affect or be affected by anything else. For the space between them is a sheer emptiness. *Nothing* can happen in this conception of the world except by means of the transmission of force between one particle and another. And that means when we ask after the *reason* why anything happens, we find that *everything* depends upon and can be reduced to this transmission of force between particles.

Later, of course, we find that these atomic particles are not the ultimate 'building blocks of nature'. For it appears they can be broken down into even smaller *sub*-atomic particles. And then quantum mechanics shows that this entire 'bits of matter extended in space' understanding of physical being is no longer consistent with our actual observations of quantum phenomena. But by then it is too late. For quantum mechanics presents us with something we cannot even picture because it contradicts our basic perceptual notions of physicality. And so, by and large, we stick with the old idea of atoms and sub-atomic particles, and think of the quantum level as some other dimension from out of which the reality of our physical particles can still emerge. At least, such was the basic picture I was presented with at school. It is this picture that still lies in the background of my everyday understanding, projecting itself into the world in much the same way (I imagine) that being brought up in a devout Catholic culture would have populated my world with saints and angels and fears of falling into sinfulness, even *after* my rejection of the faith.

But what has this to do with Darwin and the theory of evolution? Well, the task of physical science is to provide a physical explanation of the physical phenomena that can be observed in the physical universe. And once nineteenth century science had decided, partly on the basis of empirical inquiry, and partly on the basis of an ontological presupposition, that *atoms* are the ultimately real physical constituents of the universe (because if you take away the empty space, only atoms remain, so only they have real physical being), it follows that only *they* can have any real causative powers (because empty space cannot be the cause of anything), then the task of physical science becomes one of explaining how these atoms, purely on the basis of their simple causative powers, could combine together, without the interference of any non-physical agency, to produce the physical universe that is revealed to our scientific measuring instruments.

That does not mean we *reject* the idea of a non-physical agency (not yet), only that, insofar as we are engaged in an *objective* inquiry into *physical* being, a non-physical agency is not going to arise for us directly (because we are *only* observing what is *physical*). It is like a game, to see how far we can get without having to call in such an agency.

It is the principle of evolution that enables physical science to cross the boundary from the inorganic to the organic. It is like Newton crossing the boundary between the earthly and the celestial by showing how the same principles we use to explain the falling of the earth-bound apple can be used to explain the movements of the heavenly bodies. Only now we are going to show how these same essentially mechanistic principles can be used to explain the entire evolution of life on Earth. The terribly simple thing that the theory of evolution shows is how complexity can emerge from a state of lesser complexity. Once you have such a principle, you only need to keep working backwards. For instead of each complex thing having to be created out of a pre-existing intelligence that both envisages and surpasses that complexity, we can show how each complex thing can emerge, entirely mechanistically, out of a less complex thing, which itself emerged out of a less complex thing, and so on, until, ultimately we reach to something so simple it is a mere dimensionless point from out of which the entire universe explodes.

The terribly simple idea behind the theory of evolution is that of *feedback* within *circular* causation. Before the theory of evolution, Newtonian science was essentially concerned with *linear* causation, i.e. with giving reasons why one thing leads to another. But in dealing with life forms, biology encountered physical systems that were able to *cause themselves*, i.e. to make *copies* of themselves in the process of reproduction. Although Victorian science could not see into the molecular foundation of this ability to copy, it did recognise that copying was occurring. And it was clear, from our human experience of building machines, that an ability to copy is something that can be achieved entirely mechanically. For you only need to assemble together the right components in the right way and a copy will emerge. While Darwin thought it may have taken a creative intelligence to make an original form that was able to copy itself, such intelligence is not needed to make the subsequent copies—you just need the right metaphorical mould and the right metaphorical material to imprint it on. But, for *evolution*, you need one more property: that the copies you make are *not perfect*. Again, this lack of perfection can arise quite mechanically, through the effects of arbitrary outside influences on the copying process. All you need is a system that is sensitive enough to these influences that they will cause it to sometimes *miscopy* the original - just a little bit - so that, in the vast majority of cases, the copy is still able to function like the original, but is different enough to allow for some new variation in behaviour. Then there is the *feedback* from the environment. It enters into this circular process of copying by selectively *blocking* the event of copying. Again this blocking is entirely mechanical. No intelligent evaluation is needed. The mere fact that a copy occurs is all that matters. The tautological result is that copies that succeed in copying themselves will keep copying and those that don't succeed will stop copying. And in this world, to stop copying is to cease to exist.

Of course, there has to be a very fine balance in the miscopying. Too much and hardly any of our offspring will be able to function, even in an unchanging environment. Too little and there will not be enough variation to cope when the environment does change. Surely, you may think, there must at least be some intelligence behind the fine-tuning of this copying process? But again Darwin is ahead of us. For all those ways of copying that had too much or too little variation did not get to copy themselves for long. It is only those copying machines with the right degree of variation that are going to make it through. That's the wonder of feedback. It *looks* like our machines are striving to survive and evolve. But it's all being done for them by subjecting the machinery to feedback from the environment. And the environment doesn't think. It is just a name for all the events and interactions that befall our living organism.

If it just so happens that the way you were copied produces a body that can cope with the environment long enough for you to copy yourself, then those characteristics that enabled you to reproduce will tend to be passed on through the generations. Of course it may be I was just 'lucky' when the flood came and I happened to be standing on a hill. But over many generations, the unpredictable effects of such luck will tend to cancel out, whereas the more enduring effects connected with the underlying structure of the surrounding environment will tend to predominate.

Of course, there is nothing in this mechanism of circular causation and feedback to *mandate* that greater complexity is going to emerge. Consider the moon, and the rest of the solar system. The universe is not *trying* to develop complex self-replicating systems. And equally, once the complexity of such self-replication gets underway, there is nothing to mandate that it need *continue* to get more and more complex. It could be that we are the most complex self-replicators that are ever going to manifest on the earth, and that after us things will devolve into less complex structures. Our living forms are just the structures that were called forth by the physical configuration of the planet during a certain period of *its* evolution. In this mechanistic world view there can be no *intrinsic* intention, no future aim, no *teleos*. Such things require the being of a non-physical intending agency, and, as we know, such an agency had been removed from the projected objectification of physicality that formed the foundation of our nineteenth century scientific inquiries.

The Defeat of Vitalism

What we are tracing here is the collapse of a certain collective understanding that the universe is the purposeful creation of a transcendental (non-material) intelligence. For us, today, this collapse is not something of which we are directly conscious, as a culture. For we were born into the aftermath of that collapse. And so we live in the unrecognised *absence* of a collective sense of a transcendental meaning and purpose. It remains unrecognised because, for most of us in the developed West, this absence is all we have ever known. Our journey back into the origins of our scientific way of thinking is a journey into the origins of this collapse.

Our concern is not with any particular philosophical position or point of view. It is with the collective background understanding that has replaced our former sense of the being of a transcendental intelligence. This background understanding is not something we can necessarily articulate. It is something we have absorbed from everyone else in the process of our upbringing. It is present in the way we behave, the way we go about our business, in what we take for granted. It only becomes conscious insofar as we care to make it conscious. It is not as if we were sat down one day at school and told: actually there is no ultimate meaning or purpose in the events we see occurring in front of us. Neither were we positively told that there is a God in heaven in whose hands our destiny lies. What I encountered was a profound not knowing and uncertainty and evasion of such questions. This not knowing expresses itself in a kind of agnostic resignation that says: perhaps these deeper philosophical questions have no answer, or at least no answer that is accessible to our human intelligence. And so we turn away and get on with our lives the best we can.

But in such turning away we do not escape our situation. Instead we fall into the collective background understanding of our scientific culture. This understanding is not some carefully worked out position developed by those best qualified to decide such matters. It is simply what has managed to survive the catastrophes of the twentieth century. It is something we have inherited by default. It is called scientific materialism. It has survived because it still works, it still delivers the discoveries and understandings that fuel the technological and material progress that have become the justification for the continuation of our way of life. Scientific materialism is what remains of Descartes' dualism when you deny both the existence of God and the independent existence of subjective consciousness. It is an understanding that says: unless something can be verified by means of objective science, then it can only be granted a kind of secondary pseudo-subjective appearance of being, a being that ultimately depends upon the one true reality, i.e. reality as it is revealed by objective science.

As we have seen, scientific materialism, taken as an all-encompassing explanation of our being here in the universe, is a kind of absurd, obviously false, fiction. And yet, as a culture, we are in its grip. That is because it is imbued with a collective authority. To go against it, publicly, is to challenge that authority. It is like challenging the Church in medieval times, only without the inquisition. The inquisition is not needed, because scientific materialism is largely immune to the damaging effects of dissent. Its hold is not that of a positive faith in something transcendental. It is the hold of an absence of such faith. Its faith is a faith in *human* reason. That is why we are able to reasonably criticise objective science without being locked up. For science is not fixed in the way the Church was fixed. It can respond to criticism and thereby neutralise it. And it is not that all scientists are scientific materialists. Quantum mechanics has already shown that our collective, simplistic idea of materiality cannot be ultimately true. But we are like the medievals with their Aristotelian understanding of earthly and celestial being: we can't simply let our current paradigm collapse into chaos. We can only let it go when a sufficiently compelling alternative emerges, something that can coherently reorganise our entire conception of reality.

The reason we are now looking in such detail into the theory of evolution is because this is the pedestal on which scientific materialism stands and upon which its overarching coherence depends. And it is here, in the arguments that developed both within biology and between science and religion, that we collectively believe scientific materialism to have triumphed. The great debate within biology was between *vitalism* and *mechanism*. And this is the ground where Descartes' dualism of mind and matter was finally rejected. Vitalism is the idea that life has a non-material reality that governs or influences the behaviour of living organisms in a way that cannot be reduced to the mechanistic interactions of molecular components. Of course, Descartes himself was *not* a vitalist: he thought plant and animal behaviour could be explained entirely mechanistically and that only the human soul could escape this net of determinism. What vitalism did was to project Descartes' idea of an unextended soul substance from humans into the broader domain of plant and animal being. The basic idea was the same: that there is a non-material reality that can influence material events.

Within biology itself, vitalism attempted to find objectively physical biological phenomena that could not be explained in purely mechanistic terms. So, for example, Hans Driesch[4] performed experiments on the embryos of sea urchins, splitting the cells after their first division, and observing that both cells developed into complete organisms instead of each developing into half a sea urchin. His conclusion was that there must be a non-physical life principle that guides the development of these embryonic cells to maturity. This life principle would be a property of the organism as a whole, as it guides it to wholeness, and seems to 'know' what to do when the embryo is split. In contrast, if there were a mechanistic program in the first cell, Driesch expected that it would be split into two when the cell divided, so the instructions for the left half of the body would go into the left-hand cell and those for the right would go into the right-hand cell.

Now it could have been that Driesch was right, and that objective science had encountered an inexplicable mystery that no molecular interaction could explain. But it turned out here, as for many other supposedly inexplicable biological phenomena, that if you do look closely enough at the molecular interactions and reactions going on in a living cell, that some form of plausible mechanistic explanation can be found for the previously inexplicable behaviour. What is remarkable is not the macro-scale phenomenon, but the extraordinary order and complexity of the micro-scale events occurring in the cell itself.

Once molecular biology had drilled down to discover the double helix structure of DNA, and so came to understand how the genetic material of the parents is passed on to their offspring, it seemed that all the phenomena proposed by the vitalists to show there was some non-material aspect to life, were just examples of a lack of molecular imagination. And so vitalism fell into scientific disrepute. This in turn was taken to be a refutation of dualism itself, i.e. the idea that *any* non-physical agency could have *any* effect on the objectively physical behaviour of *any* physical entity. It wasn't that vitalism or dualism were finally and comprehensively proven to be false. It was more that the project of biological mechanism had proved more successful than vitalism, in terms of making discoveries that led to actual outcomes.

At the same time I think we should acknowledge that the discoveries of molecular biology have been extraordinary. For we could conceivably have lived in a universe where the interior of each living cell was the site of an essentially magical process, where the cell material was moved according to entirely new principles that only operated where there was life. But this is not what was discovered. The same kind of chemical processes occur in both organic and non-organic material. The only thing that seems to distinguish organic material is the complexity of those chemical processes. Hydrogen is still hydrogen and it still binds with carbon inside a living cell in the same way it combines outside a living cell. And the behaviour of a living cell still appears to be explicable entirely in terms of the interactions of its molecular parts. No reliably repeatable exceptions to this have been found, despite all the research. So it really does seem that vitalism, at least as it was originally conceived, is false. For if there were such a life force, and a benevolent God were behind it, then why would He attempt to so effectively hide its presence in all this extraordinary molecular complexity?

But this still leaves us with the question of the origin of the living cell itself. For, finally, biological vitalism, in thinking that the processes of morphogenesis were too incredible to be explained mechanistically, had simply highlighted the miracle of the creation of a living cell in the first place. This is where Darwin handed things over to God. And this is where modern synthetic biology is working away in the quest to build a living cell out of components of inanimate matter. If this can be achieved, it would represent the final vindication of mechanism over vitalism.

Of course, whether such synthetic life is possible is still a matter of speculation. But objective science has moved a long way since the discovery of DNA. We now have complexity theory and the idea of self-organisation to explain how chains of chemical reactions that assemble self-replicating hydrocarbon molecules could have emerged billions of years ago from the primordial chaos of the oceans. In all those trillions of trillions of molecular combinations occurring in that primordial molecular nursery, we only needed a handful of self-replicators to emerge, for, of course, each molecule that *they* assembled would then assemble *another copy* of itself, and then another, and then another, and so on, until all the available suitable material became self-replicating. And how long would it take for one of these molecules to find itself surrounded by a membrane of other molecules, and for a circular chain of reactions to occur that was able to sustain that membrane? Once we have a self-replicating chain of chemical reactions enclosed in a membrane that is also sustained by these reactions, then we already have a living cell.

To our modern understanding, that ocean becomes an inconceivably complex parallel search algorithm. It doesn't need to 'look' for chains of reactions that reproduce themselves. They simply arise by 'chance' and then by the simple act of reproduction, transform their surrounding environment into a new stable system that is more 'complex' than the one that came before. Was this effect somehow 'contained' in the primordial chaos as an already existing idea? Hardly. If a simple disordered system is able to move through many different states of which the vast majority are unstable and a tiny minority show some form of stability, then, given enough time, the stable states will predominate. Why? Because an unstable state will quickly change into another state (for that is what it means to be unstable), whereas a stable state will tend to maintain itself (for that is what it means to be stable). So, despite what Descartes thought, in certain circumstances, when the conditions are right, we should *expect* stable complex forms to emerge from the chaotic activity of less complex forms. That circumstance, in all the billions and trillions of planets in the universe, certainly appears to have occurred here on Earth. That doesn't mean there was a hidden cause that had the idea of self-replicating cells already in mind. It is simply something that will naturally emerge, given a big enough search space. And the universe certainly looks like a big enough search space.

Do you begin to see what a mighty idea it is that lies behind this ostensibly simple theory of evolution? It takes away all our notions of the need for a transcendental agency to guide the course of our physical existence. By means of such thinking we have cast the net of our reasoned explanation back to the moment of the Big Bang. Now God is only needed to set up the initial conditions. Here He is supposed to have fine tuned those initial conditions in just such a way as to ensure that complex living systems like us could emerge. But why should we start with just one universe? What if there are trillions and trillions of different universes, just as there are trillions and trillions of different planets, with each universe having slightly different initial conditions? Then we no longer need an intelligent agency to pick just the right conditions for life to appear. In fact we don't need an intelligent agency at all. It is equally as conceivable, and much simpler, to think that all these trillions of universes just created themselves. Why should there be something more? Isn't that just the way *we* think? Always searching for intelligible reasons?

THE CAUSAL CLOSURE OF THE PHYSICAL

What we are missing here is a consideration of the background within which our evolutionary thinking is moving. In the clash between mechanism and vitalism that background was still essentially Cartesian. What was being put to the test was the idea of an absolutely non-material causal agency having effects on an absolutely material substrate. If this were to occur in ordinary life, we would think of it as a kind of miracle, like Jesus feeding the five thousand. The conclusion that objective science draws from its failure to reliably detect the effects of such a non-material agency, is not that there is a problem with the original division between the absolutely material and absolutely immaterial. Instead it concludes that everything that happens in this material universe is entirely determined by the interactions of microphysical particles. In philosophy this idea came to be known as the *causal closure of the physical.*

The belief that everything is finally explicable in terms of purely physical interactions is *terribly significant.* For while the great majority of people remain unaware of exactly what Descartes and Darwin had to say, this basic message of scientific materialism has still percolated down into the day-to-day life of the ordinary citizen. That is not to say that everyone agrees with this message. It simply sits there in the background. When there is a diagnosis of cancer, you are free to seek help from alternative therapies that work on your immaterial 'life force'. But inside the hospital, in the medical trials, in the laboratories that are studying the behaviour of the cancer cells, it is scientific materialism that rules. Of course it may help if you develop a positive attitude concerning your survival and recovery. It may even help if you believe in a non-material agency that has the power to reverse the progress of the disease. But finally, from a medical perspective, your positive attitude is only going to be effective insofar as it leads your brain to release chemicals that have real physical effects on the disease process.

Once upon a time, and not so long ago, it was widely accepted that diseases could be healed by faith alone, by the power of holy relics, or the touch of a saint's hand. But we no longer believe that. Doctors do not take electives in faith healing at medical school, and there is no office for faith healing in our large metropolitan hospitals. If we, as a society, truly believed in the power of an immaterial agency to alter the course of events, then we would pay for it with our taxes, as we did when we built the great cathedrals. But we don't do that any more. Before we lay out our money, we expect to see hard clinical evidence. And the hard clinical evidence tells us that our diseases have physical causes that respond to physical interventions.

And yet, collectively, we still believe there is a small corner of the universe where an immaterial agency really does have the power to change the otherwise fixed course of material causation. That is the place of the freedom of my personal will to determine my own bodily actions. It is here that we are allowed our Cartesian illusion. However, when objective science concluded that vitalism cannot be true, that did not just apply to microbes or liver cells. It also applied to *brain cells*. What follows from the causal closure of the physical is the idea of *epiphenomenalism*. This says that if everything that occurs in the universe is determined by the physical interactions of microphysical particles, then the fact of my being conscious can have no effect whatsoever on those microphysical interactions. And that means my behaviours and inner states, including all I say and think and feel, are completely determined by these microphysical events. On this view, the feeling I have that I am directing the course of my life is just that, a *feeling*. In 'reality' the microphysical particles in my brain 'decide' what I am to do and think, entirely according to the logic of their mechanistic molecular interactions. When I feel I am choosing which shirt to wear in the morning, I am simply having a conscious experience of a certain brain process that is *calculating* which shirt to wear. There is no immaterial 'me' that is directing the events occurring in my brain. 'I' am simply passively undergoing those events. This is experienced as a freedom of will because it is a consciousness of the process that is *actually* determining what will happen. It is the process itself that is 'free' in the limited sense that it has several courses of action open to it, and, until it finishes its calculation, it is unclear which one will be actualised. If someone had a gun and ordered me to wear my checked shirt then no calculation would be needed, and I would lose the feeling of being 'free'. Such 'freedom' is not that of an immaterial Cartesian being who can step in and somehow interfere with the mechanical unfolding of my neurological machinery. It is simply the feeling the machinery experiences while it is making up its mind. And for an ideal observer, who knows exactly how the machinery works, this is no freedom whatsoever. For it is pre-ordained in the wiring that on this particular day, in these particular circumstances, my calculating brain has no choice whatsoever but to 'choose' the checked shirt.

Taken in isolation, this story of choosing a shirt, may seen quite inconsequential. But what about the broader context of your entire life? What scientific materialism is saying is that *everything* you do is determined by the mechanism of your physical body in its interactions with the equally mechanistic environment in which it moves. Your consciousness is like the consciousness of a spectator at the movies. In 'reality' there is nothing you can do to change the way this movie is going to play out. The entire stream of your experience is completely fixed, and all your thoughts, your finest feelings, your moral anguish, are just the calculations of an extremely sophisticated matrix of microphysical events.

It is here that we encounter the edge of the net of our objective physicalistic reasoning. Everything we experience is now caught in that net except our *consciousness* of that experience. That is the last remaining vestige of inexplicable immateriality. Our being has been reduced to that of a causally inert witness that experiences a complete identification with the pre-ordained events occurring in the mechanism of a human brain. In this dream of scientific materialism, we are no more than conscious automatons experiencing the execution of a neurological program. Here there is no God. At least not for us. We are abandoned to our fate, to endure, to suffer, to enjoy, to die, all according to the roll of the micro-physical dice. *→ Also God!*

This is where our Cartesian dualism has reached. Brave reason embarks on its quest to construct a universal science guided only by its own natural light. This is to be achieved by dividing up the territory of our immediate experience into the mutually exclusive regions of immaterial mind and physical matter. But the sciences of the mind begin to founder. And philosophy falls into disorder. Only the objective sciences of physical matter are able to make sustained progress. Soon their explanations reach from the birth of the universe in the Big Bang to the birth of self-consciousness in the human brain. These objective physical sciences are not arrogant or dogmatic. If they had found evidence of a divine creator, or of an immaterial life force, or of human consciousness freely determining the course of physical events on Earth, then this would have been declared as a great discovery. But no such (unambiguous) discoveries have been made. What has been discovered is remarkable enough. Just look around you, at the world, and how it has changed in the last four hundred years.

Now, pure consciousness, seen through the eyes of the objective sciences, appears as a kind of incomprehensible enigma. According to its objective observations, if consciousness were somehow eliminated from the universe, it would make no difference whatsoever to the manner in which physical events unfold. *Excellent!* There would still be physical human bodies, and they would still walk and talk and build computers and write books like this. The only difference is that there would be no one there to experience what is happening. From an objectively physical point of view, it seems there is no *reason* for consciousness to exist. For if it has no observable physical effects, it could not have been advantageous for survival. That means we cannot invoke Darwinian evolution as an explanation. And if we can find no physical reasons for the development of consciousness, that suggests it was an inherent potentiality in physical matter from the very beginning: something co-given along with the potentiality of matter to form into discrete particles, that was only waiting for the right circumstances in which to manifest itself. But if physical matter always already had this inherent possibility within itself, then we can no longer think of it as purely extended physical stuff. For if mind-as-consciousness is potentially present in all matter, then mind and matter form an essential unity that we cannot subsequently separate into two mutually exclusive independent regions of being.

At this point, to make things absolutely clear, we should remind ourselves that objective science's assertion that consciousness is causally inert is *obviously* incorrect. The demonstration is our ability to communicate about consciousness. Such communication is clearly having an effect in the objectively physical world of atoms and molecules. The question is, how did those microphysical particles in our brain manage to self-organise themselves in such a way as to form an idea that their collective activity is being experienced as something conscious? We can explain how a brain could form a concept of a physical object from correlating the activity of the various sensory input streams. But which input stream is it that informs the brain it is conscious? There is no such stream. For objective science has already decreed that consciousness is causally inert. And yet, if it were really causally inert, then it could have no effect whatsoever on the behaviour of any neuron in any brain. So how could the brain form an idea that it is conscious? It couldn't. But my brain has definitely been able to form such an idea. And not just on the basis of someone telling me about it. I can actually verify that I am conscious in my direct experience of being conscious and I can talk about this experience as an experience that is directly dependent on my actually being conscious. That means, for *me* it is certainly the case that my being conscious makes a difference to my behaviour. How about you? MYU conversational reality

So what are we to conclude from this brief history of scientific materialism? Firstly, it is important to see that it is something we have inherited. It started out as a dualism of mind and matter whose coherence was guaranteed by the intelligence of a benevolent creator God. But the Age of Reason demanded that everything have a reason. So it was not enough to call on God to act as the intermediary between the world of physical objects and our conscious experience of that world. It required empirical demonstration. And then attempts at such a demonstration ran into difficulties. It turned out that without faith in God, we could have no real certainty about the existence of a physical world outside of our immediate consciousness. And yet common sense told us differently. Finally Kant began to see into the background from out of which our experience of objectivity is projected into space and time. Again common sense held its ground. For the independent reality of the physical world was not to be seriously doubted. And so our dualistic understanding soldiered on, philosophically battered, but defiant. For a while, following Locke, we kept to our faith in God *and* the natural light of reason. But Christianity was in retreat. The more reasonable we became, the less reasonable seemed our faith. And then came Darwin and the birth of evolutionary thinking. With this tool, objective reason began to piece together an entirely different narrative to the one found in the Bible, a narrative where God was no longer needed for the creation of the plants and the animals and the human race. This net of physical cause-effect explanation slowly replaced what remained of our Biblical inheritance. The entire drama of sin and redemption and divine purpose for humanity was transformed into an essentially nihilistic vision of a struggle for survival, where any sense of meaning was something human intelligence had to create for itself within a universe that seemed entirely indifferent to our existence. This message was then driven home with some force during the course of two world wars. With no collective living faith, and no reasonable explanation for the being of our consciousness here on Earth, the developed world fell into a state of indifference concerning the ultimate questions of meaning and turned its attention to the task of immediate material self-gratification. In so doing we made gods out of science and money and material and technological progress. As individuals we no longer possessed a coherent understanding of what we are doing here on Earth. If questioned we would probably express some form of the old dualism, claiming to have a free will that we exercise in an otherwise material world. But underlying this lies a ruthless scientific materialism that understands human beings as little more than conscious automatons to be manipulated and ordered according to the self-interests of collective corporate entities and national governments.

It is this scientific materialism that is now embedded in our culture and holds sway over our lives in a way that remains strangely concealed. It operates as if it possessed a kind of self-evident truth. But this is not the case. It is a purely human invention, a manner of thinking, passed down to us from Galileo and Descartes, that has become sedimented into those very institutions of health and education and commerce and government that now direct the course of our lives. Despite the extreme sophistication of our technologies, we remain unable, as a culture, to seriously reflect on the situation in which we find ourselves. We don't even recognise the manner in which this understanding operates. Instead we dream of a life we imagine we are living, a continual projection of a future fulfilment that lies just around the corner. Meanwhile, back in 'reality', we live increasingly robotic lives, driving robotic cars down robotic highways, watching robotic news and entertainment programs, exchanging robotic messages via robotic social media platforms, electing robotic governments and performing robotic tasks in robotic organisations, whose robotic aim is to make us so robotic that we can eventually be replaced by actual robots.

THE CHURCH OF REASON

Of course, our lives are not entirely robotic. The previous tirade is a kind of caricature that covers over the true complexity of our situation. It is an attempt to indicate something that is occurring under the surface of our day-to-day consciousness. It is here that scientific materialism operates less as a philosophical position and more as a psychological condition. It is because it lies at the margins of consciousness that it exercises such profound effects on our behaviour. For if it were brought fully into the light of our *collective* consciousness it would lose its compelling character. Why? Because it is based on an obviously false notion of absolute physical being; a left-over remnant of an incoherent dualism that was only held together by our faith in a creator god.

It was the idea of Darwinian evolution that precipitated the collapse of this dualism of mind and matter. Now we stand in the wreckage of that former conception of being. The collapse itself is not something to be regretted. For our dualism was based on a story of creation that no longer corresponds with the facts as we understand them. What is to be regretted is our collective inability to arrive at a coherent understanding of being that *does* correspond with these facts. At a certain point, as a civilisation, we simply gave up on this search for coherence. I only have a vague notion as to why this happened. Certainly the two world wars were pivotal. Before the final catastrophes of the 1940s there were clear signs in the development of phenomenological philosophy in Germany of the emergence of just such a coherent understanding of being, an understanding that had evolved out of the transcendental idealism of Kant.

Atomic/nuclear weapons/power?

Back then there was still a coherent European intellectual culture, a kind of leading edge of consciousness that was connected up with the rest of society in such a way as to influence its path of development. But the world wars destroyed all that. Since then all we have known are counter-cultures. It is not that we have lacked the intelligence or creativity to develop new and more coherent understandings. It is that the underlying mass culture has become impervious to fundamental change. Of course our counter-cultures have had significant effects on the way we treat each other and on our taste in music and on our choice of lifestyles. But these effects have been assimilated into the mass culture in a way that has left its essential form unaltered. What is that form? It is the scientific materialism of corporate capitalism.

What this scientific materialism is attempting to do is to regulate the world in such a way as to ensure the survival of the corporate and governmental entities that are attempting the regulation. Behind the gloss of the surface of this technological world, there lies a logic of manipulation of the human race in order to co-opt them into the service of these abstract organisational entities. This logic of manipulation sees each human being as a kind of input-output machine. The ongoing task is to discover the rules that determine the behaviour of these machines so we can cause them to buy our products or work more effectively in our organisations or vote for our parties in the election. The idea that each of us is an end in ourselves does not enter into the equation. As purely material, mechanical entities, our being is our behaving, and our behaving is determined by the inputs we receive and how these inputs interact with the machinery of our bodies. To influence one of these machines to buy one of our products is not some kind of denial of intrinsic worth. For each person simply *is* a machine, completely determined by microphysical events. If it is not our advertising that determines things it will be some other equally mechanical impulse that sends you to the beach and makes you buy an ice cream.

At this point you may be thinking of counter-examples. Like the provision of social health care or social education. Don't these endeavours respect the intrinsic self-worth of the individuals concerned? Here we should remember that it is only a conscious human being who is capable of respecting the self-worth of another human being. So it is not that there is no such respect being shown to us by the people we meet, who

+ twelve context

care for us, who teach us, or who serve us at the till in the supermarket. We are speaking here of the *organisations* that collectively make up the structure of the society that implement strategies that seek to determine the way we behave toward each other. For example, consider my experience in the Australian university system. In order for the government to transfer the money to the university that paid my salary, the university had to show it qualified for that funding. That first meant attracting enough students (for funding is tied to student numbers). And in order to attract those students we had to compete in a marketplace against other universities and 'degree providers'. And that meant creating a corporate 'image' and advertising that image on billboards and television and social media. To back this up we had to provide 'objective' evidence of the quality of our education. And that meant making sure our students provided good feedback in questionnaires and did not drop out before finishing the degree. And that required us to attend courses to learn techniques to ensure that students did indeed give good feedback and did not decide to drop out. Inside the classroom, my basic concern was for the human individuals before me, with their understanding, and with their enjoyment of being in the class. There it was still a matter of our respect for each other. But outside of class, in my interactions with the management system, things were different. It was there that scientific materialism took over. Now, if someone had gained something of intrinsic value from what they had learnt it counted for nothing. What counted was the averaged student satisfaction score that was obtained from the end of semester questionnaire for that class. Of course there would be a link between that score and the intrinsic value for the students. But the system of regulation is only interested in the score, in the behaviour of the students, in their placing of ticks on questionnaires, and not their inner experience. How could it be? It is a mechanistic system of rules, not a conscious human being. But it is this mechanistic system that determines the functioning of the university as a whole. It is no longer run according to the collective experience and agreement of the individuals concerned. Instead our jobs become a matter of meeting objective targets set by the rules of this objective system. People even think that it is *fairer* to develop such a set of rules because it eliminates 'personal bias' in decision making. But this is not simply the elimination of personal bias, it is the elimination of our conscious human intelligence from the process of deciding how we are to live our lives together. It is here that we hand ourselves over to a scientific materialism that, from then on, treats us as the robots we have allowed ourselves to become.

This transformation of the university systems of the developed world is typically understood to be a form of semi-privatisation, where formerly inefficient government run enterprises are subjected to market forces in order to make them more competitive. And that is thought to be a *good* thing. But behind this surface explanation there lies the projection of a scientific materialism that understands the university, and all the people involved with it, as purely objective physical entities whose behaviours can measured, predicted and controlled in just the way that any other physical system can be controlled. In such a projection, the subjective experiences of the people involved are also understood objectively. The ideal would be to measure the actual events occurring in the brains of students and staff and work out how to manage things from there. But for now there are technological and legal barriers that rule out such an approach. Instead we have to indirectly measure these 'inner' events by means of questionnaires and observations of behaviour. Once we have a sufficiently detailed model of the university and the environment in which it operates we can develop strategies to enable it to compete more effectively to secure the staff and students and funding needed for its continued survival. The terrible logic of this situation is that unless the university adopts this process of objectification and control it will *fail* to compete effectively, it will lose students, staff will lose their jobs, and finally it will cease to exist. For funding depends entirely on the ability to control the objectified measures of performance that the university is obliged to make public.

There is a certain irony at work here. For, within the context of our global civilisation, the university system represents the very principle of reason itself. It is here where science in all its purity is to uncover the truth concerning the being of all aspects of the universe. The idea is to be guided by nothing but the empirical fact of what presents itself to our immediate experience and the natural light of reason that enables us to construct theoretical understandings to account for those facts. It was well understood from the outset that the progress of such science depended on institutions that were relatively free from the control of the powerful self-interests that have more or less free-rein in the rest of society. Such freedom was granted because it was clear that science and the technological development it facilitates are the new engines of economic survival and progress. It was from out of this citadel of reason that the scientific materialism of Darwinian evolution emerged. But now, almost as if it were decreed by some retributive cosmic law, it is the university system itself that is to be re-organised according to those very principles of materialistic evolution that it gifted into an unsuspecting world.

The irony is to do with the idea of scientific truth. For once we submit to the logic of scientific materialism as a principle of organisation, then truth loses its position as the ultimate value around which a university is constellated. Robert Pirsig saw this when he described the university as a *Church of Reason.* Scientific materialism is unable to grant truth any *intrinsic* value, because intrinsic value is *immaterial.* As a pure materialism, it can only grant truth a *pragmatic* value. For if our understandings are true, that means they *work* in enabling us to predict and control events that we would otherwise not be able to predict and control. True ideas or true understandings now become *products* that the university is paid to produce. It is no longer truth, as an end in itself, that is the paramount value, it is the *survival* of the organisation. This is how the Church of Reason becomes *deconsecrated.* The transcendent intrinsic value that it once served is no longer present. Of course, there are still individual researchers whose inspiration is the pure pursuit of truth. But they are now strangers in an institution that serves another end.

This deconsecration of the Church of Reason is shown in the way it now presents itself to the world. Instead of factual prospectuses that list the subjects that are on offer and the factual publication of the grades that are expected for entry into each course, the task of attracting students is handed over to the public relations and advertising departments. Here we meet with a very different idea of truth. Rather than uncovering the deep structure that lies behind our subjective impressions of the world, these departments are concerned with *managing* these impressions. This generally means creating an imaginary picture of the kind of life you will have if you enrol in one of the university's courses. Truth only matters here in the sense of not being caught out in telling an obvious lie. For that would create a *bad* impression. Unambiguous lies are easily avoided once you leave the world of facts behind and enter into a possible world of how your life *could* be. This art of creating a good impression also extends into the world of research, only here there is less room for manoeuvre. What

happens is that external bodies come and measure how you are performing. Generally these bodies publish the metrics they use in advance. Now
the task is to create the impression you are doing good research by scoring
well on those metrics. Promotion comes to those who score well, and not
to those who spend three years on a problem that so engages them that
they can think of nothing else.

DARWIN'S CHILDREN

The underlying technology that has facilitated the spread of scientific materialism into nearly every aspect of our lives is the computerised digital
information system. The connection here is so close that we can even understand our global computer networks to be physical manifestations of
the idea of scientific materialism. This goes back to our consideration in
Chapter 5 of the mathematically ideal structure of scientific theory and
the scientific models we derive from those theories. As we saw, these models 'exist' in a domain of pure ideality that is abstracted from the ground
of our immediate perceptual experience. Scientific materialism then proposes the existence of another reality 'outside' our experience where these
mathematically derived abstractions (i.e. microphysical particles and energy fields) have their 'real' being.

With the advent of digital computer technology this world of mathematical ideality has started to manifest 'out here' in the technologies with
which we interact. For the digital computer is essentially a *mathematical
machine*. It is mathematical because its function is to manifest the ideality of a binary state in the physical actuality of an electrical charge. Computer input and output, and the programs that operate on the input to
produce the output, are all forms of ideality expressed in a mathematical language the computer is physically designed to implement. Finally
we can think of each stream of input and output, and each program as
a *number*. For a computer program has a completely unambiguous and
precise mathematical form. Because of this it can be fed into a 'compiler'
and transformed into a pattern of 'bits', i.e. an extremely long sequence
of two alternating states. We think of these states as representing 'zero'
and 'one', or 'on' and 'off', but in actuality they can represent anything
we want them to represent. In themselves they are simply an expression of
a relation of *difference*. For all that distinguishes one state from the other

is that the first state is different from the second state. Once we have such a relation of difference we can encode anything we care to express within our ideal mathematical universe. So when we say a computer program is a number, that simply means we can take the bit pattern the compiler has produced and *interpret it as a number*.

For example, if we call our first state α and our second state β, then we can let $\alpha = 0$ and $\beta = 1$, and use binary (base 2) arithmetic to turn our bit patterns into numbers, i.e., $\alpha\alpha\alpha = 000 = 0$, $\alpha\alpha\beta = 001 = 1$, $\alpha\beta\alpha = 010 = 2$, $\alpha\beta\beta = 011 = 3$, $\beta\alpha\alpha = 100 = 4$, $\beta\alpha\beta = 101 = 5$, and so on. In this way each and every digital computer program can be transformed into an unambiguous number, as can its input, and its output. And if we let the number of the program $= c$ and the number of its input $= i$ and the number of its output $= o$, we can represent the running of any computer program by the expression $c(i) = o$, which means: program c is exactly equivalent to a mathematical function that transforms i into o.

Now this reasoning does not just apply to scientific programs performing calculations on the equations of quantum physics. It applies to WhatsApp and Skype and Zoom and Facebook and every email program and every website. What our computer technology is doing is creating a secondary ideal mathematical world model that represents the actual experiential world in which we live. It is here that those aspects of our lives that are relevant to the functioning of the organisations that make up our global economic systems, are systematically measured and recorded by the computerised information systems of those organisations. Each such information system contains an abstract mathematical model of that part of the world it is measuring, made up of abstract entities that represent the actual people and things and events that are being measured. These entities are specified exactly and unambiguously in the logic of our α β binary states. It is *we* who assign meaning to these states, just as we assign meaning to these strange squiggles we call written words. For, despite what we have been led to think, an information system *in itself* does not contain any information about the world. It is our human understanding that first creates the information system and then bestows meaning on the patterns of binary states it contains. Even the idea that it contains patterns of binary states is a meaning we are bestowing.

What is happening in all this frenzy of recording is that huge swathes of our experiential life are being captured and transposed into a parallel mathematical representation of the world we live in. Once such a parallel representation is available, our information systems can perform calculations to detect patterns or regularities that would otherwise lie hidden. These regularities allow us to predict the future behaviour of the worldly objects and events that are represented in these patterns. We can even predict what *would* happen *if* we were to change that world in some way, such as how our profit margins would alter if we were to reduce the wages of our employees.

This entire process of measurement, mathematical representation, detection of hidden regularity and deduction of future consequences is *exactly* founded on the template of Galilean science. Now, instead of studying the mathematical regularities in the trajectories of falling bodies, we use the same approach to study human behaviour in the relentless pursuit of greater efficiency and potential profit. And the more our ability to capture digital information grows (through cameras and microphones and sensors and cookies), the greater becomes the reach and ability of these digital information systems to influence and ultimately control the course of events on Earth.

Of course, it still appears that we, the human race, are in charge of these systems, and that it is up to us how they are used. For it is we who use our information systems to monitor the processes of climate change and so recognise the actions that are needed to maintain climate stability. What counts is whether or not we take these actions, not the recommendations of the information systems themselves. And it is equally the case that without our corporate information systems it would be impossible to coordinate and maintain our current ways of doing business together. It seems it is up to us whether we use these systems to exploit each other further, or to create a more even distribution of wealth.

But thinking this technology is neutral only hides the reality of the situation in which we find ourselves. For, in our use of information technology as a means to monitor and control the world we live in, we are *already* acting out of a scientific materialistic understanding of being. It is the technology itself that embodies this understanding, and our using of it to order our human world actually causes the world to conform with this same understanding.

We could say that the technology was neutral, *once.* For it was we, the people who first developed the information systems, who transferred our own scientific materialistic understanding of being into the manner of operation of these systems. But once the systems are in place, they come to embody the very understanding of their creators, just like any other human social organisation. The difference now is that these understandings are mechanistically fixed in the way we are required to interact with the system, i.e. according to rules that are embedded in the interface by the person who programmed the machine.

Here we are no longer dealing with a conscious individual. We are dealing with a digital machine which controls the manner in which we can interact with it, and, insofar as our interaction with the rest of the world is through this machine, it also determines the manner in which we act on the world. Pretty soon, that machine starts to replace actions that were once a matter of our discretion. Now it works out your salary and transfers the money to your bank. It decides which shifts you are going to work, who is going to work with you, where you are to go, who you are to see. At the end of the shift, the only information that counts is the information you input back into the machine, and again it is the machine that decides the form of that information. More and more the space within which we can operate with any freedom of discretion is narrowed down. Once that freedom reaches zero that means the machine is doing everything and you no longer have a job. And once the machine has replaced you, we have one more sector of the world that is functioning according to a pure scientific materialistic understanding of being, where the events that formerly required the presence of human consciousness are now entirely determined by processes of physical cause and effect that involve no consciousness whatsoever.

As a society we do not see the underlying character of what is occurring. Slowly, systematically and inexorably we are developing technologies that are designed to remove us from having anything to do with the actual running of the world we live in. It began with the machinery of the factory and the field that removed us from the land and from the occupations of our creative handiwork. Now it is concerned with mechanising our perceptual and judgmental capacities. Each new step along this path unfolds according to a certain irresistible logic. Just as the army of dockers was never going to stop the advance of the shipping container, neither are the guards and train drivers going to stop the advance of the totally automated train. Each innovation comes with the same justifications: it is more efficient, it is safer, it is more reliable, it is going to make our lives easier, it is going to save us money. In the end, the human beings whose way of life is being destroyed are seen as being selfish. For why should we, in the rest of society, be expected to subsidise their way of life by paying more than we need to for the service they are providing? And if these arguments do not work, there is the more drastic mechanism of extinction: for if an organisation exists in any kind of competitive environment, and it fails to deploy a technology that will make it more competitive, it will simply be destroyed or taken over by an organisation that does deploy that technology.

Here we see, enacted in the world before us, the very process of mechanistic Darwinian evolution that we theorise was responsible for the evolution of the species. Indeed, the fact that our technological systems appear to evolve in this way is seen as a further justification for continuing on this path of technological development, as if it were our destiny to complete what nature imperfectly started. This thinking then moves to another level when we consider the possibilities of genetic engineering. For it is here that the scientific materialism of Darwinian evolution stops being a mere theory and turns into a technology for the mechanistic development of new kinds of biological automatons.

Behind this mechanistic transformation of our way of life there lies an understanding of being that we have unthinkingly inherited from a past that could not possibly foresee the consequences of that understanding. *We* are the consequences of that understanding, and it is up to us to make that understanding conscious. For the way we are living is *absurd*. Instead of organising our society to serve us, the *human beings*, we organise our society to serve our *organisations*. And so we sacrifice ourselves, our capacity to live meaningful, creative and cooperative lives together, lives where the *quality* of our actions comes first, to the service of the very organisations that are systematically replacing us with robotic information systems that destroy the meaning and significance and value of our working lives.

To repeat: the understanding of being that lies behind our way of life is the scientific materialism of corporate capitalism. Scientific materialism is an understanding of being that defines what is real in terms of what can be objectively measured. What can be objectively measured are purely physical things and events. What scientific materialism leaves out of consideration is human consciousness itself, as something that has an

No! Given in to dualism have!

independent existence and value that is distinct from the purely physical processes that are associated with our verbal reports of being conscious. It is not that scientific materialism denies there is such a thing as consciousness. What it denies is that it has any intrinsic value in itself or any independent causative power that cannot finally be reduced to the objectively physical interactions of physical particles. What corporate capitalism adds to this scientific materialistic understanding of being is the idea of *money*. Money is the objectification of human valuing. Just as scientific materialism says that all different forms of being can be reduced to one ultimate form of being: the objectively physical, so corporate capitalism says that all the different forms of human valuing can be reduced to one ultimate objective mathematical form of value: *money*.

This is not some kind of philosophical argument, as if there were a large group of people who are attempting to prove that only the physical is real and the only form of value is money. We are concerned here with a certain understanding of being that is embedded in the technologies and organisations that determine the way we *actually* live together in society. These organisations hide their underlying form because they are in the business of managing *us*. Such management typically involves the creation of a false image, as if the organisation were a friendly person who really cares about you and your thoughts and feelings and opinions, as if your subjectivity really mattered. But underneath, of course, your subjectivity is irrelevant. The intention is to persuade you into certain courses of action. And yet how easily we collude in this deception. For it is often the case that a corporate representative really is being friendly and feels themselves to be positively identified with their organisation. Such positive identification is itself managed in training days designed to foster a 'team spirit'. Here it is not a matter of some sinister intent or conspiracy. Each organisation, in order to survive, has to manage the impressions it creates in the human beings on which it depends, i.e. on its employees and its customers, or, in the case of governments, on the people who vote for them. This is done quite shamelessly and openly. The idea that an organisation that must compete for its survival should tell the truth about how it operates and about the real reasons for its actions is not even recognised as practical possibility. That truth is generally only spoken by people within the organisation, meeting in private, in those spaces where the intrinsic value of our human consciousness is still recognised. Every now

and then a truth teller will crash out into the open. But our corporate organisations know how to deal with such eventualities. They simply create
an alternative plausible version of events, one that steers carefully around
the known facts in such a way that it cannot obviously be refuted. Once
such an alternative explanation is in place, the seed of doubt is sown. Cu-
mulatively, in our wider society, this creates such a field of disinformation
that it becomes impossible to tell what is really going on. It is this field
that neutralises the democratic power of the people to change the underlying direction of progress that the scientific materialism of our society is
determining.

One hardly needs to be a psychic to see where things are heading (if left
unchecked). The scientific materialism that rules in the research establishments of the deconsecrated Church of Reason is systematically working
toward uncovering the mechanistic principles that underlie the physical
functioning of the human brain. Once these principles are understood
they will inevitably be used to construct machines that perform brain-like
functions. This is the very area in which I worked during my career as a
university academic. Indeed you may think of this book as a kind of debt
I am repaying to humanity for having assisted in this endeavour. Once we
have machines that can think like the human managers of our corporate
organisations, those managers, like the dockers and train drivers before
them, will be replaced. Here it is not likely that we will greatly miss their
human touch or human understanding, for they were already required,
by the logic of the organisation itself, to implement the same principles
of financial scientific materialism that will now be programmed into their
robotic successors. These new artificial managers will just be better than
the old ones: faster, fairer, more efficient, more rational, making our lives
easier and saving us money. They will outcompete the remaining humans
by making superior decisions. To begin with it will only be lower-level
managers that are replaced, those concerned with the day-to-day running
of things. But, if our new artificially intelligent managers really do think
like humans, only better, and faster, directly accessing vast memory stores,
seeing deeper patterns, and predicting further ahead, then this process of
substitution is inevitably going to reach to the top. For it is a dog eat dog
world, as they say.

Do you begin to see what kind of madness this is? Our entire global
economy is heading towards a form of total robotic organisation. The

artificial 'intelligences' behind this won't just be controlling economic production, they will be directing the public relations and advertising departments. They will be running the media and entertainment industries. They will be controlling the lobbying groups that are seeking to influence government policy, policies that themselves are being implemented by robotic government agencies. And they will be launching and defending legal cases designed to further each organisation's self-interests. What about the judges? Do you think that they will still be human? And yet this is now the trajectory of our 'civilisation': to replace ourselves with machines. This trajectory reflects an unconsciousness in which we have left ourselves out of our own future, just as scientific materialism leaves our consciousness out of its understanding of what truly exists. For what are we supposed to do once these artificial intelligences have taken over the running of the world?

Psychologically, I think we are caught in a kind of regression back to an infantile state. Our collective intention is to become completely idle selfish consumers of the goods and services supplied to us by our completely robotic technological systems. Our machines will be like our mothers and fathers, providing us with everything we need, while we will have nothing in particular to do, but enjoy ourselves. Hooray! This is the dream of a child: to be on some kind of continuous summer holiday, without having to think or take responsibility for anything or anyone else. Somewhere in this dystopian future I see the final perfection of a totally convincing form of virtual reality. It's already there in our contemporary science fiction movies, like *The Matrix* and *Avatar*. Our artificial intelligences will only have to provide us with little cubicles and sufficient nutrition and muscular stimulation, and then pipe us into the new virtual collective world of humanity, where everything is possible. Instead of carving up the physical earth, you can simply imagine the house you would like to live in, the person you would like to be with, and lo! there it is! You'd like to fly over this virtual earth like an eagle? Lo! there we are! You have some more sinister fantasies that you would like to explore? No problem! Of course, it doesn't have to be completely imaginary. If you want to, you can still interact with the avatars of 'real' human beings. But after a while, as the software develops, I wonder if you will be able to tell the difference between a 'real' human transmitting from their cubicle and a pure software entity. For, after all, in a world of pure scientific materialism, it's all the same whether the process that is generating the experience is situated in a living brain or in an electronic circuit. And if we assume that our software is going to stimulate those regions of the brain that 'real' experience stimulates, then our virtual experience is not going to be just like 'real' experience, it's going to be *even* better, *even* richer. For if there is a perfect or ultimate state of pleasure, then these machines are going to find it, and they are going to recreate it in your brain. Do you begin to see that the idea of leaving the cubicle is going to become less and less attractive?

Of course, one problem is that if you don't leave the cubicle, then how are we to make babies? Well, perhaps our robotic helpers can arrange that too. Once they have harvested an egg and some sperm, they could place baby in an artificial womb, while we experience an artificial pregnancy, along with all the hormonal dimensionalities. Even the father could find out what that is like. Once mature enough, we could place baby in their own cubicle. And yet, after all, wouldn't it be easier to have a virtual baby? For the whole process of growing a real physical body is very expensive. We could read your DNA and the DNA of your chosen partner and make a *perfect* baby. And then, of course, with birth, there is the problem of death. The wonderful thing about the virtual world is that your virtual body does not age, or get sick. Back in the cubicle, as the technology develops, even a serious illness can be hidden from you. We just block the signals to the brain. And if we can block the experience of physical illness, what about the experience of mental distress? What if you started thinking about the ultimate meaning of your cubicle life and this made you depressed? Shouldn't our benevolent artificial intelligences intervene here as well, and block such negative thoughts? In fact, do you think perhaps we would prefer not to know anything at all about our having a physical body back in the cubicle?

I think you can see where this is going. If our intent is simply to experience one long dream where every wish we can have comes true and where there is no hardship or suffering or disappointment, and we program our artificial intelligences to deliver this for us, then, the most logical outcome is for those machines to protect us from the reality of our mortal situation. They will simply engineer our dream for as long as our bodies can support a normally functioning brain, and then, one day, when the physical brain starts to fail, we are (compassionately) switched off. Of course, in order to protect those who continue to dream, our virtual presence continues to appear in *their* dream. But finally, the day is going to come when the last body in the last cubicle dies. And that will be that, as they say. Perhaps the rest of life on Earth will have survived this experiment. And perhaps the robots will switch themselves off as well. Or perhaps the whole virtual world will continue on with virtual versions of ourselves continuing their dreams without there being any physical bodies involved. A kind of meaningless mechanism, devoid of consciousness and purpose. In fact, just the kind of meaningless mechanism that scientific materialism understands our universe to be already. Only this time there really will be no consciousness in the machine.

This picture of the demise of human consciousness is not intended as a prediction of what will *actually* happen. It is an attempt to bring to light the kind of future that is *already* implied in the way we are living. My more realistic prediction is that our current scientific materialistic understanding of being is going to collapse long before we can develop the kinds of technologies I have envisaged. For scientific materialism is only a passing phase in the development of human consciousness on Earth. It has certainly proved useful as a framework for the development of a mathematical understanding of the underlying objective structure of the universe. But as an ultimate understanding of being, it is certainly false. This falsity can be demonstrated in any moment we care to become conscious of being conscious. And yet such insight does not transform our *collective* belief in the ultimate and obvious materiality of things. Such belief can be partially dispelled if we live for a time in a different kind of culture, like an ashram, or if we go into a solitary retreat. But in our everyday lives, going to work, watching the media, doing the shopping, talking to friends, it seems we cannot help but share in the fundamental outlook of the culture in which we live. Of course we can consciously disidentify and separate from such collective phenomena, but they nevertheless remain active, and influence the way the world seems to us. We could think of this as a kind of *spell*. Despite one's conscious rejection of a collective belief, it still has an influence in the background. That is why we need to chase it down together, and consciously examine how it operates in the fabric of our everyday experience. And we especially need to see the pernicious effect it is having on our way of being together and on our relationship with the living planet. For it has turned the earth and all its living creatures into a collection of resources to be exploited in the competitive evolutionary survival of the organisationally fittest. Here it is not a matter of government legislation. It is a matter of a fundamental change in our understanding of ourselves, of the meaning of our being here, and of the being of the animals and the plants and the earth and the oceans and the skies and the manner in which they are manifest to us.

#themoonbelongstoeveryone

NOTES

1. This is a reference to the first lines of William Blake's poem *The Tyger*: "Tyger Tyger, burning bright, In the forests of the night; What immortal hand or eye, Could frame thy fearful symmetry?" (1789/2019).

2. Charles Darwin (1809-1882) published *On the Origin of Species By Means of Natural Selection* in 1859. In it he proposed a theory of evolution that today forms the foundation of modern evolutionary biology.

3. "The force that through the green fuse drives the flower" is the opening line of a Dylan Thomas poem of the same name. See *Selected Poems* (1974, p. 34).

4. Hans Driesch (1867-1941) was a German biologist and philosopher, who explained and defended his version of vitalism in *The History and Theory of Vitalism* (1914).

PART 2

The Resonance of Intelligence

CHAPTER 8

THE HUMAN CONDITION

THE MISSING MANUAL

IT was only after my experiences at Warwick University that I came to find it strange that no one had ever sat me down and explained exactly what it means to be a human being. Of course objective science already knows about the evolution and history of the human race. We even have some rudimentary psychological theories and therapies that are designed to help if we start showing symptoms of mental distress. But the underlying 'philosophy' of my upbringing was that human children are much the same as any other mammal. All that is needed is for us to be thrown into the world as we find it, and, with the help of our instinct for imitation, we will quite naturally learn to adapt and adjust ourselves to that world. If there is a task, it is to successfully navigate this process of adaptation, and to find some niche where we can enjoy a healthy and prosperous and fulfilling life.

Now, *God knows*, to achieve that, in this modern world, is not an easy task. For what does it mean to lead a fulfilling life? To live in a state where all your desires are continuously satisfied? Is that even possible? For our desires conflict, and what seemed desirable from a distance can dramatically change its character when we come to hold it in our hands. After a while, when it becomes clear that it is impossible to satisfy all of our desires all of the time, we start to become more realistic. Perhaps it is enough to simply not be *unhappy.* And yet, to manage that, we need to inquire into what it is that we *really* need and not just what we *think* we want. For there are certain basic needs that underlie the play of our surface desires, such as our enjoying reasonable health, sufficient food, comfortable shelter, human fellowship, loving sexual intimacy, engagement in creative and meaningful activity, and a connection with the natural beauty of the earth. To be without any of these for long is going to create a lack that naturally turns into an intention to do something about it. But, for those of us fortunate enough to live in a modern affluent society, who have our health, these needs are by no means impossible to satisfy.

And yet many (most?) of us are unhappy. Perhaps not all of the time. But there will be some aspect of the life that is persistently causing us trouble. Typically we get caught in a conflictual situation from which we find ourselves unwilling or unable to escape, such as a difficult job or relationship, or a persistent addiction. And when we go into the roots of these conflicts we find that what is playing out in the external world also has roots in our inner lives. It is now that we discover the presence of the *human psyche.* What do we know of this place? Well, to begin with, it is where we *actually* live. It is the origin from out of which our entire experience of living here on Earth emerges. When I said earlier that no one ever sat me down and explained what it means to be a human being here on Earth, *this* is what I meant. For me, in my childhood upbringing, as far as the adults around me were concerned, it was as if there were no such thing as the human psyche. Everything was oriented around what happens 'out there' in the space of the external world. As for my actual home, my being here in the midst of this experience of being human in a humanly projected world, it was as if this psychic ground did not exist.

Of course it was recognised that I had thoughts and feelings occurring in an inner world. But these were treated as another kind of 'thing', as private possessions that only I could see, but which made themselves known by various signs and symptoms. And when these thoughts and feelings manifested themselves in the external world, it was my job to manage them so as to adapt myself to everyone else. How I was able to do this, I did not know, and neither, it appeared, did anyone else. It all worked on a system of approval and disapproval and of reward and punishment. These external checks and restraints seemed to work well enough, at least in my case. So long as I behaved well, then all was considered to be well.

'Adapt to everyone else' is only concept of relationality so far.

What was missing was any kind of understanding of what was really going on. We were all caught, mother, father, teachers, friends, in a great stream of tradition. We did as the others did. As children, if we were asked why, then we would say we have to behave this way, otherwise we are told off. And in this system of behaving, we assumed the adults knew what they were doing. But as we grew up, we found that the adults did *not* know what they were doing. They too were doing as the others do.

What strikes me now as being strange, is our almost complete lack of collective interest in understanding the principles of the functioning of the human psyche. For it is not that we lack the capacity to develop such an understanding. Think of all the training and learning we engage in, the years at school, then at university or at work, all the books and talks and manuals, the essays, the homework, the vast stream of information we absorb, and how all this is considered so important for our success and happiness. And yet, in all this learning that is designed to prepare us for our future lives, where is the course and the manual that tells us about the functioning of the human psyche? I don't mean the self-help book you pick up at the local bookstore. This issue is something fundamental for our entire culture, something that involves everyone, and that can't be fixed by a few people reading a few books. It concerns the basic ground and being of our lives together—such as whether we spend decades stuck in a pattern of suffering that was passed on to us by the ignorance and unconsciousness of the people who managed our upbringing. Or was I again asleep in the class at school where they taught us all about the effects that our childhood relationships with our parents were going to have on our future lives? I don't think so. As teenagers we were simply thrown together and left to try and work things out for ourselves.

The trouble here is that because no one ever sat us down and explained what it is to live in the human psyche, we do not even know what that would be like, and we do not even miss the knowledge we do not have. We simply take for granted that to live here on Earth is to live like everyone else, and everyone else is, by and large, simply getting on with their lives. We do not question the foundations of our being here. There's no time for that.

At the same time, there is the almost complete failure of psychological science to arrive at any kind of settled or unified understanding of our situation here. So even if we feel a compelling need to investigate this ground of being, we are again confronted with the labyrinth of human opinion. Before the Second World War there was Jung, and Freud, and Adler. Now these various schools of psychology and psychotherapy have fractured, like the protestant Church, as each individual theoretician and practitioner has headed off in their own direction, according to their own inspiration and experience. Added to that are all kinds of new techniques of psychic access and control, like cognitive behavioural therapy and mindfulness training. Recognising this almost complete lack of coherence in our mainstream culture, there are now the many hundreds if not thousands of self-proclaimed spiritual teachers, who lack even a basic foundation of experience in dealing with the psyches of other human beings. Instead they base their practice on the direct experience of their own journey of descent.

Of course, as we have seen, this situation is an understandable consequence of the scientific materialism that lies at the heart of our technological civilisation. For, in an essentially materialistic world, the human psyche becomes a kind of fiction, an immaterial epiphenomenon whose 'emanations' are entirely determined by the physical functioning of the human nervous system. As a consequence, all the forms of psychology that take the psyche seriously, by granting it its own kind of non-mechanistic being and inheritance, are bound to become marginalised. Anything that strays outside the orthodoxy like this is generally labelled as a 'pseudo-science' on the basis that it cannot be 'falsified'—which simply means its essential subject matter cannot be measured and objectified. In this way, mainstream applied psychology becomes concerned with the development of techniques that can produce measurable and objective outcomes in the behaviour of its 'subjects'. To support this endeavour, theoretical psychology becomes neuropsychology—the study of the physical processes in the brain that supposedly determine our psychological experiences. In all this objectification, the human psyche, as it is in itself, as we actually know it to be in our immediate experience, *disappears.*

And so, after all, I should not have been surprised that no one was speaking to me of the reality of the human psyche in my early life at school and university. Like everyone else, I had to pick up what clues I could from my own experiences and from the writings and teachings and songs and artistic creations of those few others who had turned their intelligence inwards.

THE RETENTION OF NOW

So how are we to proceed? For the human psyche is not an object. It is not something we can get outside of and study at our leisure. In some paradoxical way 'I' emanate from this place, while at the same time remaining unified within it. And yet my consciousness does not encompass the entirety of the psyche. In any one moment, there is only consciousness of the aspect that is manifesting now— my consciousness *is* that aspect. But the psyche is more than a collection of aspects. It is that from out of which this immediate content of consciousness has emerged. It is, if you like, our interface with the unthinkable source of experience itself.

Our task is to ask after this human psyche—a psyche that remains essentially concealed to us. We are asking after our situation here, our 'thrownness' into the world.[1] We are asking after this experience of being human. And yet, in order to phenomenologically engage in this inquiry, we only have this Archimedean point of consciousness, this ground of our immediate contact with what *is*. Here we only know what it is to be conscious. That does not directly reveal the structure of our being in the psyche. For we are *being* that being. Do you see the situation? We can easily speak of 'egos' and 'ego death' and relate that to some experience that occurred in the past. But what about now? This moment. The sound of the birds singing. A slight feeling of hunger. The clicking of the keyboard. Insofar as I am immediately present then 'I' have disappeared. Only when I reflect on this immediate experience do I regain some sense of separation. If we stick with what is revealed in each moment of consciousness then I can certainly ask after the phenomenon of a feeling of hunger. But what of the structure of the phenomenon of the being of my egoic self and its relation to this place of immediate consciousness? For, in this moment of immediate consciousness, my egoic self does not appear before me as something I could directly investigate. And if I attempt to *reflect* on this egoic self, then it still fails to appear. For, in ordinary reflection, it is the egoic self that is doing the reflecting. This leads to the same infinite regress of reflection reflecting on itself that we encountered in Chapter 2 (page 28). Are we then to conclude that the ego is some kind of illusion, that it does not *really* exist because it cannot be directly observed? No. Our ego cannot become an immediate object of investigation for us, not because it does not exist, but because it is a *way of being*. As such, it can only become manifest insofar as we are *being* that way of being, and our being that way means we cannot, at the same time, observe that way of being, because, to engage in a pure phenomenological witnessing is to be in *another* way of being.

But all is not lost. For even though we cannot directly encounter this egoic way of being, we can at least inspect the traces it leaves behind. Here we meet with Husserl's notion of *retention*. This is like a 'buffer' where the immediate past of experience remains present as it fades away. As a demonstration, try listening to one of the sounds occurring in the environment around you now. It is best to pick something you cannot perfectly predict, that lasts a few seconds, and that is being repeated with some regularity—such as the sound of a bird singing. The task is to listen to the way the sound endures in each moment. In this enduring we hold on to the beginning of the sound as something we still hear, for a while, as being present, even though the moment of that beginning has past. This is not like a memory, where we reconstruct an experience that has already ended. The beginning of the sound remains *sensorily* present as the sound continues. And then, as the beginning fades away, what follows it also *sensorily* endures and fades away. If our experience of sound were not extended in this way, we would literally hear nothing. It is only by means of such retention that the form of the sound is able to be 'stretched out' so as to be capable of manifesting as a direct sensory experience. William James[2] described the structure of this experience of temporality as a kind of *saddle* (see Figure 8.1).

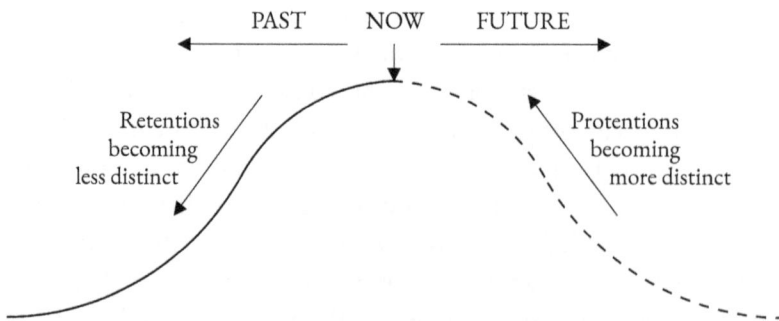

Figure 8.1: William James' saddle model of time perception.

At the top of the saddle is the immediate now point from out of which the leading edge of the sound is emerging. This is like a stream of water emerging from a hidden spring. In the instant the sound is experienced it becomes part of this stream of past that flows away down the saddle in strict time sequence, so that the further away in time a particular sound form moves, the more it decays, until it slips out of immediate sensory consciousness altogether and joins with the intuition we have of the 'full form' of whatever it is we are hearing. This form also includes the front of James' saddle, which is our protention or expectation of the sound that is to appear next. And just as the already emerging sound decays to make way for the ever arising source, our future protentions decay in the direction of the future, becoming more and more indistinct the further away they are from the source of emergence 'now'. In this way they flow 'up' the saddle toward now as they become increasingly more definite and distinct until their moment of actualisation, at which point they are immediately transformed into an element in the retreating flow of retention.

Of course, we do not actually 'hear' these protentions in the way we hear the retentions. They are sensorily empty expectations that only properly reveal themselves in the moment of their sensory actualisation (or in our surprise when they do not actualise as we protended). Finally, as soon as the sound and our protending of it ceases, the entire structure of retained sensory sound experience also ceases—for we do not hear the end of the sound decay in the way we hold on to the decay of the beginning. Once the sound ends we have our complete intuition of its form. And that means we no longer *need* to hold onto it, as it no longer contains any uncertainty for us.

At the same time, we must recognise, in the immediacy of temporal experience, that there are no actual elements or points of protention or retention in the stream of time consciousness—the stream is the streaming of a flowing unity. It is only when we stop and reflect and take the experience apart that we freeze this flow into particular elements. In actuality there are no future now points coming toward me or past nows flowing away. There is just this extraordinary stillness and stasis of the ever present now through which all experience is continuously passing. It is only on the basis of this stasis of now that there can be anything like an experience of flow, for the flow is only flowing relative to that which does not flow, which is now, which is my being conscious now.

catalytic integration

form = retention/harmony

It is the phenomenon of retention that allows us to perceive our flowing experience as it slips away into the past, for just as long as it is retained in the present of a present retention. It is here that we can catch the *phenomenon* of the functioning of our egoic self. In fact we have already been employing this approach in understanding what it means to think. For each moment of thought leaves its own retentive trace in the time stream of consciousness so that when we stop thinking and enter a state of pure witnessing, we can still catch the retreating presence of our former thinking state. It is this retreating presence that informs us we are no longer thinking without our having to think "I am no longer thinking". Instead we can directly 'see' the distinction that lies before us, as the retreating thought is swallowed up in stillness.

Our task, then, is to turn this witnessing consciousness onto the retained traces of our egoic activity, and so uncover the manner of being of the ego itself. But, in order to do that, we must already know what it means to be engaged in egoic activity. And that means we must already have some idea of what the ego *is*. But if we look back on the preceding chapters, and all the references to being in an 'egoic state', and having an 'egoic self', and experiencing an 'ego death', there too it was assumed that we already knew what the ego is. This was based on the further assumption that each of us have, in some way, already experienced a cessation of the egoic state, and that we came to know this egoic state by means of such cessation. But nowhere, as yet, have we explicitly come to examine exactly what it is that we mean by this word 'ego'.

So how are we to recognise the functioning of this 'something' that we only become aware of when it ceases to function? And, indeed, what does it even mean that it ceases to function? Do you see how language works here? How we use words with great certainty whose meaning remains quite obscure to us, especially those words that refer to our inner states. We may have an implicit sense of what we mean, but if we are asked to 'spell it out' then we find, perhaps with some embarrassment, that we do not explicitly know what we are talking about. For that we need to engage in a further inquiry. And yet here a straightforward witnessing will not do. For it is not as if we have this ego directly in front of us, like the cat in the garden. We first have to delineate what we mean by the word 'ego' by looking into the way we actually use it, collectively, to refer to something that is manifest to us all, both inwardly and outwardly.

THE EGOIC SELF

Originally, in English usage, 'ego' was the Latin word for 'I'. But its meaning altered after it was used to translate Sigmund Freud's[3] concept of 'das Ich' (literally 'the I'). After that, it came to represent a personal centre of agency existing within the greater horizon of the human psyche, a horizon which also included (for Freud) the Id (das Es or 'the It') and the super-ego (das Über-ich or 'the Over-I'). Freud's Id was a repository for the basic (unconscious) instincts or drives and for hidden or repressed memories, and the super-ego was his repository for the moral conscience. This model, and its further development in the various schools of psychoanalysis, has had a significant influence on our collective background understanding of ourselves. It spread through our intellectual culture via books and literature and film and via the opinions and pronouncements of experts, even when those experts disagreed with Freud and his intellectual descendants. This is analogous to the way that all scientific ideas influence our background understanding, even though we may only hear of them indirectly. The key idea that psychoanalysis introduced to the world is that our egoic self exists within a human psyche of which we are largely *unconscious*.

As a society, we have slowly come to recognise how this personal and collective unconscious can influence and even determine our behaviour in ways we can generally only recognise after the event, by means of analysis and reflection. For instance, when we meet an authority figure who behaves in any way like our father, we will tend to unconsciously project the situation of our childhood onto the events occurring in front of us. We will see motivations and judgments and character traits that are not necessarily present in that authority figure. We will become emotional and accusatory. Then later, if we are fortunate, someone we trust will point out that we were behaving strangely, even childishly, and that perhaps we had mistaken this person for someone in our past. In such a recognition we can have a direct insight into how unconscious processes are able to take over our normal state of consciousness and project stereotypical judgments that generally only partially correspond with the actual situation. Such automatic judgments are perhaps useful when being attacked by a tiger, but not so much in the complexity of modern civilised society.

In the midst of the recognition of an unconscious projection, there is a certain 'someone' who is 'taken over' by the projection, who later realises it was a projection, and who subsequently adjusts their behaviour in the light of that realisation. This 'someone' is my *conscious ego*, or, as we would normally say, my *self*. From the outside, looking at someone else, this self appears as the person who is living the life of that person's body. It is not the entirety of the psyche, with all its unconscious processes and motives. Neither is it the momentary state of that person as they present themselves to us now. It is their enduring, psychological, egoic self. And when the person says "I feel unwell" or "I will meet you tomorrow" or "Do you remember when I mistook that man for my father?", this 'I' refers to that egoic self, just as the Latin translation suggests.

The being of this egoic self becomes more clearly delineated when we recall it is the same entity that the law holds to be responsible for its actions. For if we are sleepwalking, or under hypnosis, or in the involuntary grip of a strong emotion, then the law will recognise we have *diminished responsibility*. This implies our egoic self has a domain of agency that (assuming we are *not* under some involuntary or unconscious compulsion) allows us to freely determine what we say or do, so that we *can* be held responsible for what we *actually* do. And again, according to the law, it is the same enduring self who committed a crime in the past (even decades ago) who is to be held responsible for it in the present. This implies my freely deciding egoic self is essentially who I am, and that, even though my life experience will have changed the way I behave and think, it is still the same 'me' who acts now, and who is responsible for how I acted in the past. This inheritance of responsibility is based on the further idea that we are responsible for the person we have become, because we have become that person largely on the basis of the free decisions we have made in the past.

Of course the idea that we can freely determine what we do and become is highly questionable. It certainly does not agree with our objective understanding of how the brain is determined by purely physical events. And yet the idea that we are free, autonomous, self-determining individuals is part of our collective understanding of what it means to be a human being. It is not just a matter of law. We directly *feel* another person is responsible for what they have done long after they have done it, especially if they have harmed us or someone we care about. At the same time we feel an enduring *personal* responsibility for our actions, born of the conviction that we are, when circumstances allow, capable of determining these actions for ourselves. This implies the presence of a certain kind of *agency* that *holds sway* over our actions, that has at its disposal an array of habitualities and capacities that it is free to enact or not to enact according to its own private calculation or preference. If we take all this together: the capacity to freely reflect and decide on what we do, combined with all the various capacities we have at our disposal, in their sheer potentiality, and we recognise that each of us possesses our own manner and style of reflecting and deciding and acting, we begin to see the outline of our egoic self. We can think of it as a kind of ongoing process that is present during large portions of our waking lives. If it is functioning coherently then it is generally the case that our living life is also functioning coherently, insofar as external circumstances allow. We find we are able to hold down a job, enter into enduring relationships, procure adequate shelter and food, and so on. And these are not unimportant achievements. If the egoic self starts to break down, then so does our ability to function in society. For it acts as a kind of gatekeeper in relation to the rest of the psyche. It understands what needs to be done to keep the current life functioning and it understands that certain drives and impulses arising in the psyche cannot be 'acted out' without serious consequences.

So what was all this talk in Part One about ego death and moving beyond the egoic state, as if the ego were something we had to get rid of? For the death of the ego would presumably mean its disintegration, and that hardly seems desirable. To see this one only needs to read the psychiatric literature, or to have been close to someone who has experienced a psychological breakdown. After witnessing such an event, you would not say that the ego is something illusory or fictional. It is an actual process, or way of being, that holds our otherwise disordered psychic states in some kind of purposeful unity, and connects us with the egoic processes of the people around us. It may be that there is some illusion in our understanding of this ego, or in the ego's understanding of itself. But as a way of being, it simply is what it is.

The Ground of Resolution and Preference
But is this ego who *I* am? Let us examine this question. It is something we already touched on in Chapter 2 (pages 24-29). There, as an example,

we looked at the interplay of the various intentionalities that were arising as I attempted to observe myself breathing during the Vipassana meditation. But who was this 'I' who was attempting to meditate? Do you see how easily we assume the presence of an agency that directs the course of our lives? It is built in to the way we use language. And yet, in this case, there certainly seemed to be *someone* who was interested in watching the breath and who was needed in order for the meditation to continue. For there were several other 'agencies' operating in me who did not want to meditate. Despite their interruptions, this someone was resolved to continue, and keep those other intentionalities in check. It is this *overarching* intentionality that we generally identify as our egoic self.

The first thing to observe is that this self is not a particular, active intention. Active intentions come and go, while the egoic self and its view of the world endure. I may get angry, I may get depressed, I may hate my job today, and feel that I can't go on. But, as with the Vipassana meditation, or my writing of this book, *something* keeps us going in a certain direction, as least for longer than the duration of the next impulse. The Vipassana experience was particularly instructive because it set up a situation of *conflict*. Ordinarily we allow one thought to follow another to follow another while maintaining a general sense that 'I' am the one who is thinking these thoughts and that it is up to 'me' whether I think them or not. In the Vipassana experience this 'I' decided to watch the breath and to *stop* thinking. In such a situation it soon becomes clear that 'I' am not in charge of my thinking processes, at least not for more than a few seconds or minutes. And yet this resolve to meditate was still effective. For despite my thinking about girlfriends and motorbikes, I did not *act* on those thoughts. My *body* remained seated and as soon as the current daydream ran out of steam 'I' again started to watch my breath. Of course, this all relied on my *resolve* to carry on. I could have 'decided' at any time to have got up and quit the entire project, but I did not. So what can we say about this resolve?

Firstly, it is clear that you cannot simply decide to have sufficient resolve to stick to a certain course of action. The resolve, its strength and tenacity, is a capacity that is already present beforehand, that you bring to whatever it is you intend to do. It is partly determined by how meaningful or valuable or necessary you find the task to be, and partly by the strength of your already existing general capacity to 'stick things out'. And yet things are not quite so simple. For you can influence how attracted you are to a goal by using your imagination, or by looking at advertising material, or by talking to other people (but only if you are resolved enough to do so). Or you can develop your general level of resolve by boasting to people of what you will achieve, or by undergoing some form of training (but again, only if you are resolved enough to do so). And yet, even if we take these avenues of increasing our level of resolve into account, 'you' are still not really in charge. For there is an infinite regress at work: you can only increase your resolve if you have the resolve to do so, or if something happens that independently acts to increase your resolve (such as your standing to lose a lot of money).

OK, you may say, perhaps I cannot directly determine how resolved I am, but I can at least determine the action or the goal that I am committing myself to. But here again, you are only going to commit to something if you are sufficiently attracted or compelled by other reasons to do so. And like your level of resolve, your level of attraction is not something that is under your control. It is more like an oracle that issues its decree. I remember the last time I was looking to buy a car. I would research all the vehicles on the internet that suited my fussy criteria, pick what seemed to be the best one, and look at all the photographs, stoking up my feeling that this indeed was a car I wanted to buy. And then I would call the owner and make an appointment, full of hope. But it was only when I saw the car and owner in person that the oracle would silently speak its yes or no. Often I would try and ignore the no, which generally came in the first few seconds, so that I could prolong my dream, go on a test drive, and generally justify the hours I had wasted on another fruitless journey.

But hang on, you may say, perhaps these background capacities and preferences lie beyond my control, *in this moment*. But where have they come from? Haven't they been formed on the basis of all my past experiences of choosing, and deciding, and preferring, and resolving and sticking things out? Isn't it the case that I have *learnt* to stick things out because of all the *little* decisions I have made, all my little successes and failures, and because the consequences of these actions have directly shown me that it is better to stick to what I resolve than to follow my momentary impulses? And likewise with my preferences. Isn't what I prefer really the distilled wisdom of all my previous experiences of gaining and losing? Take the example of buying a car. The reason I didn't want to buy the car is that I could see, underneath the polish, that it had not been well cared for, and that the internet pictures were deceptive. In my foreground consciousness I was trying to ignore this obvious fact because I wanted to avoid a feeling of disappointment. This was not the operation of a mysterious oracle, it was the operation of an experiential capacity to read the signs without having to explicitly think about what they were. I simply *felt* this was not a good car. In the *background* I knew exactly what was going on.

So, once again 'I' emerge as my egoic self. Perhaps I am not completely in charge of my life. For there are impulses that get past me, that make me do and think and say things that go against 'my better judgment'. But, by and large, I am living the life I choose to live. I can see that in any particular situation my level of resolve and my underlying preferences are relatively fixed and that my actions are fairly predictable. But, the point is, that this resolve and these preferences are *my own creations*—they have come out of all that I have done and experienced in the past—they are the living embodiment of that past, that expresses the person I am. They are like a work of art, an expression of my inner character, something I have grown over the years and that now, quite naturally form and guide my actions.

The Free Agency of the Ego
And yet, if we are to save this idea of the ego as a free agency, there must be a domain in which it is *genuinely* free to determine the course of its own life. For we have only put this moment of freedom back into the past by saying that the ego created our resolve and our preferences out of its previous actions. The question is where, in all these actions, does this domain of egoic freedom lie? If we look carefully, it is not present in the everyday actions of our lives. For in each such everyday moment we find our actions are simply performing themselves. We can demonstrate this by again slipping back into a state of immediate consciousness and observing our behaviour. If you wait and don't 'decide' to act, you will find, in the moment that an action occurs, that there is no egoic self involved. The body just moves—it does what it does. Even if the egoic 'you' decides to do something beforehand, in the moment of the action, it is still the body that 'does' it. There is no other entity standing over it, guiding how the leg is raised, or calculating where the foot is to be placed. It all

happens by itself. This is especially obvious when it comes to touch typing. It's even worth learning to type just to be amazed at what the body is capable of, how it knows how the words are spelt, how the sentence is going to end, without the slightest intervention of any supervisory self. Or consider playing tennis—how the body knows how to place a ball on the other side of the court—and how our egoic thoughts and concerns about winning can actually interfere with this bodily skill.

But again, all these bodily skills that we so effortlessly enact, were once unavailable to us. We had to *learn* to walk, and to type, and to play tennis. These are now capacities we have at our disposal. We no longer have to guide them—but we did once, when we were learning. We had to practise. We had to *think* about what we were doing. Surely it is here that we finally come across the domain of our real egoic self? Yes. The place of the ego is the place where we *think*. It is here that we consider our actions, watch over our behaviours, and come to our decisions. To do this we have to engage in a special form of thinking called *self-reflection*. This is where, instead of simply being the being that is enacting whatever it is enacting, we turn our thoughtful attention on *ourselves*. In so doing we *objectify* ourselves. This objectified self is something that is both me and not me at the same time. For my being is the being of immediate consciousness now. And yet, within this immediate experience of being it is possible to think about the being that I am. Such thinking objectifies my being in a way that is quite natural for us, because that is the way we think about everyone else. What we do is to make ourself into an *other*. And yet, at the same time that I am thinking of my objectified self like this, I am also subjectively present as the one who is doing the objectifying. So I am not thinking about myself as I am now. I am thinking about myself as I would look if I were someone else. We are in a hall of mirrors here, in which the possibility of deception and error starts to loom large.

We need to look at this form of self-objectifying thinking very carefully. When we inhabit this place, we obtain a sense of being separate from the person that we are, the person who actually performs the actions of our lives. We find we can think about this person and judge them, and make plans and decisions on their behalf. In this separation from our self and the immediate circumstances of our life, we obtain a sense of *freedom*. We are no longer condemned to simply live each impulse as it arises. We can think about each situation of our lives and we can imagine it turning out differently. We now have this place of reflection where we feel we can make things better for the person we are thinking about. At the same time, the extraordinary fact is that this person who we are thinking about, from whom we have separated, is actually who we are, and we have not really separated, for it is this self-same person who is doing the reflecting.

Now, it seems to us, in the moments of our reflection, that this ob-jectified person who actually lives our life, is a poor kind of figure, who is pushed around by inner impulses and external circumstances, and who is in need of our help—the help that only we can provide from this place of separation and freedom from compulsion where we see the bigger picture of both the inner and outer life of our objectified doppelgänger. And, of course, in a way this is true. For we do possess this valuable capacity to ob-jectively reflect on our lives. But, in another way, it is not true at all. The error comes when we think that the place we inhabit when we reflect on ourselves betokens the independent existence of a separate self, possessed of its own free will and power to decide the destiny of the life upon which it reflects. For the reality is that it is the same poor schmuck who lives the life and thinks about living the life. To see this we must examine exactly what transpires in our acts of self-reflection. The question is, from where do our thoughts arise?

A classic example is the demonstration of our free will in relation to our bodily actions. Let's say someone asks you to prove that you have free will, and you say: "OK, in one second from now I am going to raise my arm", and then, right on cue, you raise your arm. Here we encounter the amazing phenomenon of the 'I can'. All I need to do is to *think* of a certain action in the right way, and that action actually occurs. There is no more to it than that. I predict it will occur and lo! it occurs. Here it is entirely up to me what action I choose, whether I stop it half way through, or whether I do nothing at all. All paths are open to me, none is more attractive than another, for there is no external compulsion or reward.

Surely here I am completely free in what I decide to do? And if I am free here, then I am free in relation to when I go to the supermarket, and in relation to what I say or do next. Perhaps in the bigger events of life things are more or less predetermined, but at this micro-level, it seems I am freely deciding thousands of little actions all day long. And as these ac-tions accumulate from childhood onwards, as a result of my freely playing and amusing myself, I learn to do things and form the habits and capacities that grow into that very work of art that is the adult life we spoke of earlier. But we still have the question: where did the thought-intention that caused me to raise my arm come from? Did 'I' decide that it should arise?

If we look carefully enough, from a place of immediate consciousness, at the retained trace of such a thought-intention, we will see that it simply arose, by itself. There is no other reflecting self that decided the thought should arise. The entire idea of my personal freedom resides in the idea that I *could have* intended something else. But the truth is, in that moment when you did intend to raise your arm you *could not* have intended anything else. The thought-intention simply arose and the action started to unfold.

Ah, you might think, ask me again, and this time I will do something else—that proves I am free. But that only proves you are free from any internal or external *coercion.* The next time you demonstrate your free will by doing something else, the same process will occur: a thought-intention will arise to do some action or other and lo! that action will occur. And yet in the midst of all this, you have no idea what caused that *particular* thought-intention to arise rather than another.

But now, of course, I feel the neuroscientist on my shoulder insistently saying: "Exactly right! The reason you cannot say what caused that particular thought intention is because it was caused by physical processes in your brain of which you are unconscious. These processes are not at all mysterious—they are the result of low-level microphysical interactions that determine all your supposedly free actions according to the physical regularities of your nervous system. Your behaviour only looks free and undetermined because of the extraordinary complexity of this underlying physical system." From here it is but a short step to understand all of our finest philosophical thoughts and transcendental intuitions as being caused by the mindless interactions of this same extraordinarily complex physical system. Again we have to pause and say to the neuroscientist within: "I know, immediately and directly that I am conscious and that this knowledge of being conscious cannot be explained in terms of the interactions of purely physical neurons." That means we can only accept the neuroscience as partial truth, valid within the context of its physicalistic assumptions. And yet, at the same time, we must recognise that both the neuroscience and phenomenological experience agree that the notion of there being a separate, truly independent egoic self, that exercises its free choice over our actions, is simply false.

What occurs, as we can verify when we consciously observe the phenomenon, is that thought-intentions continually arise into consciousness independently of the willing or not willing of any egoic self. That does not mean these thought-intentions are causally inert. For they clearly have the capacity to produce bodily physical actions, or emotional reactions, or to call other related thought-intentions to mind. At the same time, the egoic self thinks that *it* is the cause of at least some of these thought-intentions—particularly those that arise when it is calmly and rationally reflecting on itself and the events of its life. And even though we can clearly demonstrate that the ego is not the causal power it thinks itself to be, this idea of egoic self-determination is hard to dislodge. For we are dealing with something more than just a logical assertion. There is also a *feeling* of free agency that pervades the egoic state.

THE ACT OF IDENTIFICATION AND SEPARATION

So let us pause and examine again what we are doing here. We are asking after the human ego. At least we are asking after something we *call* the human ego. Our way of proceeding is to look into our immediate experience for the traces of activity that this ego leaves behind. These traces provide *clues* as to the *form* of the ego. But the form itself resides somewhere else. We are looking at the traces because the ego cannot become an object for us, like a table or a chair. We have to 'catch it in the act' while at the same time, in a certain sense, *being* this ego *while* it acts. We are in a hall of mirrors and we must tread carefully. For the hallmark of the ego is its activity of reflection. And what are we doing here if not reflecting? Could it be that we are fooling ourselves? That the ego is up to its old tricks, thinking that it is seeing the ego, when, in fact, it cannot see itself, because it is the one doing the seeing. It can only see an objectified *model* of itself. And there is some truth here. For the capacity we are using to reflect on experience is the same capacity that is producing our experience of being an ego. The difference is that we are not reflecting *egoically*. Instead we are looking and seeing from the place of immediate consciousness. There is no attachment in this looking and seeing. We take the model of the ego

? ⓇⒷ brain relatedness of self non-central in wholeness;
Ⓔ brain - egoic 'I' - reflective

that is emerging from our investigation of egoic experience and we *put it to the test.* That means looking into the human psyche in a way that is analogous to our looking into the world. Of course, the human psyche is not at all like the external world. It is the place in which the external world appears. But just as we project a model of perceptual form into the stream of sensory experience in each moment of perceptual experience, we can also project a model of the ego into the encompassing psyche. Then, remaining in this place of disinterested immediate consciousness, we await the psyche's feedback . . . .

What we are waiting for is a kind of *resonance.* What this resonance says, if it arises, is: "Yes, this model you are holding in mind has captured an aspect of that which it is attempting to envisage." This is the same resonance we experience when we perceive an object in the world—the very resonance that bestows actuality on what we perceive, that says the dynamical model you are projecting really does correspond with 'something' here in the unthinkable source of experience. Now, however, in projecting a model of the ego, we do not get an experience of sensory actuality, we get a sense of the actuality of the form of our own manner of being. Of course, the problem with such inner resonance is that it is impossible to verify objectively. The best we can do is to confer together on what it is that we find in such inner inquiry, searching for the right language, to see if there is a correspondence between us. For, even though I am asking after the *impersonal* form of the human ego, it could be that I am only seeing aspects that are peculiar to me alone. Or it could be that my experience of resonance means nothing more than my having a feeling of affinity for the model I have created. Do you see how much depends on the correspondence of our seeing into this matter? It is not enough just to imagine what I am seeing. We have to test this out together.

So how do things stand with this model? As we saw earlier, the ego is not *nothing.* We cannot simply dismiss it as some kind of illusion. For there is *something* that reflects on the life and our behaviour, that makes plans and forms resolutions and experiences regret and vetoes certain impulses while allowing and encouraging others. This 'something' certainly appears to 'hold sway' over our voluntary capacities for speech and behaviour, exhibiting a certain style or manner of being that we can recognise in ourselves and in others. And it was this same something that was present when we acquired these capacities, that learnt to ride a b and learnt that falling off again is painful, and learnt to keep getting back on again until we were able to cycle all over town and so learnt that it was worth persisting. In the midst of all these capacities there resides something quite extraordinary. It is the entire unified coherence of our experience of life to date. It is how we have learnt to survive in this modern world. If we consider each capacity that we have acquired since birth as a certain form of behaviour that we have at our disposal, such as walking, reading, writing, talking, driving, dressing, making conversation, delivering a lecture, using a chisel, performing mental arithmetic, reasoning logically, and so on, then we can see how all these capacities are connected together, in such a way that each is unfolded as and when it is needed, and is then enfolded back into the unity of our overall capacity to deal with the situation of our living life.

action focus

I should like to say that this unity of potential capacity is our *real* ego. It is a *process* that is operational in us and not a kind of person. It is not the *only* process, but, if all goes well, it is the *predominating* process that keeps our lives from falling apart, both inwardly and outwardly. It is the *form* of our normal habituality. And it is the means by which we acquire new habits and discard old ones. These habits are the form of our manner of living in the world and with ourselves. And like the monk's habit, it is what we wear in order to deal with our embodiment here on Earth.

So, this is our model: Firstly we are proposing that the ego is a certain manner or way of being that we have inherited and developed according to the way we have lived our lives. Here we immediately encounter a circularity: for the way we lead our lives is at least partly determined by the form of our ego, while what we actually do *changes* the form of that ego. Just think of yourself as you were in earlier life. Look at how your interests have changed, and your self-understanding and your behaviour, and how the egoic self you once were had its hand in setting the course that has led to where you are now. For the ego is not something fixed—its development involves processes of feedback and circular causation like those we encountered in our consideration of Darwinian evolution in Chapter 7. And yet neither does the ego change completely (except, perhaps, in certain extreme circumstances). For we are, at least partly, the consequence of what we intended to be in the past, so that those early intentions not only live on, they can, in the right circumstances, come to a kind of fruition.

Secondly, the being of the ego is centred on our ability to *reflect* on our experience. It is in this space of reflection that the ego manifests itself to consciousness as an activity that is attempting to control the overall direction of our behaviour. If we were to think of this in terms of our current understanding of the brain, the ego would be the psychic counterpart of the executive control function that is thought to be centred in the prefrontal cortex. This control function is something that holds sway over our 'voluntary' capacities to think and act and pay attention. And these capacities can in turn be used to develop new capacities to think and act. So the ego holds sway not just over our actual capacities, but over a field of potential capacities that we could develop if our resolve and circumstances allow.

But what is perhaps most important for us here, is that this idea of the ego also holds sway over what we could call our *intellectual centre*. This is the place where we develop our conscious understanding of the world and our place within it. And it is here that our egoic process can come to think of itself as being the one who is in charge of our lives. Now this way of thinking is certainly understandable. For there are processes occurring in us all the time that appear to be determining what we do. Like my seeing the shopping list in the kitchen and thinking that I really must go to the supermarket. But suddenly I remember the telephone call I promised to make this morning and so I decide to go to the supermarket this afternoon. And then, later in the afternoon I find I do actually go shopping just as I intended to this morning. What happens here is the development of a conviction that there is a person 'behind' all these egoic processes, who actually creates the intentions that are expressed in the thoughts that appear to determine our actions. This 'person' thinks of itself as the one who appears or emerges in each act of reflection. It is the 'I' who is looking at 'me'. If we are to give it a name, we could call it *the thinker*.

What happens in an act of reflection is that the formerly unified world of my immediate experience becomes *fractured*. Instead of my being at one with whatever is happening, I become a separate entity who *looks on* at what is happening. It is in this position of looking on that I find I can think. I can refer to my memory and I can form chains of reasoning that lead to conclusions that are also laid down in memory. From this position of separation it appears to me that I am the one who is determining what happens to me according to the chains of reasoning and association and wishing that are occurring in me. I forget that in the actual moment of action, such as getting into the car and driving to the supermarket, I, the thinker, am no longer there, there is just the experience of getting into the car and driving. I do not notice this because I only notice the things that I think about, and when I think about my getting into the car and going to the supermarket, I immediately insert myself into that situation, and think that 'I' got into the car and that 'I' drove it to the supermarket. But this I of reflection is only there in the moment of reflection. The great illusion that 'I' am in charge of my life is perpetrated in the projection of my 'I' of reflection into every moment of my life. I simply assume that I was present in all those former moments because I am present now when I reflect on those former moments. The one who was *actually* living those moments—lets call that *me*—remains silent, because the I-that-is-me does not separate or reflect—it *is* me—and it remains being-me even in the apparent separation of reflection. For, as we have already seen, reflection is only an *appearance* of separation. The all encompassing being of consciousness is still holding the experience of reflection together, manifesting it to be the experience that it is. All that has happened is that my centre of consciousness has become *identified* with the thought processes that are occurring.

THE FEELING OF BEING ME

Now does this imply that there is *no one* who is in charge of my life, that I am, in fact, *not* responsible for what I do, and that I am a kind of robot controlled by processes occurring 'behind my back'? Here again, like all good philosophical questions, the answer is yes *and* no. It all depends on what you mean by 'I' and 'me' and 'my'. I remember, many years ago, listening to Barry Long saying "I am me", and not fully understanding what he meant. Now it seems much clearer. Insofar as I think of myself as being the thinking reflecting I that emerges when I separate myself from the immediacy of experience, then it is true that 'I' am neither in charge of, nor responsible for my life, and that 'I' am indeed a kind of robot controlled by processes occurring behind my back. This is the great irony of our modern objective scientific understanding of human intelligence. For insofar as the scientists and philosophers who think that we are nothing more than biological automata are identified with their thinking selves, then it is true that they are thinking the thoughts of a biological automaton. But once we withdraw from this identification with the being of a disembodied thinker, the situation changes. For it now becomes apparent there is a *source* from out of which my thoughts and impulses are *emerging.* That source is what we are calling the *human psyche.* And if I stop thinking and stop projecting and become present to how it is in each moment, then there is no longer any activity of a reflecting 'I' looking back on a separate 'me', instead there is the direct experience of this source from out of which all my thoughts and impulses and feelings are emerging, actually *being* me. This is my being. I am that. As Barry said: I *am* me.

We are not talking here about some kind of enlightenment. We are talking about something that emerges quite simply and directly out of a careful and sincere inquiry into how things actually stand for us in our immediate experience. For 'I', as my reflective self, am not the author of the thoughts and intentions that are arising in me. They are just arising. But from where? From within 'me'. So they are 'my' thoughts in the sense that they have come 'out of' me. The entire context that informs them is the context of the life that is also 'in' me, the life that 'I'—the being I am—am living each moment. What we are encountering here is the distinction between the I of reflection and the I of being. The I of being is who I am. I cannot reflect on this I. I cannot create an inventory of its contents. I am this I in every moment that I am, not just in those moments of immediate consciousness, but also when 'I' have apparently separated from this state of unity. For there is still a unity of consciousness that holds this state of separation together. This state of unity can never become an object of reflection for this reflective intelligence because it is the ground from out of which the reflection emerges and, at the same time, it is the encompassing consciousness that illuminates whatever it is that is being reflected.

What is at issue here is the egoic feeling that I am the author of my actions and thoughts and that I am in some way responsible for these actions and thoughts. Not, perhaps, for the thoughts themselves, for there can be many strange and contradictory impulses arising in me. But I do feel a responsibility over whether or not these impulses are *translated* into actions. The question is whether this feeling of responsibility is something that only arises when 'I' reflect on 'my' experience, or whether it is a phenomenon of experience itself. The task is to remain completely present and witness the experience of being a body performing actions in the world. Until now we have been primarily focussing on the experience of perceiving things and events and not the actual experience of being a body—even though this body has been intimately involved in all these perceptions. For I can only see what I see because my eyes are directed toward what I see. We could say that the events occurring that are centred in this body, such as my looking here and looking there, my breathing, my fingers typing these words, are also perceived quite impersonally, just as I perceive the movement of someone else's body, or the movement of an inanimate object. It's simply a matter of the focus of the attention. But there is a difference. For I not only perceive the actions of my body externally, I *inhabit* these actions so that they have a quite different quality in comparison with perceiving the actions of someone else.

For instance, in this moment, there is not simply an anonymous typing producing these words. The typing is related to a certain looking that is occurring in *me*. The words are arising in consciousness in the very moment that they are being externally manifested in the bodily action of my typing. The book itself lies as a kind of potentiality in 'me' and the language, the typing, the movements of my fingers, are emerging from out of that place. Each moment is a projection of the being that I am 'into' the potentiality of experience. And the shape or form of this potentiality is expressed in the shape or form of my body and all the capacities for action that it enfolds. Within this experience there is the place of pure witnessing where I can *bracket* the phenomenon of my inhabiting this body so that I can *see* the way this experience of being a body is manifesting. It is through such witnessing that we can first see the phenomenon of actuality that we bestow on the being of the things we perceive in the world around us.

What we are looking at now is another such core phenomenon: the phenomenon of my experience of being this body. We know this is a phenomenon because it can be disturbed. For example, people who have suffered a severe stroke can have the experience that a paralysed part of their body (typically an arm) belongs to someone else.[4] It is this feeling of ownership that is of interest. It is not something that arises only when we reflect on experience. It is our constant companion, something that clearly distinguishes our body as a special sphere of agency that is objectively distinct from the rest of the world. This feeling of ownership *is* a feeling of agency. It is the feeling of an 'I can' that protends the potentialities of all the capacities that can be brought to bear on the current situation. These capacities are the bodily mirrors of all the hidden horizonal aspects of the world that lie in front of us. For these aspects are only accessible to me because of my bodily capacity to unlock those aspects by, for example, picking something up with my hands or walking over to the other side of the room with my legs and then looking with my eyes at the aspect of the room that was previously hidden to me. It is in this way that every moment of experience is shot through with the protending expectations of the potential bodily movements that lie at my disposal. This entire capacity to freely open up the horizon of the world is experienced as a pure sense of agency—an agency which expresses the sheer potentiality of experience that stands open before me as my immediate future.

It is this phenomenon of agency that the 'I' of reflection appropriates to itself. But agency does not 'belong' to the 'I' of reflection. It does not belong to anyone. It simply expresses, as a *feeling of potentiality*, the degrees of freedom that lie open to the intentionalities that experience themselves as being centred on this body. These intentionalities are not just concerned with bodily actions. They are also concerned with my degrees of inner freedom to think and imagine and remember, to plot and plan, to wish, to cast spells, to pray, and so on. There is something entirely objective about this. It concerns the domain of experience over which these intentionalities *directly* hold sway. Such holding sway means that intending something is enough for it to actually happen. It is this distinction that defines the being of the external world as that over which my intending does *not* have immediate effect. So, for instance, I cannot make these words appear on the screen by simply forming an intention that they appear. I *indirectly* make these words appear by *directly* intending to type them by means of my holding sway of my capacity to move my fingers.

So what of the feeling of responsibility for my actions? Again, there is something entirely objective here. For it is simply a fact that what is directly intended by 'me' has consequences that return to me. So, for example, if I say something harsh that results in my losing the friendship of someone I care about, then I will feel I was responsible for losing that friend. That is because I am responsible. It was my action that did it. I may even feel regret. But that again is something quite objective. It expresses the feeling of the loss of the potential of that friendship that was once open to me but now is closed.

But what of the intentionalities themselves? The various impulses in me that, once they cross a certain threshold, come to hold sway over what this body does. How is it decided which of these intentionalities gets to have its day in the sunlight of consciousness? Surely it is here that the real question of agency and responsibility lies? And surely it is here that we once again encounter the ego and the necessity that we reflect on our experience so that only the most suitable intentions get to be enacted? For in each one of us, during the normal course of our lives, there is 'something' that is attempting to regulate our lives both inwardly and outwardly, that attempts to curb our excesses and strengthen our resolve in order that we find some kind of stability and fulfilment in our lives. Isn't this something my ego, and isn't its proper role to reflect on my experience and decide what is best for me? Yes! Of course! What have we been thinking?

So let us look again at this ego. Carl Jung called it an *ego complex*. For the ego is not just a single intention or intentionality. As we saw earlier, it is an enduring, coherent and coordinated way of being that we have developed on the basis of our inherited predispositions and conditionings, and according to our actual experience of living and the consequences of that living that have fallen upon us. It is, in effect, the way we have learnt to adapt ourselves to the conditions of our lives. It is this 'entity' that generally 'decides', when all is going well, and we are calm, and not agitated by a strong emotion, which intention it is that is going to hold sway over our actions. Now, according to Jung, and according to our own experience, there are a number of other complexes residing within us, that have very different outlooks on the world to that of our calm and rational ego. These complexes embody different ways of being in the world, ways that are triggered by various kinds of archetypal situation, like being physically threatened, or experiencing strong sexual attraction, or needing to assert your psychological independence from your parents. As those who are in the business of manipulating public opinion know only too well, all that is needed is to provoke a strong emotion, and start speaking the language of one of these complexes, and the rational egoic self is soon supplanted with something much more easily controlled. A brief consideration of the history of the twentieth century should soon convince us that far from undermining this ego and its sense of agency, we should be bolstering it up. For it is certainly a better master than the mass psychoses that are waiting in the wings.

But that is not our concern here. We are interested in how it actually stands with our egoic process in its normal everyday operation as the gate-keeper of our psychic lives. The question we are asking now is: what would happen to this ego if we disidentified with the 'I' of reflection? If, instead of thinking that it is 'I', this thinking self that emerges in acts of re-flection, who decides what course of action to take in any given situation, we stopped and looked again at what actually occurs. Let's say you have to decide whether or not to buy the car I was looking at earlier. Here the process involves collecting together all the information you can about its age, mileage, service history, number previous owners, MOT history, etc. and then looking at the actual condition of the vehicle, and assessing the owner, their sincerity, their resolve, looking at the documentation, and then considering the *price*. What happens in me is that I simply absorb all this information and inwardly check whether there is the intention in me to buy this car. This is simply a matter of *looking*, the intention is either there or it is not. If it is there, there is also the question of whether I pay the asking price or make an offer. Again, if I look, I will generally find there is a figure that comes to mind. So I make the offer. Either it is ac-cepted or it is declined. If it is declined, I look again. Same process: is there an intention to offer more, if so, how much? In this entire sequence of events I do not find any reflecting 'I' who is making any decision. There is just the reading of my inner states. Generally I find this reflecting 'I' only emerges when I am not actually making a decision. It is there, looking at the car on the internet, thinking "I really like this car" or "I wonder what they would take for it" or I start imagining what I would say in order to bring the price down, or I start picturing how great it would be to own this car and drive it down some idyllic country road in the summertime in England.[5]

The thing to see here is that this ego that we have developed out of our own experience of buying and selling and driving and dealing with money and other people, knows perfectly well how to behave when buy-ing a car without my having to reflect on anything. For once again, like our looking out into the garden to see if there is a cat, if there is the possibility of doing something that is standing immediately in front of me, 'I' do not need to 'decide' anything, I simply need look and see whether the intention to act is in me—*after* I have looked at all the relevant information that could have a bearing on that intention. This is still a form of reflection—let's call it *deliberation*—but it does not involve the illusion of there being someone standing over the process who is somehow freely making a decision. The decision, if we must call it that, emerges out of the oracle of the condensed wisdom of all my former experience. It emerges out of *me*—the being that I am, not 'I' the thinker who *thinks* him or herself into being.

THE EGOIC STATE

An egoic state arises when there is a splitting and identification of consciousness with the egoic processes that are otherwise quite naturally taking care of the life of the body. As a result a new and habitual form of thinking and self-understanding emerges that features an egoic 'I' as the agency that is in charge of the body. Now, instead of acting as an intermediary between the body and the psyche, the ego takes itself to be the sole authority that decides, according to its own power of reflection, what actions the body is to take. This occurs because, in the act of identification, consciousness bestows upon the ego the very feeling of agency and responsibility that properly belongs to the original disidentified state of immediate conscious unity. It is only on the basis of understanding this event of identification that the earlier experience of 'ego death' at Warwick (page 2) can be fully articulated. For the ego did not die. What died was the egoic 'I' that believed itself to be in charge of my life. And what replaced it was the immediate consciousness of the underlying unity that I am. This immediate consciousness does not need to think or refer to memory in order to understand or make sense of what is going on in its life. It simply needs to inquire of itself and await whatever response arises. It was this ability to access the profound self-knowledge of consciousness that my 19-year-old self found so amazing. For this is not the knowledge of a personal egoic self gleaned from personal life experience, it is the impersonal self-knowledge that consciousness always already has of itself but can only explicitly realise when it comes back to itself. In being immediately conscious (like now) we have direct access to the very intelli-

TG – Unity is conversational!

gence that is manifesting experience in every moment, that understands space and time and sensory quality because these are the very mediums in which experience is expressed. Once we withdraw our identification with the egoic processes that themselves are attached to worldly existence, we can start to examine the transcendental dimension within which the experience of the objectivity of the world is formed—as we have already been doing in Part One—and, of course, we can also examine the form of our previous egoic identification.

The essential stance of the egoic thinker is that it is *turned away* from the unthinkable source of experience. It is interested in the fate of its own objectified self, and only sees that objectified self existing in an objectified world. It thinks that self has control over its own destiny because of the capacity it has to objectify itself and think about itself and so appear to decide its own actions. The thinker thinks it is free and undetermined in what it thinks and decides because it does not see itself, as it is; it only sees its *objectified* self. And it seems to the thinker that it has complete freedom to determine what this objectified self is going to do. It just has to think "I will raise my arm" and up goes the arm. Do you see that this thinker is only aware of what it is thinking of? And it thinks it is thinking of itself, of the being it actually is, and that it is directly determining what this being is going to do. But it is not thinking of itself, of the being it actually is, it is thinking of an objectified self, a *model* of itself. The being that it actually is, is the being who is doing the thinking and the reflecting. And these processes of thinking and reflecting are not determining what I am actually doing, for I am actually thinking and reflecting. And this thinking and reflecting is emerging from a place over which 'I', this thinking and reflecting self, have no direct control whatsoever. So, do you see, 'I' only have the illusion of control? It is a trick done with mirrors.

And yet, what is going on here? For am I not reflecting right now? Have I not also objectified myself and made a model of myself? Isn't the only difference that this is just a more sophisticated model? For instead of objectifying my worldly self and thinking about that self, I have now objectified the thinker, and am now thinking about that thinker. Yes! That's exactly what is going on. But it is not our *reflecting* that is the issue here. It is our *identification* with this act of reflection that creates the unconsciousness of the source from out of which this stream of reflection is emerging. If we remain in a state of disidentification then it becomes ob-

vious that there is no independent, disembodied, self-determining thinking self who is in control of my actions. There is just this stream of reflection. But, at the same time, it matters whether or not consciousness
becomes identified with this stream of reflection. For such identification
creates a sense of self in that stream that then determines the form of the
reflection, i.e. it becomes a reflection that remains unconscious of the fact
that it is emerging from a deeper source. The subtle point here is that it is
the deeper source itself that bestows this sense of autonomy on the stream
of reflection, by the very act of identification.

But this raises another question: why should consciousness allow itself to become identified in this way? For surely it, from within itself,
should know what it is doing? And yet, if we look at how things stand
with humanity today, it is this state of identification with one's thinking,
egoic self that seems to quite naturally predominate. The reasons for this
remain obscure. Perhaps the best we can do is to remember our history.
For this identification with our thinking processes has not happened by
accident. On the contrary, our Western objective scientific culture has
quite deliberately taken this act of identification and turned it into the
ground of our understanding of being and of our way of life. For we
were not always turned away from the source of being and convinced of
our lonely egoic isolation in a desolate and uncaring universe. In relation
to the entire history of humanity, this emergence of the sovereignty of
human reason is a very recent occurrence. Before that, of course, there
was *God*. And with God there was a very lively sense of the being of the
human psyche. This was a place populated with all kinds of entities: angels, saints, deities, devils, demons, sprites and faeries. Good spirits and
bad. These were personifications of those very powers that we are now
considering as lying behind the emergence of our thoughts and feelings
and impulses. In our modern myth of objectivity, the spirits are understood as material processes occurring in our brains and our human self is
identified with a certain class of process centred in the prefrontal cortex.
Here too, the idea of a free and autonomous rational self deciding for itself what the body is to do, is recognised as a kind of illusion, that cannot
survive the scrutiny of a purely objective rationality. It is this rationality
that sits on our shoulder and says: "It is not objective science that holds
this absurd idea of the sovereignty of the egoic self—you cannot hold *us*
responsible for *that*—that was just a throwback to the old religious idea

of the soul—no, we see through all that now—it is pure *impersonal* rationality that is our guide."

And yet I do hold science responsible for setting up and maintaining our egoic identification. For scientific thought *is* the rationally purified thinking of this egoic self and it still holds to the idea that such thinking is the final authority in deciding what is to be true and real. Moreover, it is still the case that each purely rational thinker identifies the immediate being of their consciousness with their process of rational thought. The only difference, for those who think this through to the end, is that it becomes inconsistent to identify the ultimate being of the source with their egoic self, because that self appears as just another physical process occurring in the physical universe. Instead, a purely objective rationality is forced to identify the ultimate being of the source with the objectified being of matter itself, for only matter has real, causally effective being. And, as we know, such thinking is absurd, because it remains unconscious of consciousness. But now we can understand that the reason it remains unconscious is because of the original splitting of consciousness and the identification that occurs in the processes of rational self-reflection.

This shows the *inner* trajectory of the history of modern science. It began with Descartes' identification of his immediate consciousness with the processes of his rational, objectifying self-reflection. From then on it was to be human objective reason, already unconscious of its source in the human psyche, that was to attempt to build an entirely objectified understanding of itself and the soon-to-be objectified world it encountered. This objectifying understanding of being was then passed on to us, the people of the earth, as we were told about the true objective being of the sun and the moon and the planets and gravity and molecules and atoms and electrons and cells and DNA and evolution by random mutation and natural selection. And as a form of proof we were given televisions and radios and computers and antibiotics and heart transplants and cars and jet aircraft and spacecraft and weapons of mass destruction. In the process we, more or less, *lost our feeling for the source*. We no longer thought in terms of there being a God or a higher power that was responsible for us and for the creation of the universe. Instead we began to experience ourselves as being rational autonomous individuals who are capable of determining the course of our own lives according to the light of our own reason.

And yet it only takes a few moments of serious self-reflection to recognise that this supposed rational autonomy of ours is a complete fiction. Each thought, each feeling, each action that we claim as ours is emerging out of a place, a source, upon which we are totally dependent and of which we are virtually unconscious. It is this place that composes the sentences we speak, that effects the movements of our bodies, and that produces the feelings we experience in relation to the events that it already expects and protends. If we were able to drop the obviously inadequate idea that this entire stream of meaning was somehow the result of the physical interactions of the microphysical particles that supposedly make up our physical bodies, then the sheer wonder of what is occurring may start to dawn on us.

Returning to the Light

So what became of the missing manual mentioned at the beginning of the chapter?[6] Not so long ago we had the Holy Bible. But the scientific revolution has more or less swept away the medieval form of consciousness to which the Bible was speaking. Now we are in need of another kind of manual, one that can make sense of our being here within the context of all that has happened in the last four hundred years. For it is not that we have outgrown our need for a guiding myth, it is that we have fallen into a kind of unconsciousness that has ceased to care or even notice the wasteland[7] we have created for ourselves.

The first part of the book provided the beginnings of an understanding as to how we arrived in our current predicament. Our task now is to find a way out of this labyrinth, to find a living myth that is vast enough to contain all that has occurred and all that has been discovered since the advent of modern mathematical science, to *discover* a meaning that is profound enough to make sense of this situation we have made for ourselves. For we cannot simply invent this myth. It has to grow out of the soil of the psyche that is, for us, the source of all meaning and experience. So it cannot be the invention of an individual consciousness—something put together out of personal choice and preference. That would be a mere entertainment. It must needs be the response of the psyche itself to what it has already manifested. And yet, at the same time, the psyche *needs* our individual consciousness. For a myth cannot become a myth until some consciousness, in some individual, is able to project and experience the meaning it contains.

What we are asking here, is whether there is not, already, just such a myth, available to us—not something we must construct, but something that, in a sense, must necessarily arise, as the hidden meaning of what we have already experienced. We are not to make this meaning, we are to inquire after it.

To begin then, in Part One of the book, we have taken as our clue the historical and philosophical origins of objective mathematical science. For it is our science and our scientific thinking that have brought us to the collective situation in which we find ourselves. What I am suggesting is that our path does not lie in rejecting this science, but in revealing the essence of the impulse that lies behind it. That essence is expressed in the idea of the *natural light*. It says that each of us is a light unto ourselves. It says that I should accept nothing as being true until that truth can be demonstrated in my own immediate experience. It says that 'I' am the one who is the ultimate bearer of this power to discern truth from falsity, and that, as such, I am *responsible* for the truth. In such responsibility there lies *freedom*. For if I am to discern the truth in any meaningful sense, there must be the possibility of falling into error. Otherwise I am a mere automaton, for whom there is not even the possibility of distinguishing between truth and error. So how am I, in the freedom of my being, to avoid falling into error? Only by remaining *true* to what the natural light reveals. This act of remaining true, of being (in Heidegger's language) *bound* to what that light reveals, is an act of *freedom*—for if I am free to be bound by what is true, then I am equally free to fall into the freedom of error, where I can determine the 'truth' in any way I 'choose'.

This is the drama of our European scientific civilisation. We set out on this path of discovery, illuminated by the natural light of reason, and we have ended up here, in a desert of meaningless technological progress. Either this meaninglessness is the truth, or we have fallen into error—as we were free to do. We are the result of an impulse to create an entirely new form of civilisation, where each of us is to be responsible for the truth, and where no external authority is going to determine or impose their truth upon us. And, at a deeper level, in such devotion to the truth, we were to become gods in our own right, with ultimate power over ourselves and over the world we were to create in the light of our enlightened reason. A world where true justice and equality were to be realised for the first time on Earth. For, as far as the natural light is concerned, we are all equal in our direct access to the light and to the truth and error and freedom it bestows. And that means we are each an end in ourselves, not some tool to be used by another, to be manipulated and deceived, for, deceive me as you may, I still have the light within me!

And yet it seems to us now that all this talk about the natural light was just a form of seventeenth century naïveté. Something that unravelled into the labyrinth of contemporary philosophical confusion and that we can now reject in the sophisticated irony of our post-modern scepticism. But in such rejection, we reject the founding and guiding impulse of our culture. It is not a sign of sophistication, it is a sign that we have lost our way, that we have failed to develop or even understand what has been passed on to us. For we don't get to choose what we inherit. As a civil-isation, this is all we have. That is why we have been working our way back into the clearing of this natural light—tracing out for ourselves the errors of our ancestors. Where we have arrived is at the error of our false sense of egoic independence. It is not the natural light that is responsi-ble here. It is we who commit this error, insofar as we identify our con-sciousness with our ability to rationally reflect. For the natural light is not an individual capacity to reason. It is that which reveals what is ac-tual to us. This identification of consciousness is our *habit*. It did not start with Descartes. He just inaugurated a new form of identification with objective reason. As a result, our collective conscious connection with the human psyche, that was once lived out in our Christian faith, has become *severed*. This former faith, when lived sincerely, still allowed us a certain commerce with the psyche via the symbolic forms of Chris-tian worship. But such faith stands directly in the path of our impulse toward egoic self-determination. And so we no longer hear Jesus saying: "I can of mine own self do nothing: as I hear, I judge: and my judgment is just; because I seek not mine own will, but the will of the Father which hath sent me."[8]

This severance from the psyche[9] takes the form of a rigid separation of the subjectivity of our inner lives from the objectivity of the physical world, culminating in the enthronement of objectivity as the criterion of ultimate reality, because only what is objective is causally effective. This turns the human psyche into a kind of fantasy land that is generated by the automatic discharging of the action potentials of the neurons in our brains. So much for the Father which hath sent me. And yet, we know that this psyche is the very source of the stream of meaning that we experience as our living life here on Earth. And we know that discharging neurons are not meaningful. Our identification has led us into error. We have become unconscious of what it is to be conscious, of what it is to experience this streaming of meaning. Instead we have become fixated on the objective world that appears in this stream and the projection of our lives being enacted in that world. We have forgotten the psyche, the source from out of which this streaming of meaning is emerging each and every moment . . . .

At the same time, we should not think of this as a kind of mistake. For this identification of consciousness with our thinking self has enabled the extraordinary development of our capacity to reason and plan and organise and objectively inquire into the objective structure of the world around us. It was not as if our medieval ancestors were *disidentified*. They were in another kind of error of collective belief in whatever the Church decreed to be the truth. Perhaps it was even *necessary* for us to follow this path of rational reflection in order (one day) to *collectively* recognise the situation of our own identification. But we should be careful here. There is a tendency in us to escape the stark reality of the knife-edge of the present by imagining that it will all work out for the best in the end. But, the truth is, we do not know how it will work out in the end, and whether or not humanity will collectively 'come to its senses'. How it actually stands, and has stood for a while now, is that we are continually collectively poised on the brink of possible self-destruction. This is the collective human condition as far as we know it, and not our dream of a future mass enlightenment. As for how it will turn out for each of us individually, that is much easier to read: for we are heading toward the inevitable death of this human body and all that connects us to the worldly fate of this human collective.

And yet here we are, in this space and place of reflective consciousness. It is a place that has been made for us—it is a *gift* from those who came before. Now we are to learn how to be here *consciously*. That, if you like, is the historical task that has been laid upon us—for we are the outcome of the Western scientific impulse to *see things as they really are*. It is this impulse that has mercilessly stripped away our projection of the psyche onto the objectivity of the external world. But this process is not complete. We still project the being of mother matter as the blind and mindless source of all there is. This new deity is a cruel goddess who has abandoned us in a loveless universe. She cares not for us, and neither does she value us. She cannot care or value, she cannot even be cruel, for that would imply she knows what she is doing. No, it is we, her abandoned children, who project an idea of cruelty into the bleak silence of our predicament. For our goddess does not speak, she simply calculates the consequences of the laws of physics.

Do you see, of all the creation myths of all the peoples of all the earth, how inadequate this is? For it is asking us to accept that our foundational source, that from out of which we have emerged, and that which is producing our meaningful experience of being conscious in every moment, is herself without mind and consciousness—as if consciousness and meaning came into being one day, by accident, because of a series of random mutations in the material substance of our unconscious mother . . . . It is both tragic and comic that we should have come to believe in such a ridiculous myth. And yet it expresses a truth about us and our culture. For we collectively live in the denial and denigration of our own intrinsic being. In this way we reap what our ancestors have sown. We bring down on ourselves the mother we deserve. She consigns us to the oblivion of a meaningless materiality. We think we have no myth any more, that myths are for primitive peoples, and that we are facing reality as it is. But that is not true. Every culture has its myth, its understanding of being, its story of creation. It is only that our myth has become unconscious. It is the *myth of mindless matter.* In our identification with our unconscious objectifying reflective thinking processes we have turned our backs on the very source of meaning that is continuously creating our world. And so we come to think we live in a universe without meaning. Of course we do!

So what does it mean to become conscious in this place of reflection? Well, once again, it means that we withdraw our identification with the thinking processes that are ceaselessly clamouring for our attention. We become present and remain present. We do not allow these processes to 'carry us away' into another time—the time of the world that they are thinking of—which is not the time of my *true being.* For the time of my true being is always *now.* That is all really. It's so *terribly simple.* It is only

↳ Can be pushed to an extreme – but held in balance by conversation . . .

now, in this place of immediate presence, that we can begin to survey how it actually stands with our being here. Not how I think it stands, or how I imagine it stands, or how I theorise it stands. For here I am. I am the being of this place of consciousness. But if I am to know this place as something explicit, as something I can speak of, then I need a *mirror* that can reflect back to me the meanings that are emerging in me. For, to begin with, I *am* those meanings. There is no space in me to know the meaning I am in each moment, for each moment that meaning changes and slips away. It is only by means of this mirror that I can come to contemplate the form of the meanings that express the being that I am. But in this moment of reflection, in order that I can register what it is that is being reflected back to me, my consciousness must undergo a *modification*. I am no longer simply being the stream of meaning that I am, I have to separate myself from myself so that I come to possess a certain focus of attention that I can use to look into the mirror of reflection. This is the marvellous capacity of human consciousness—the capacity to reflect on the stream of meaning that it is. If all goes well, and I remain poised in the present moment of consciousness, then I, the being I am, by means of this mirror of reflection, can become conscious that I am conscious. This is the great revelation that is open for us as human beings here on Earth. Of course the animals also appear to us as being conscious. We can even transpose ourselves into their experience to a certain extent. But are they conscious that they are conscious? I think not. Just look into an animal's eyes, and then look into the eyes of a conscious human being. What do *you* think?

But, as we have seen, there is a danger in this capacity of reflection. The danger is that the separation of consciousness that is needed in order for the reflection to be a reflection, itself becomes unconscious. For this capacity of reflection opens up a space in which it is possible to *think*. Instead of simply observing the stream of meaning that I am, I can also allow intentionalities to emerge into this place of reflection and to play themselves out, rather than being directly enacted in the external world of experience. Such playing out involves the creation of an imaginary world where these intentions can envisage their own fulfilment in an imaginary time. In such imaginings, the intention can intend an image of myself, as the bearer of the intention, a self who is taking imaginary actions, in a world of images of other people and things that more or less correspond with the actual world of experience. After a while I get engrossed in the shadow play of these intentionalities, much as I get engrossed in a movie. But here the movie features an image of myself in the leading role, an *avatar* I appear to control and with which I can identify. As soon as I do that, I lose my poise, I become unconscious of my being conscious now, and I enter into the experience of being an imaginary self. The signature of this identification is my starting to experience *actual feelings* that correspond with the imagined situation. I become tense, my heart rate goes up, and my palms become sticky. Now I am no longer conscious of myself as the consciousness I am in this moment. I have become split. The part of me I need to maintain my consciousness of being conscious has been carried off into another time, an imaginary time.

This state of thought-imagining can begin to take over my entire inner life. I start to think about everything I do. I start to think that my thinking about what I do actually determines what I do. And then I start to think about thinking itself. In such thinking I enter into a new and higher level of intentionality. It is now that the possibility of *philosophy* arises. Instead of my desires and fears and anticipations playing out in my imagination, a new kind of disinterested intentionality can emerge that asks after the ultimate meaning and purpose of my being here. This is something unprecedented. For even though 'I' am identified with these thought processes, it is not 'I' who is developing this capacity for disinterested inquiry. It is emerging from out of the psyche according to an impulse that is already *in* the psyche. 'I' think 'I' am doing it, but 'I' am not (*not I but the Father within me*). And yet, if 'I' do not identify with this reflected thinking self, then *I am that*, I am that place that has developed language, that has developed this capacity for disinterested inquiry, and that now wishes to express itself in this space of reflective consciousness. I can now *consciously* dwell in this clearing of reflection where the intelligence that lies behind this manifestation of being can come to know itself for the first time in the mirror of its own experience . . . .

NOTES

1. "Thrownness" is a term Heidegger introduced in *Being and Time* (1962) to express the manner in which we inherit our particular situation in the world.
2. William James (1842-1910) was an American psychologist and philosopher who wrote about the saddle-like structure of time perception in Volume One of his *Principles of Psychology* (1890, p. 609).

3. Sigmund Freud (1856-1939) was an Austrian neurologist and founder of psycho-analysis. He first published his model of the ego and the Id and the super-ego in *The Ego and the Id* (1923/1975).
4. For more detail on feeling someone else is controlling your body, see the note on alien hand syndrome on page 383.
5. This is a reference to the Van Morrison song *Summertime England* ("Will you meet me in the country in the summertime in England, will you meet me?").
6. I remember now that I must have picked up the idea of the missing manual from Timothy Leary (1920-1996). Originally a Harvard psychologist, he studied the therapeutic use of psychedelic drugs, before losing his job and becoming a leader in the counter-culture of the 1960s. During my time at Warwick I read most of Leary's books, one of which had the subtitle: *A manual on the use of the human nervous system according to the instructions of the manufacturers* (1977).
7. I am thinking here of T. S. Eliot's poem *The Wasteland* (1963, p. 61).
8. John 5:30
9. This severance from the psyche is symbolically expressed in Philip Pullman's image of children being severed from their daemons in his *Dark Materials* trilogy (2011).

CHAPTER 9

REMEMBERING YOURSELF

SPIRITUAL TEACHINGS

IN 2015 after living for twenty-five years in Australia, I returned to England, partly to spend time with my mum in her final days, and partly to break up the habit of my former life as a university academic. One day, soon after my arrival, I was browsing through a second-hand bookshop in Worthing, when I came across a rather extensive collection of works concerned with the spiritual teachings of George Gurdjieff.[1] Clearly someone with a deep interest in Gurdjieff had recently died or had had enough of reading what he had to say. In the 1980s, before moving to Australia, I too had become interested in Gurdjieff's teachings, reading most of his work, and also that of his interpreter, Pyotr Ouspensky. But as soon as I became involved with Barry Long, I more or less forgot about Gurdjieff— for why should I concern myself with someone who died in 1949 when I could spend time with a living teacher in the here and now? Then Barry too died in 2003, and my impulse to seek the truth in the wisdom of another died along with him. Now, in the Worthing bookshop, out of an almost nostalgic curiosity, I bought a copy of Ouspensky's *Psychology of Man's Possible Evolution*.[2] Here, in the span of 85 pages and over a course of five lectures, Ouspensky provides a basic account of the core of Gurdjieff's teaching.

On reading this book I became more and more surprised at how familiar it all seemed to me. For, in the intervening thirty years, what had once been a mere collection of interesting ideas and possibilities, now had elements I recognised as being true in my own living experience. This seemed significant—for I had not deliberately followed any of Gurdjieff's methods, and yet I seemed to have arrived at a place at least part the way along the road he was indicating—and that by having taken a quite different route. For there is very little in common between the teachings of Gurdjieff and the way of phenomenology we are exploring here. This led me to think, after all, that Gurdjieff did possess some genuine objective knowledge of the human condition. My opinion before was that he was probably a bit of a rogue, who took advantage of his gullible students in order to lead the grand life of a great teacher. He certainly thought nothing of relieving those students of their money, and of subjecting them to ridicule and humiliation—all in the name of bringing them to greater consciousness, of course.

Figure 9.1: George Ivanovitch Gurdjieff 1866(?)-1949.

And yet now it seemed to me that there was a jewel hidden in all the obfuscation of Gurdjieff's mythical talk of celestial beings and cosmic laws—this jewel being the aforementioned genuine objective knowledge of the human condition. This knowledge begins with the assertion that our normal state of waking consciousness is a form of *sleep*. What makes it a form of sleep is the possibility of our being awake in other states of higher consciousness. These higher states have a relation to normal waking consciousness that is analogous to the relation waking consciousness has to the dream states we experience while actually sleeping. What defines us as being asleep while we are awake is that we entertain false ideas about the situation we are in. Firstly we think we are in a continuous state of self-consciousness, whereas, in fact, we are only properly self-conscious in those moments when we become conscious that we are conscious. Secondly, we think we have a will that determines what we do, whereas, in fact, we simply follow whatever impulse it is that currently has control of our bodily system. Sometimes outside events, or an inner impulse, will lead us back to a state of self-consciousness where we get a sense of being awake and in charge of what we are doing. But these states only last momentarily and, having arisen by chance association, are soon extinguished as the next chance association, bearing the next impulse, takes us over. Thirdly, we think that behind these impulses there is a permanent ego or self that is guiding our life and that all the various impulses we experience are an expression of this ego, whereas, in fact, our inner life is a kind of anarchy, where the impulses we enact each have their own 'I' and their own understanding of the world, and where there is no one 'centre' that unifies their expression. Instead there are several such centres, in which various impulses are brought together purely on the basis of their intrinsic similarity of outlook and their temporary agreement concerning what they are intending. What really grounds our sense of having a permanent ego is our experience of the permanence of our underlying bodily system and not any experience of psychological coherence.

The picture that emerges is that the ordinary state of a human being on Earth is that of a virtual *automaton*. And what is worse, we are automatons who believe we are not automatons, but conscious individuals with a coherent psychological identity and an autonomous and self-directed capacity to determine the course of our own lives. Gurdjieff's teaching then goes on to say that all is not lost, and that we do possess the capacity to be conscious, and to develop a coherent psychological identity, and to autonomously direct the course of our lives. All that is needed is for us to *make an effort* to attain these things. The first and most fundamental form of such effort is to expand those moments of self-consciousness that are already occurring by developing an intent to engage in what Gurdjieff called *self-remembering*. It is in this place that Gurdjieff and Barry Long

⤷ relatedness to self. What about others? and God?

and Edmund Husserl and you and I can all dissolve into the immediacy of being conscious now. So long as we don't start thinking about it.

What we are facing here is the rather delicate question of the use and value of spiritual teachings. For, in what I said earlier about the absence of a manual that explains what it means to be a human being, I was speaking of our *collective* situation—of the fact that our former manual, the Holy Bible, no longer has any collective authority and that what has taken its place—scientific materialism—simply fails to address the questions that matter to us. But when we move outside the collective, we are faced with an almost bewildering array of spiritual teachings that each, in their own way, provide a manual for those who are willing to step into the alternative world that the teaching presents. Amongst these teachings there are some, like Gurdjieff's, that present a picture of our human condition that, at least to some extent, corresponds with the phenomenology. For it is true, that in our ordinary human egoic state we are only rarely and accidentally conscious that we are, in fact, conscious. It is the way of phenomenology, just as it is with Gurdjieff's notion of self-remembering, to cultivate this state of being conscious of being conscious, in order to inquire into how it actually stands for us, in the ordinary everydayness of our human condition. And just as Gurdjieff foretells, what we discover is that our everyday notion of egoic autonomy does not stand up to careful scrutiny.

But the parallels go further: aside from being an end in itself, the discipline of self-remembering is concerned with forming a *magnetic centre* of intentionality and habit that can remind one to re-enter and maintain a state of pure consciousness, which in turn can be focussed on the aims and intentions of the teaching. In the same way, as phenomenologists, we cultivate this place of pure consciousness, as our mode of inquiry into whatever it is we wish to question. And so we too develop a magnetic centre of interest and intention and habit that reminds us to return to this place of pure consciousness.

It now starts to look as if phenomenology contains a hidden teaching, or, alternatively, that Gurdjieff's teaching is a form of phenomenology. But neither of these views is correct. What finally distinguishes a teaching is the presence of a teacher, even if that teacher is dead. A teacher, in order to be a teacher, is someone who possesses or who has realised a superior form of consciousness and knowledge to that of their students.

That is certainly how Gurdjieff and Barry Long presented themselves. In contrast, phenomenology is not the teaching of a teacher. It is a mode of inquiry. It may be that someone teaches that mode of inquiry, or demonstrates it. But the person doing the demonstration is not claiming to possess a superior state of consciousness or knowledge. They are merely acting as a colleague and co-inquirer. And yet there is still a deep commonality between phenomenology and all genuine spiritual teachings—for both are concerned with uncovering the truth that lies behind the surface of our human condition. The essential difference is that phenomenology is a way of inquiry that explicitly rules out the presence of an external teacher or teaching. Its very *ethos* is the ethos of our Western inheritance which says: I will not accept the authority of anyone who attempts to tell me what is true and what is false—I will only accept what I can see according to my own inner light. Phenomenology contains this ethos because it *is* our inheritance—it is the refined essence of the philosophical impulse that gave birth to our Western scientific culture.

Saying that is not to dismiss the value of teachers and spiritual teachings. For it may well be that there are individuals who permanently dwell in higher states of consciousness and knowledge. And if you were such an individual, then you would surely wish to impart to everyone else (who is willing to listen) what it is that you are seeing. And yet, we must ask, as phenomenologists, what is occurring here, when one person goes to listen to another, in order for that second person to inform them about the truth of their own life.

This question, of course, relates directly to my own experience of being with the teachings of Barry Long. For Barry was quite aware of what we are saying here about our Western inheritance. He even described himself as the *Master of the West*. As a result, he frequently insisted that everyone who came to listen to him should not believe him, and that they should test everything he said in their own experience. And then he said, if there is something that goes beyond your direct experience, then you should listen for the 'ring of truth', which we would call the resonance of the psyche confirming that what is being said corresponds with some inner psychic form. Barry did not want disciples or believers. He disapproved of Eastern practices of devotion and worship. He didn't take on an air of superiority, and he called himself an ordinary man. He even used to say "There is only one 'I' in the universe and it is sitting on that chair looking out of your eyes". And if he thought someone was forgetting that, he would say: "Don't put your 'I' on me!"

But what Barry was saying and asking, in relation to his being an ordinary man and not believing him, was only reaching to the *surface* of the psyches of the majority of people who came to see him. At least that was my observation. And I suggest it is the same for all the other teachings where the teacher asks the audience to remain conscious and take responsibility for what they are hearing. For if we generalise our initial understanding of the human condition, it follows that most of the people coming to see the teacher are going to be in that very place of unconscious automatism that we recognise as our normal egoic state—that is *why* they are seeking a teacher, to find a way out of that automatism.

OK, you may say, it may start out like that, but if the teaching is effective, then people will start waking up and taking responsibility for themselves and testing what the teacher is saying before accepting it as being true. But, again according to my personal experience, by the time the teaching is consciously understood, it is already *too late*. For what we are dealing with here is an example of collective *unconscious projection*. Such projection is recognised as something fundamental and unavoidable in the practice of psychoanalysis, even possessing its own technical term: *transference*. But outside of psychoanalysis, people remain largely unaware of this phenomenon, and so continuously get caught up in experiences that they do not consciously understand.

The Phenomenon of Transference

I don't know if you have ever been involved with a spiritual teacher. Most people are immunised against such an experience by the illusory stance of egoic independence and individualism that pervades our culture. It is this stance, and our ignorance of the being that stands behind it, that makes us so open to the possibility of transference when we first recognise the presence of a genuine teacher—i.e. someone who has knowledge that comes from beyond the egoic state we collectively inhabit. What happens in transference, as far as I can make out, is that I project an image of my true being onto someone else, and then take that person to possess all the qualities of the being that really inheres in me. What makes the transference so powerful is that I am usually unconscious of these qualities. I do not know who I am, I do not know of the wisdom and love and integrity that lie within me. Of course I don't, because, as yet, I have been unable to properly manifest these qualities in the external world, except, perhaps, in a few extraordinary moments. But now, through this other, I glimpse the possibility of becoming the being that I am, I get a 'taste' of what this means—the projection actually manifests my being as if in a mirror. But it is in the nature of the transference that I do not consciously recognise what has occurred. Instead I experience a 'special' connection with the teacher. They are not like other people. It becomes the most important thing in life to be with them, to absorb what they have to say. It is as if the teacher is the custodian of my true being and that through the teacher I am going to realise that true being in me. But, of course, that true being already *is* me. What happens, as a result of this projection, is that I make the teacher *responsible* for my being the being I am, even though I have not done this *consciously*. The projection just happens. That is what makes it so *tricky*. You really do see these marvellous qualities in the teacher and you are completely unaware of the mechanism that is doing the projecting—for if you were aware, then the projection would collapse. And when you are projecting in this way, it becomes almost impossible to see the other person as they would be if you weren't projecting. For the projecting not only changes the way you see the other person and the way you behave toward them, it also changes their perception and behaviour toward you. And here we must remember that the teacher, if they are famous, is receiving these being-transferring projections from hundreds if not thousands of people. This must have a profound effect on the teacher and perhaps not entirely for the good.

In practice, when you first come across a teacher with whom you resonate, you do not usually immediately enter into a transference and surrender responsibility for your life—although, of course, there can be cases of 'love at first sight'. Instead you start by listening and trying out the various methods that the teaching suggests. But then it turns out that something actually produces a result. You discover another state of consciousness, one from which it seems that your normal state really is one of being asleep. That ordinary state is now in abeyance and you see there is a greater truth of which you were formerly ignorant. It is now that a deep desire is going to emerge that wants to be free of this state of egoic consciousness. The desire can see that we have only come to this 'higher state' because we are in contact with someone who appears to live in that state, who is free of this burden of self, and who has the knowledge and power to lead us to the same freedom. It is now that the transference occurs. I come to accept the teaching and the teacher as knowing better than I what I have to do, in order to become free. There is a great relief. I can lay down the burden of uncertainty and responsibility for my life, and simply follow the teaching to the best of my ability. It doesn't matter if the teaching tells me not to do this, if it says "You must not believe, you must test everything for yourself". For I can easily say to myself: "OK then, I must not believe, I must test everything for myself". But if I am simply testing because the teaching tells me to, and not because I am already in the place of being responsible for my life, then this is all hopeless. It is like the scene from *The Life of Brian*[3] where Brian shouts to the multitude: "You've all got to work it out for yourselves!" and the multitude replies, as one voice: "Yes! We've got to work it out for ourselves!" and Brian says: "Exactly!" and the multitude replies "Tell us more!" . . . .

But life is paradoxical. If you do persevere with a certain method, like self-remembering, even if only because a teacher has told you to, then you start to develop a certain configuration of habit that is concerned with the teaching, and with 'waking up'. Now, instead of being caught up with intentionalities whose directionalities are entirely focussed on one's egoic self and the external world, there arises another centre of interest, that joins in the inner conflict, and every now and then wins through and reminds you to withdraw, to disidentify, and to remember yourself, your being conscious. Such a state of being conscious of being conscious is no longer robotic. There is a gap, a pause, a space where one's consciousness is no longer compelled to identify with the next intention that automatically emerges. It is from here we can start to observe how it actually stands, how we are placed, how we are determined, and yet how we have this *leeway* of consciousness.

And so it seems, just as Gurdjieff thought, that you have to form a habit of remembering in order to break the habit of identification. And yet who is it that makes this habit? Who is it that is behind this impulse to be conscious of being conscious, who is *capable* of such consciousness? Of course, it is *me*, my own true being. The very being I have projected onto the teacher. It is (I am) still here. I am even responsible for the transference. For it is this I, the I of being, who is *actually* here in each moment, living this *actual* life, who bears whatever it is that *actually* happens—the consequences of the former actions that *I* enacted—and all this quite independently of my thinking self-identity, who thinks it makes the decisions, who is trying to follow the teaching, who then disappears in the moment of crisis, when reality breaks through and crushes this projected thought-self and its fantasy of an imaginary enlightenment.

And yet if you think you need a teacher then it seems you will have to have a teacher—the irony being that the job of the teacher is to teach you that you don't need a teacher. To do that they first need to prove they are a teacher by demonstrating some real knowledge. Once that occurs you can happily project the being of your own true self onto them and absolve yourself from the anxiety of the responsibility of being that self. Then the teacher can tell you what you already know but were unable to make explicit—and so relieve you of the effort of finding out for yourself. Finally, the teacher's job is to show you the error of your ways. This is generally done by *letting you down*. For no human being can live up to what you are projecting. Out here, we are human, all too human.[4] I think Gurdjieff understood all this.

So things are paradoxical. It seems to be a law of life that if you do realise something, you are required to express that realisation to *someone*. And yet it also seems that a kind of error occurs when you set yourself up as an authority, as one who 'knows', who has the power to bring others into a state of realisation. For then you are *inviting* the projection of someone else's responsibility for their own state of being. I understand this now as a *moral* problem. The question is, is it right to take on the responsibility for another person's inquiry into the truth? And the answer, I think, is no. Equally, we could ask, is it right to give up the responsibility for your inquiry into the truth to someone else? And again, the answer must be no. But out here, in the 'real' world, the situation is not so clear cut. For many teachers will say that they are not asking for, or taking on responsibility for the people that come to see them, that they are just telling it how they see it, and that if you project something onto them, that is not *their* responsibility, but *yours*.

But, as with all situations in life, we should not only listen to the language, we should also look at what is actually happening. Generally it turns out that we, the people coming to see the teacher, all start acting rather strangely, as if we are under a kind of spell. We hang on the teacher's every word, we feel elevated and honoured if the teacher pays us special attention and, if others are looking on, we feel even more elevated in their eyes. There starts to be a semi-conscious kind of assessment as to who has 'got' the teaching and who has not, a kind of ranking of consciousness, where the teacher stands at the apex. These are all signs that a collective transference is occurring. And then there is the question of *money*. It is almost always the case that the teacher is making their living out of their students, and no longer needs to go out each day and drive a taxi.[5] For being a spiritual teacher is not like holding down a job in the outside world. There is no boss putting on the pressure, no disaffected, competitive colleagues, no targets being set, no new corporate strategy being implemented, and no having to apply for leave. One is emperor of one's own private empire. At the same time, everyone else is (more or less) pleased to serve you and take on the worldly strain of supporting you. And, if they are less than pleased, they will generally simply go away. Essentially you *bask* in the light of the role that everyone is projecting onto you. This just happens automatically, because you are actually fulfilling that role, whether you repudiate it or not.

And, for the egoic self, the whole situation is very pleasing. I know this because I was a teacher—not a spiritual teacher, but a university professor. The day I started teaching I felt this change in my status, in the way I was being seen and treated by everyone else. Before that I was a student, and before that a gardener, and I was treated as 'nobody special'. But once you have a PhD, an associate professor title, and a research lab with PhD students of your own, then people simply cannot help but treat you differently. You find that their pausing and paying more attention to what you say is OK with you, is only right, only what you *deserve*.

So, the situation is, as far as I can see it, that when you set yourself up as a teacher, and you have some genuine knowledge, and people come to see you because they recognise that knowledge, then it is simply going to happen that the people are going to project the status of being a teacher onto you. And that is for the good reason that you are actually being a teacher and you are actually there to teach something. That means, whether you admit it or not, that you *are* taking responsibility for passing on that knowledge to the people who come to see you, and the people *are* surrendering their responsibility to you, so that the transmission can happen. If, in addition, you are setting yourself up as a *spiritual* teacher, and you have something genuine to say, then something more serious is going to happen, for people are going to project their *real being* onto you, not through choice, but in much the same way as we fall in love. This is going to be a deeply meaningful experience for the one doing the projecting. They are going to see great profundity in you, and you are going to acquire significant power and influence over them. They will give you their money in order to be with you. They will make love with you if you wish. They will adore you. And this will test your character. In an *ideal* world this would not happen, for we would all be in the place of immediate consciousness, where it no longer matters whether you speak the truth or I speak the truth, for it is all emerging out of the same unthinkable source. But in such a world, there would be no more teachers, because there would be no more need for any teachers. For we would all already be in the place of our real being.

But back here in the 'real' world, what are we to do, when we have this inner sense that there is something more to life than is being presented to us in our culture—and yet we have no idea how to proceed, how to reveal this 'something more'? Indeed, what was *I* to do, all those years ago, after my Vipassana experience? Like many people I had had a glimpse, but I had no clear understanding as to what had happened. Outside the Vipassana state what would have become of me if I had simply slipped back into an ordinary life with ordinary people living in the unbroken habituality of their egoic states? I think perhaps I would have 'fallen in line' and gone back to sleep.

It now seems to me that the being I am knew exactly what it was doing when it prompted me to travel to Oregon and become a sannyasin of Bhagwan Shree Rajneesh. That was what was right for me then. I had no magnetic centre to speak of. I was adrift in the world. 'Something' had made me stick out that meditation in my Brighton bedroom, and now 'something' was sending me to Oregon and projecting itself onto the form of this charismatic Indian guru. Did it matter what his motivations were? Well, yes. But that was OK as well. It caused the entire sannyasin world to come crashing down a few years later,[6] and that meant I could become free of a teaching that was no longer serving me. How can we possibly judge all this? All we can do is describe it. And yet there does seem to be a kind of progression involved. For there has to come the day when one says, *enough is enough.*

That day came for me some time after Barry Long died. I was involved in the posthumous publication of his autobiography,[7] when something happened as a result of my reading it. I started to see, behind his realisations and his deep insight into the human psyche, that Barry was also a man, just like me, and that he had lived this remarkable life, entirely according to his own lights, whereas I had spent a good part of mine listening to him and trying to follow what he was telling me to do. I had done this because I thought him a kind of god. But now, seeing the events of his life laid out before me, and seeing the consequences of his teaching working out in the people around me, I began to see the outline of a different man, a man who was again, human, all too human. I could almost see the glamour of my transference begin to fade and the reality of what I had done begin to dawn on me. For the being I am had known what was happening all along, and yet somehow I had chosen not to see the reality of my situation. Only now was I ready to take responsibility for my life—despite my having thought for years that I was indeed taking such responsibility. But my former idea had been a simple matter of not blaming fate, or other people, like my father or mother, for the situation of my life. Now I again understood what I had chosen to forget, that my true responsibility is to *manifest* the being I am—to *listen* and to *act* from this place of being, rather than from the place of an egoic self that seeks to ignore that being. It was the death of my father that finally made this clear. For Barry Long had been a kind of spiritual father to me. He was there because my upbringing had not provided me with any kind of explicit spiritual instruction. But in the end, in his final days, my *real* father, the man who had actually been there during my formative years, whose genetic system I had inherited, was able to manifest his true being to me, and I was able to do the same in return. In this manifestation something profound occurred. After his death I felt that my father had passed his authority on to me. This all occurred in a symbolic realm on the 'other side' of death. I can't explain that any further. But I know in *that* transference that the entire drama of my needing a father outside of me had ended.

THE QUESTION OF FREEDOM

In all we have said about the ego and the egoic state, and the self that lies behind this state, we have still not clearly addressed the question of

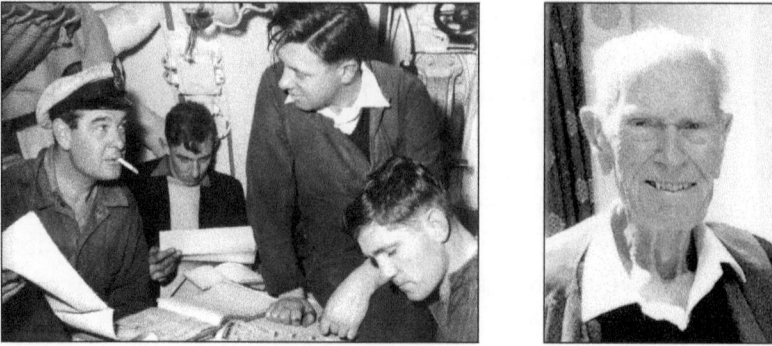

Figure 9.2: Dad: Leslie Thornton 1926-2013. (a) With the cigarette, during his sub-mariner days in the Royal Navy and (b) Shortly before his death.

freedom. For, if there is freedom, freedom to be bound to what the natural light reveals, then *who* is free? Our first and most natural answer is to say that *I* am free: I, the being I experience myself to be in each moment of my being inseparably present in the immediacy of what is happening now. It is only when we 'take things apart' in the mirror of reflection that a different 'I' can emerge who appears to be separate from this immediate experience. Such separation is always after the fact. It is an artefact of our ability to imagine ourself behaving differently. And then we come to think that we are that self who could have behaved differently. But we can never behave differently. We can only behave exactly as we do behave. At the same time our thinking about behaving differently influences how we subsequently behave. But that does not mean that this reflected image of our being is the author of our thoughts or of our subsequent actions. It just means that what we think has an effect on our actions.

So what has this to do with my freedom? As far as my thinking processes are concerned, 'being' is just a name I give to the unthinkable source of experience. But then, when I withdraw from such thinking, I find I *become* this source, I find it *is* me, that 'I' am no longer separate from 'it'. What can we make of this being-at-one-with whatever is occurring? What kind of freedom is there in such a going-along-with? If this is my being, what kind of being is it? Does it possess any kind of unity or intelligibility beyond the unity and intelligibility of its being consciously experienced? For if we *reflect* on what emerges from out of this unthinkable source, as our everyday experience, we find very little evidence of the presence of a

Celtic

'higher self', of a being who could, in some way, be responsible for this ca-cophony of intentionality that is my actual life. It is only in the moment of withdrawal, of self-remembering, that anything like the presence of the being of my being can again come to itself. And it is only then, in the poise of immediate consciousness, that the disordered flow of my worldly intentionality loses its grip, and relinquishes its otherwise undisputed ac-cess to this clearing of consciousness where all becomes manifest.

Clearly we cannot even have the possibility of freedom until this iden-tification with our conflicting intentionalities is dissolved. For they are not conscious in and of themselves, they are only conscious of what they intend, and what they intend always refers *away* from now. The freedom of consciousness is the freedom of not being determined by these *sub-personalities*. It is the freedom to intend the magical intention that only intends what is, that intends this moment, that intends *now*. Such intend-ing intends nothing. It is what returns us to the presence of being. Only from *here* can we engage in a *conscious* reflection on our actual situation. Only from here can we inquire and question and attempt to catch the mechanism of our intentionality 'in the act'.

And yet there is still something puzzling at work. Something, I think, that has puzzled everyone who has attempted to make some logical sense out of our human situation. For if I, the being I am, have this freedom to come back to myself in this state of immediate consciousness, then I must also have the freedom to *not* come back to myself, to instead get caught up in all these intentionalities that lead away from myself, away from the consciousness I am. Now why would I do that? Why, in my freedom, would I *allow* myself to forget myself? And yet that is what happens. I am continuously forgetting myself and getting caught up in various thought processes that lead me into fictitious worlds. And it is not as if there were a necessity for that to happen. I can still function in the world without being identified with what I am doing. But still I get caught. I get 'pulled' back. There is an allure, a possible future that beckons. And after all, it is all very well being present for a while, but being present all the time? There is something in me that cannot bear it. I don't bear it. I *want* to slip away. It all takes too much energy, holding to this place of consciousness. I positively seek to escape into my other world, to entertain myself, to dream. And then, when I have had my fill, I am reminded again about being conscious, I remember myself, I come back to now . . . .

So where is the freedom in this? Am I not still an automaton, who gets a certain 'kick' out of being present, a certain enjoyment, so that this practice of becoming present, combined with this feeling of enjoyment, creates a habit that reminds me to become present again—but only occasionally and only for a little while, for my feeling of enjoyment is soon replaced by a feeling of boredom, that again, quite automatically, causes me to start thinking? What is going on here? Am I not simply being determined by quite predictable feeling states that automatically transform themselves from the positive to the negative and back again? Where am I in all this, the being I am, the one who is supposedly responsible for this life?

Here, again, the labyrinth beckons. And once again we have to stop, and examine the actual situation. For it is true that I can and do enjoy being immediately present. I hear the sweetness of the blackbird singing. I see the beauty of the way the leaves in the tree are fluttering in the wind. And as everyone knows who promotes meditation as a means to relieve stress and improve mental health, there are lasting benefits associated with disidentifying with our thinking processes. But we are not here to become more relaxed. The point is that once I become immediately present as the consciousness I am, then *the robot is no longer in charge.* I have not become present in order to enjoy a particular feeling. It doesn't matter that I may get bored in a minute or two, or that I get distracted and caught up in a particularly compelling thought. In the moment I am immediately present, then I am freely and consciously being the being I am. This being is not 'trying' to do anything. It is not engaged in some struggle to remain conscious. It does not fight against the other processes that are trying to get back into the 'driver's seat'. It is simply being. Its power lies in its ability to reveal how things actually stand in this life, in this psyche. This capacity to 'see' alters my understanding, which alters my behaviour. Such alterations change the structure of the *system* that I am. It is this system, that, in a way, *resists* my being conscious. It is what I have *inherited.* It is my past and the past of all those who came before me.

WHAT STANDS IN THE WAY

What stands in the way of my being immediately conscious is this *past.* Here I do not mean the historical past, the collection of events that have already occurred, and our memories of those events. I mean the past that is present to us now, as the entire system of *habit* within which we live. What distinguishes our being conscious in the midst of these habitualities is the presence of a leeway of uncertainty or indeterminacy, of something still being in question, still not being *completely* habitual. That is why we are able to stay conscious in our perceptual experience, because, *if we pay attention,* there is always an element of uncertainty concerning exactly how things are going to turn out. Once an event becomes determined, once we can consistently predict how it is going to unfold, then we tend to become unconscious of it, unless we make some explicit effort of observation. That's just the way our system works. We can observe this each time we acquire a new habit.

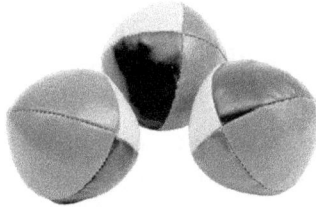

For example, let's say we are learning to juggle using three juggling balls. We start by throwing the balls into the air one at a time and we predict we are going to keep them there. We send this prediction through our arms and hands, and they respond, as best they can, according to the habitual capacities that they can already manifest. We don't perform any mathematical calculations. We simply wait to see how it turns out. In this waiting, our seeing remains conscious. We are in a state of impersonal witnessing, just as we were when looking for the cat in the garden (page 29). And, of course, unless we have already learnt a similar skill, our predicting soon fails. The balls fall on the ground. And so we try again. Only this time, because we were conscious, we remember something of our previous attempt. This allows us to send a slightly different prediction through our system and again observe the result. We predict in more detail how our left hand is going to pass the ball to our right hand each time it performs a catch—for it already has the habit of catching and passing, just not of doing it with the required speed and regularity. Then we predict how our right hand is going to throw each ball higher than before, and do so immediately it receives it from the left hand. And so we try again, and this time it works, but only twice, because our consciously predicting what the right hand is to do meant we stopped consciously predicting what the left hand is to do. For it turns out it is very hard for us to consciously predict more than one thing at a time. But after a while, the more we repeat it, the right hand starts to 'get the idea' about how high to throw each ball. It starts to 'do it by itself' so it no longer needs a conscious prediction to keep it in motion. Now I can consciously predict that my left hand is going to pass on each ball as quickly as it can. And lo! after a while my predictions come to pass. I stop dropping the balls. I simply predict that I am going to juggle, and lo! I am juggling! To begin with I find I still have to remain conscious, and predict myself juggling, otherwise I start dropping the balls again. But then, after practising over a period of days, I find I do not have to pay attention any longer. The juggling is simply happening by itself. Perhaps now I can start to learn how to do something else *while* I am juggling.

The point here is that more or less everything we do, all day long, is like this juggling example. Once upon a time we had to learn to predict the forms of the objects in front of us, by consciously predicting how these forms would change in relation to the changing position of our bodies. And we had to learn how to change the position of our bodies by predicting how that would feel and how the objects around us would change. These habits of perception and action have become our ability to live in the world. Now they are almost completely *unconscious*. We simply see the world in front of us. We are no longer aware that this seeing is an achievement of our predicting what is to appear and of our registering how our experience matches those predictions. All this is occurring just as juggling occurs for a professional juggler. Instead of experiencing the streaming sensory qualities that actually comprise our various sensory fields, we experience ourselves as existing in a unified world of form that these sensory qualities manifest. In this mode of being in the world we remain unconscious that it is we who are projecting the form of the world and that this is a skill we have learnt to deploy automatically. There is a kind of law here that says: as soon as we are able to successfully and reliably predict the form of something, we become unconscious of ourselves as the one who is doing the predicting, and we come to directly experience that form either as something potentially or actually existing in the world, or as a capacity that I can bring to bear on that world. It is here

that we meet with our fundamental unconsciousness of the world forming intelligence that lies behind our experience of embodiment.

This capacity to predictively form the world into which we project ourselves is a basic *habit* that we learn during the early years of childhood. We
did not do this by thinking about it. We did this by making our predictions, and *consciously* observing their outcome. Once the habit is formed
we forget about it. It becomes a part of the system we are. This system
is a system of habit. A habit is a capacity whose execution has become
entirely predictable. When something is entirely predictable that means
it has become *mechanical* and its execution no longer requires or attracts
any conscious attention. As we grow into adulthood we form hierarchical
layers of habits that become our ways and means of living our lives in the
world. Eventually, our system of habit becomes so adapted to the world
in which it thinks it lives, that it more or less stops learning. That means it
has found a niche whose broad form is almost entirely predictable. Only
the perceptual surface captures its attention, where it must check before
crossing the road and where it must listen to what you are saying because
it is not quite capable of predicting how your sentence will . . . . But the
basic system of capacities are all in place. The job is done. There is no
longer any need to bring the full light of consciousness to bear on any of
the more or less predictable situations that occur in the day-to-day life.
And so we lose the external promptings that stimulated all those earlier
moments of consciousness. We enter into a twilight world where our
inner lives become colonised by habitual thought patterns and imaginings. As consciousness withdraws, it is replaced by the mechanisms of
the habits that our former moments of consciousness were able to form.
These habitual mechanisms are the embodiment of our own past. If there
is no impulse from within to maintain a state of consciousness in the face
of this edifice of habit and there are no dramatic and unpredictable situations from without that demand we change our ways, then, just as Gurdjieff asserted, we enter a state that we may as well describe as being *asleep.*

It is not that this system of habit somehow *intentionally* stands in the
way of our being conscious. It is simply what has evolved, what has proved
itself to be necessary for the survival and reproduction of our ancestors.
And so its intention, if such a word is meaningful here, is to ensure our
survival and reproduction so that we in turn can become the ancestors
of our own progeny. It does this by enacting the essential prediction that

the system it is, is going to *survive*. In order to do this, it has to predict the form of a world within which it continues to exist. It is already supplied with a set of deep orienting habits we call instincts. These instincts provide motives for behaviour. The habit forming system then has to find ways for these instincts to be satisfied that do not lead to its own disintegration. Current thinking, based on the so-called *free energy principle*,[8] suggests that our brain brings about the satisfaction of these instincts simply by *predicting* that they are satisfied and (mechanically) adjusting its neural connections until this comes to pass. It is under the supervision of these predictions that we learn the habits that enable us to survive.

If we are again to attempt to characterise our human condition, it appears that I, the consciousness I am, have somehow 'woken up' in the midst of this habit-forming system. As far as the system is concerned, consciousness is a kind of limited resource that is used to handle situations that the system is unable to fully predict. It seems to need this resource to adapt its behaviour, firstly by adjusting itself to make better predictions, and secondly by developing new habits that avoid or destroy unpredicted phenomena. Either way, by eliminating what is unpredictable, the system also acts to eliminate our episodes of consciousness. Such elimination *conserves* consciousness so it is available to meet the next unpredictable situation. It is not that our system is working against our being conscious. It has no intention one way or the other. It is simply a mechanism, developed by a process of evolution, that, in the past, proved effective in controlling the damaging effects of the inherent unpredictability of the environment on our chances of survival. We are now the inheritors of this mechanism. And, like every mechanism, if left to run by itself, it will produce mechanical results.

So what is our relation to this mechanism? Here we are not being theoretical. We are considering our actual phenomenological experience of embodiment. And we are not speaking of the body as a physical object, we are speaking of it as a *psychic system*. In our ordinary (habitual) way of thinking, we consider our body to be something we can control according to our own 'free will' and we consider our thinking, deciding self to be something quite distinct from the habits we can 'call up' and manifest in our day-to-day behaviours. But, if we observe this day-to-day behaviour, we can see that our individual actions are connected together in habit *sequences*, where our intentions to act themselves become habitual.

For example, consider the alarm going off in the morning and your behaviour as you prepare to go to work. Consider the way you get dressed,
the way you wash yourself in the shower, your pouring the cereal and eating it, your driving on route to work. Then consider the thoughts that
are running through your head as you perform these actions. Do you
see that unless something unexpected happens, or you have a moment of
self-remembering, then this entire sequence of activity is more or less entirely mechanical? It is not that you do and think exactly the same things
each day. For you will probably wake up thinking about something that
happened yesterday, or about the meeting you are having today. But the
thoughts you are having in relation to these events are still going to be
quite mechanical. Or you may have thought your cereal bowl was in the
cupboard but when you look it isn't there. And so, quite mechanically,
you look in the dishwasher, and there it is. Throughout this entire procedure of going to work, all the behaviours needed to deal with the various
situations are automatically brought into play without the need for conscious reflection on your part. Instead, you remain caught in the various
thoughts that are also automatically arising according to the chance association that was triggered by the sound of that motorbike passing by
outside . . . .

Figure 9.3: Alarm clock photo by Mohssine Chnaf on Unsplash.

The issue here is not the habituality of our behaviours, but the habituality of our *thinking*. For just as we acquire habitual ways of behaving
by first consciously observing a behaviour and then repeating it until it
becomes automatic, we likewise acquire habitual ways of thinking. Here,
instead of repeating something over and over until we get it right, we can
simply pick up a habit of thinking from coming into contact with that

habit in another person. For in our act of understanding what another person is saying we are actually thinking that person's thought, or, to put it more correctly, we are thinking our understanding of that thought. It is in this way, once we have learnt the habit of understanding language itself, that we learn to think by thinking the thoughts of the people around us. Of course some forms of thinking, like mathematical reasoning, do require training and disciplined repetition. But our everyday thinking is generally informed by habitual modes of understanding that we have simply picked up along the way.

On the surface it appears that we have consciously selected the ideas and opinions that we hold on to and defend. But, if we dig a little deeper, we find these attitudes are more or less aligned with the basic cast of our personality and can be predictably explained in terms of influences acting on us from our surrounding environment—such as the youth culture of our teenage years, our parents and our reactions to them, the people we met and fell in love with, our experience of going to work, the people we have admired, the books we have read, the traumatic events that have occurred to us, and so on. Behind this lie the basic habits of thought and behaviour that pervade our culture so thoroughly that we are barely aware of their existence, because they are shared by nearly everyone we meet. It is here we encounter the deep habits that underpin our egoic state of separation and our scientific materialistic understanding of being.

These are the unconscious habits by means of which we come to understand who we are and what the world is for us. They are habits we have picked up from everyone else, from childhood onwards, and like all habits, the more we live in them, the more ingrained and unconscious they become. We don't experience them as habits of thought, we experience them as realities. Just as our perceptual habituation projects the perceptual form of the world for us, our cultural habituation projects the way our culture understands the world and our being in that world, as if that is the way the world actually is. And so we literally experience the world as being the way we think it is. But with our perceptual projections, we have the immediate feedback of the senses to confirm whether our projections actually correspond with the source out of which all experience is emerging. In contrast, with our cultural projections, our feedback primarily consists in our agreeing with the understanding of everyone else in our culture, and not with any underlying reality. As soon as we start to test these cultural understandings, as we are doing here, we generally find they do *not* correspond with the feedback of our conscious phenomenological observations.

It is about now that it should be dawning on us that our inner mental life, insofar as we are not immediately, consciously present, is almost entirely made up of the automatic manifestations of thought habits that we have acquired from our past experience. We do not recognise the situation we are in, when we are in it, because we have developed thought habits that think we are in a completely different situation. We think we are free to think whatever we wish to think and that we are not thinking habitual thoughts, but that we are thinking brand new thoughts of our own devising. But what is brand new for us are only the *situations of our life*, the new person we just met, the new car we just saw, the new television program we just watched. In each case, we become conscious only because we have met something we cannot predict, something that needs to be classified away into our existing system of habitual thinking. And so we experience something new. We even crave such experience. But the new has come to us from the *outside*. Our response is to immediately assimilate this new experience and turn it into something past, something predictable. Now we can think about it. But such thinking is utterly mechanical. It proceeds by means of habitual associations that connect something in what we are thinking of now with something it was related to in the past, and that association brings up our next thought. These streams of association can present new *sequences* of thoughts, thoughts that you had never seen put together in that way before. But all we are witnessing is the operation of a complex association machine that is constructed out of the habits of our past experiences. It is all entirely predictable because it is our very machinery of prediction that is producing the associations.

Once you discern the operation of this machinery, you can begin to see how it is continuously transforming what is new and alive and immediately present into something that is past and dead and mechanical. We can all recognise this in our getting used to a beautiful song. After a while we are able to perfectly predict each note and nuance, and, as we do, we lose our consciousness of its beauty. Sometimes we can get it back. But we can only do that by becoming immediately conscious again. Otherwise the music slips away under a cover of predictability. It seems that you no longer hear the song itself, its pure living form, you simply experience your prediction of the song. But then again, over time your habit of prediction can start to decay—one day you go to a movie, and quite unexpectedly you hear the song again in all its originality and beauty . . . .

The assimilation of what is present into the habituality of the past continues unabated even in situations like this where we are attempting to understand the process of assimilation itself. You would think that our directly seeing into the engine of this process would enable us to break free from its effects. But that is not how it works. For the system itself is not conscious—it is a machine that is continually attempting to learn the form of whatever it is that we are consciously experiencing. Right now it is trying to learn the form of our understanding of itself. So the next time you read this, it will already be all too familiar to you. You will have more or less predicted the gist what is coming next well before you read it (you may, in fact, already be doing that). There will be no novelty in it and therefore, from the point of view of the system, no need to read any further. For you will 'know it all already'. And yet, in that place of knowing it all already, you are not present, you are not immediately conscious, you have predicted the meaning-form of the words but you have not realised their *actual* meaning. For the actual meaning of what is being said about this habit system can only be grasped from *outside* the habit system, i.e. from the place of immediate consciousness, where you hear the song in its timeless presence, no matter how many times you have heard it before.

Time is key here: our habitualities are the habitualities of the *past*, they are the crystallised forms of our past behaviours—our thinking and feeling and imagining and doing—and insofar as we lose our presence, our immediate consciousness of being now, then we *fall* into the temporal dimension of these habitualities. The temporality of these habitualities cannot refer to now (cannot *be* now) because they are not conscious in themselves. They are not conscious because of *what* they are. They are the recorded form or pattern of the activity of a *former* consciousness. But we, in losing our presence (our consciousness of *being* now), can *fall* into these habitualities, and, in a certain sense, we *lend* them our consciousness. In doing so we enter into a new dimensionality of temporal experience called *time-past*. We generally do not recognise this time-past as past because our habitualities are continually (habitually) thinking about possible (or impossible) futures. But these futures are *past*-futures, expressions of habitualities we experienced in the past, recombined and projected into a space of imagination. The *real* future is something we never see. It is the unthinkable source of this very moment from out of which the stream of experience is continually flowing. And even though the stream is present to me now, everything in the stream is already past. Only now is now, and now is the nothing from out of which the past is streaming. Everything we predict, everything we imagine is in this stream, and is already past. But as our predictions emerge from out of this unfathomable source, and as we watch them flow away, we think we are looking into the future. But we are not. We are looking into the past of habitualities we have developed that continually *imagine* the future. The future, the *potential-actual* future, *is* now, it is entirely present as the unthinkable potentiality of this moment.

This was revealed to me in Kuala Lumpur airport while travelling on one of those electric trolley carts. Instead of sitting up front and looking in the direction we were going, I was sitting on the back, facing the other way, watching where I had already been. It suddenly dawned on me that this is how it is for us. We are continually watching what has already happened streaming away from us. We think we are looking into the future, sitting up front and facing our direction of travel. But our activity of looking into the future is just another item in the stream of time that is flowing away from us. The future itself is behind us, is coming toward us from a completely invisible dimension. It doesn't matter how we turn our head (our attention), the future is always what we can't see, the source from out of which what we can see is always emerging.

Still we are asking after the situation in which we find ourselves: our being conscious in the midst of this system of habitualities. Our inheritance is of a culture that knows virtually nothing of the reality of our being conscious, and so it exists almost entirely in an experience of its own habitualities. As children and young adults we are filled with these habitualities of thought and behaviour and in our adult lives they are continually reinforced by the collective thought and behaviour of all the other people around us. If we take all this together, we can begin to see that what stands in the way of our being immediately conscious is this edifice of habituality that is our modern way of life. Our immersion in this collective habituality is like a kind of gravitational inertia that tends continuously to pull us back into a state of being identified with the streaming of our own habitualities. The way this inertia functions is part of our evolutionary inheritance. It was simply better for our ancestors to form collectives that shared the same basic habitualities. And so it has become natural for us to automatically fall into line with the collective habitualities of our culture. And if it is part of the habituality of our culture to think in ways that close off the possibility of our being immediately conscious then, of course, we will not find it easy to maintain a state of immediate consciousness. There is nothing *wrong* here. We just have to understand what we are dealing with.

Another thing we should understand is that our system of habituality is not, from within itself, going to generate habitualities that will lead us into a state of immediate consciousness. Not only will it not do that, it will tend to turn everything we do discover on the basis of our being conscious into another habituality. For as soon as we think we have found a way to enter or maintain a state of immediate consciousness, that way will become habitual and so will tend to lead us back into a state of habituality. This can go as far as our developing a habituality that *thinks* it *is* immediately conscious, and thinks that all we need to do to realise consciousness is to identify with this habituality.

In contrast, our actually being conscious is like a clearing in the forest of our habituality. Everything that we see and realise in the clearing immediately flows off into the past where it is appropriated by this system of habits. It is the *job* of this system to relieve us from the necessity of being conscious of whatever it is that we are experiencing by turning it into something habitual, so that our consciousness can be free to deal with the *next* unexpected situation. Here we must remember that the basic system evolved to deal with the extreme danger our ancestors faced living in the midst of a wild nature. But now, as we sit in our rooms, behind the walls of our civilisation, living our structured and routine lives, this system of habituality does not serve us so well. For we are collectively facing situations and crises that cannot be resolved by habituality alone. In fact, these crises are the very phenomena that are calling us to consciousness, as it is only by means of consciousness that we can transform the habitualities that are creating the crises in the first place. For it is not wild nature that threatens us now. It is our system of habituality itself.

If we were to think in terms of some greater evolutionary purpose, it would appear that the overriding task for humanity is to manifest the being of consciousness on Earth. Evolution has done its job in creating the system of habit that is our psychic human body and that now enables us to manifest ourselves to ourselves in this projection of our being here on the earth. It is now as if the system were saying: "Well, I've done all I can, after this it's up to you—I am just a system, I do not work against your being conscious, but neither do I work for it, for I am not conscious, I am just the means that enables you to be here in this existence. Now it is your responsibility to be the consciousness that you are, and that includes being conscious of this system that you inhabit, of understanding how it operates. For if you neglect that then you cannot be the consciousness that you are, not fully, for you will still be controlled by the mechanism that I am."

Now, of course, this is not a scientific or phenomenological finding. It is a form of meaning that is attempting to illuminate the transcendental situation in which we find ourselves. So our accepting or rejecting this meaning is not a matter of finding an objective proof or demonstration. It is a matter of *resonance* with our experience of being conscious. For, as I am (we are) the consciousness of this planet, and insofar as this consciousness is free and undetermined by the habituality of the past, then it lies in our own hands to project a purpose, a meaning, a reason for being into the source and await the resonance of a reply. Such a reply does not necessarily have to come from within. There can be resonances in the outer life as well. For instance, consider that Nicolette, my wife, is reading Carl Jung's autobiography at the moment. And just the other morning, she was so struck with something Jung was saying she decided to read it out to me. It went as follows:

> By virtue of his reflective faculties, man is raised out of the animal world, and by his mind he demonstrates that nature has put a high premium precisely upon the development of consciousness. Through consciousness he takes possession of nature by recognising the existence of the world and thus, as it were, confirming the Creator. The world becomes the phenomenal world, for without conscious reflection it would not be. If the Creator were conscious of Himself, He would not need conscious creatures; nor is it probable that the extremely indirect methods of creation, which squander millions of years upon the development of countless species and creatures, are the outcome of purposeful intention. Natural history tells us of a haphazard and casual transformation of species over

hundreds of millions of years of devouring and being devoured. The biological and political history of man is an elaborate repetition of the same thing. But the history of the mind offers a different picture. Here the miracle of reflecting consciousness intervenes—the second cosmogony. The importance of consciousness is so great that one cannot help suspect the element of *meaning* to be concealed somewhere within all the monstrous, apparently senseless biological turmoil, and that the road to its manifestation was ultimately found on the level of warm-blooded vertebrates possessed of a differentiated brain—found as if by chance, unintended and unforeseen, and yet somehow sensed, felt and groped for out of some dark urge (from *Memories, Dreams, Reflections,* pp. 371-372).

Figure 9.4: Carl Jung 1875-1961.

The Intelligence of Not Knowing

And so here we stand, strangely poised, in the clearing of the source. There is one here who is inquiring and there is another who is answering. And then there is the unity of the clearing and of the inquiry and of the answer, in which no one is inquiring or answering, where there is just the inquiry and the answer and the clearing and all that is manifesting in the clearing. It is as if there is a *lens* of consciousness. We can wind this lens all the way back into the unity of being conscious now, where there is no 'I' looking or inquiring any longer, where there is just the manifestation itself. But then a question arises, a question concerning how we can characterise the clearing of the manifestation of all that can have being for us. Now the consciousness I am 'takes up' the question. The previous unity is dispelled, the lens becomes focussed, and the intention in the question is no longer the empty intention of a possible question, but the active intention of a living inquiry. But here we have not become *lost* in the question. We *hold* the question. We remain conscious. We are directed toward the question and to the potential response, and we are *awaiting* that response. We cannot say from whom or from where that response is coming, except to say that it is coming from this unthinkable source. And yet there is still a sense that *I am answering myself,* only this 'I' exceeds the boundaries of the clearing. It meets itself here, in these inquiries, but only *aspects* of itself, just as we only see aspects of the things that appear in the world around us. There is division within unity. There is this 'greater' being that I am, that is the source of the questioning and of the response, that exceeds the clearing, that remains for the most part concealed. And then there is the being of the clearing itself, the being that is the consciousness of the clearing, that is the reflected intelligence of the being 'behind' the clearing. The being of the clearing *only knows what appears in the clearing*. It does not *answer* the question, it *awaits* the answer. It enters into a dialogue with what lies concealed beyond the clearing. We could say that the clearing is my *explicit* world, where the concealed being I am comes to a partial and explicit knowledge of itself through the mirror of the consciousness of the being of the clearing.

In order to engage in such a dialogue, the being I am, the one who is waiting, must already be in a place of *not knowing*. To *not* know is to have withdrawn from the place where we already know everything already. I first became aware of this distinction in listening to the discourses of Rajneesh. Over and over (it seemed) he would speak about the oracle at Delphi declaring Socrates to be the wisest man because he knew that he knew nothing.[9] At the time I did not make much of this statement. But now it seems significant. For if you withdraw and disidentify with your habitual thinking processes, then you will find you arrive in a place where you no longer know what you thought you knew. There is just an inner silence of not knowing. In this silence there is the manifestation of sensory experience, which, of course, is not silent. But it is occurring *in* silence. It is there in the nothing of the space that separates the something of everything that we see before us, and in the nothing of the silence that separates each sound we hear. It is only because of this nothing, this absence, that anything is distinguished from anything else. It is the pure receptivity of consciousness—it is the clearing of the clearing. Insofar as I am this immediate consciousness, then I am this space in which everything appears. It is not that I know nothing. For I am the conscious receptivity that knows each moment. It is that I am no longer in the place of the habituality of *knowing everything already*. I now *hear* the sound of the bird rather than semi-consciously registering something that confirms my expectations.

There is a subtle distinction here. It is not that the habitual expectation of the bird song is no longer functioning. If that were the case then I would no longer recognise the sound, I would experience a jolt of surprise, and my attention would be drawn to the sound in a new way. No, my system of habituality is functioning just as it always does, keeping an 'eye' on things for me. The only difference is that I have become conscious. I am no longer identified with the habituality that already knows what is going to happen. I am in a place where the expectation of that habituality is just another aspect of the totality of what is manifesting now. I am *allowing* the bird song to manifest, I am *hearing* its manifestation, even though it is manifesting through the habituality of my expectation. It is just a small shift of emphasis, and yet it leads me out of the captivity of my habituality into the immediacy of being now.

The 'not knowing' we are speaking of, is our not being identified with the habituality of the expectations we have concerning the form and structure of our experience. In being conscious, we are simply present as the receptivity of now. It is here that the beauty of the earth can (unexpectedly) reveal itself. And it is here that we can engage in authentic phenomenological inquiry. Such inquiry requires that we have access to our immediate intelligence, the intelligence that is able to look and see into experience as it actually presents itself, rather than as we habitually think it presents itself. For if we start to look into the deeper structure of experience, we can no longer employ the reliable sensory expectations we developed as children—the ones we use to see cats and dogs and tables and chairs. Instead we are faced with such mysterious phenomena as our identification with our thinking processes and our feeling of having a free will. It is here, in our not having a ready-made habituality to fit the situation, that confusions and difficulties can arise. For unless we come to a state of immediate consciousness, what our culture presents us with are the labyrinthine corridors of the books and theories and opinions of the various professionals who have made a career out of thinking about these things. It is there that we encounter the polar opposite of our state of not knowing.

Again we must tread carefully. We do not want to make a *problem* of this system of habituality that constitutes everything we think we know. For it is something that has arisen in us quite naturally, like the growth of a plant, according to the influences of the environment on the underlying structure of the kind of person we are. As we come to adulthood it emerges as the collection of beliefs and opinions and theoretical and factual knowledge that is expressed in our explicit thinking processes. It is here that we come to form our explicit understanding of our place in the world—our own particular 'world view'. This again is our Western way. It is our *right* to form such a world view—so long as it respects the rights of everyone else. For it would go against the spirit of the natural light for us to impose our world view on another. All we can do is point out how certain world views go against what we see to be the obvious and self-evident facts of experience.

Our situation is that we cannot help but have some kind of world view. It is the unavoidable expression of everything we have been through and how we have come to understand it. But it is still a habituality—a certain habit of thinking we have acquired in the past that we now reincarnate in the present, each time we think or tell another person what we think we know. The question here is how it stands each time that system comes to expression. Do we see it for what it is: a habituality that records and expresses the best understanding we have been able to reach up to this moment? Or do we experience ourselves as being identified with this understanding? Are we the space, the consciousness within which this understanding comes to expression, or is it that this understanding *is me*? If we are identified, then we will experience a certain emotional attachment. That means if someone disagrees with us, we will not be interested in why they disagree, not really, we will only listen to them in order to find a way to attack what they are saying, because *we know already that we are right.*

Of course, if we did not think we were correct, then, presumably, we would do something about it. So it is not that we should go around thinking we are wrong. The crucial distinction is again the distinction of our being immediately conscious. For our *not* being immediately conscious means we have *already* become identified with what we are thinking and saying. It is the identification that defines our loss of consciousness. In contrast, in immediate consciousness my belief in what my thoughts are proposing becomes *suspended*. This is not a matter of doubting. To be in a state of doubt is to be identified with thoughts that assert themselves doubtfully. The consciousness I am simply does not *go-along-with* that intending or asserting. I retain my poise. I can *see* what the thought intends *without* entering into that intention. Now all this may sound quite complicated, as if there were some fine power of distinction needed to see one's thoughts in this way. But this is not a matter of deliberately distinguishing anything. We are speaking of an intelligence that effortlessly manifests as soon as it is released from its captivity in the labyrinth of our thinking processes. The not-knowing of this intelligence is the not-believing of a not-going-along with what our thinking intends. The thoughts themselves still intend what they intend while my immediate consciousness simply witnesses that intention.

What we are faced with here are *two forms of knowing*. The first form is the *immediate* knowing of consciousness. This is our knowing of whatever it is that is happening now—for instance, my knowing of the sound of the breeze in the trees. And then there is the *propositional* knowing of my thinking. Such knowledge takes my immediate knowing and fixes it into an explicit form. For instance, it proposes that "I am hearing the sound of the breeze in the trees". But of course, I may not be hearing the sound of the breeze in the trees. I may be the subject of a strange hoax where someone is faking the sound of the breeze using a mechanical device. The point here, as Descartes saw, is that my immediate knowing can never be false in this way. For I am simply hearing what I am hearing, and not proposing *what* it is *that* I am hearing. Such immediate experience is present to me *before* I reflect, *before* I think, *before* I can formulate something in language to indicate what I am experiencing. It is only in the language of thinking that the possibility of the true and the false arises, and it is only on the ground of immediate consciousness that such questions can finally be settled.

And so, insofar as I am present as this immediacy of consciousness, I only know what is immediately present. I do not know the truth or falsity of any thought unless the object of that thought is also immediately present. For it is only in this moment that I can see whether or not there is a correspondence between the meaning of that thought (which is immediately present) and what that meaning intends (which is also immediately present). Otherwise I do not *know* anything about what the thought proposes, because it proposes something that is not immediately present. But if I *enter* the thought, then I *do* get to know what it proposes—for the thought intends the being of what it thinks. For example, if I think of New York, then I am actually thinking of New York, and my thinking refers me to New York, and as soon as I am referred in this way I lose my immediate consciousness of now and I enter the *knowing* of what the thought is proposing. Do you see, that in my not going-along-with my thinking, I do not *know* what the thought knows, even though I know what the thought *means*? For I can *hold* the meaning before me without following what it proposes into the virtual not-now of my thought world.

So, the not-knowing of immediate consciousness is not a complete not-knowing. For I do know what is immediately present to me. But I remain disidentified with that other world of thought that knows about everything else. This not-knowing of immediate consciousness is a not-knowing in relation to what we would call *worldly* knowledge. Worldly knowledge is a kind of commodity. It is something we expect our university professors to possess. And it is something, when we go to university, that we also expect to *get*. Once you become an expert in a particular area of such knowledge, and you start speaking about it to someone else, it is like entering into a kind of knowledge warehouse, where we can roam from one place to the next within an enormous structure of understanding that connects everything into a unified hierarchy. The more you read and teach and research and write and publish, the richer and more interconnected and more coherent this body of knowledge becomes. It is really something quite extraordinary. And then, one day, at the beginning of a lecture, you are standing on the podium, and you suddenly realise: but I know *nothing*! At least, all I know is what is immediately present to me now. Here are all these people expecting me to produce this specialist knowledge, and there is just this inner silence in me. I know from previous experience that when I put up the first slide that something will happen, words will arise and a lecture will emerge. But in this moment of immediate consciousness I do not know anything about all that knowledge that is supposedly inside me somewhere. I simply have to wait, hopefully, and see what happens.

Now, the strange thing is, that if you can teach from this place of immediate consciousness, your classes are likely to go much better. It is like

playing tennis. As soon as you stop making an effort, you find that the habitualities of knowledge will simply emerge harmoniously on their own.
As they do, you can start to work in another dimension, where you remain conscious of what is actually happening with the people and the
atmosphere in the room. You find you can start creating a space of consciousness that some of the people can start to enter—a place of looking
and seeing within the material that is being presented. Nothing is 'by
rote' any longer. There is the beginning of the awakening of the intelligence of now.

So it is not that our immediate consciousness somehow works against
our capacity to learn and disseminate worldly knowledge. Quite the reverse. But, if you did go to university, you may remember it was not always like I am describing. If you encounter an expert who is completely
identified with their knowing and their knowledge then they will tend
to want to turn out students in their own image. This can be a kind of
tragedy. For many students who arrive at university are still naturally in
touch with this not-knowing state of immediate consciousness.

Loss of Memory
It should be clear by now that it is no good trying to remember what is
being said here. Even *I* can't remember what is being said here. For it to
be meaningful you have to see it all again for the first time each time you
read it. Nevertheless habitualities will form. They are already making
these sentences predictable. And if you read this through again, those
habitualities will make it appear as if nothing is being said or as if the
same thing is being said over and over. That is because we are speaking
of what is *not* habitual—of what cannot be understood from habituality.
And so, if you read this within a habituality that makes it predictable,
you will leave the immediacy of the present—whereas what is being said
here is only concerned with the immediacy of the present and can only
be understood by means of our actually *being* present.

If you do try and make something memorable out of these words then
it will most likely become a *system of precepts*—things to do or not to
do, things that are true or not true, methods to follow in order to 'be
present', interpretations of Descartes, opinions about science, and so on.
We can only think about this by going into the memory, and that, of
course, means leaving the immediacy of now (unless the memory simply

provides us with what is needed in the moment—which it often does). If you do make such a system of precepts, then you have missed the meaning of phenomenology. It is part of the irony of life that this often happens with people who decide to study phenomenology professionally.

The nature of being a scholar means you have to develop a kind of encyclopaedic knowledge of your subject area. If that area is phenomenology, then you will end up developing a system of habituality that encompasses everything the key phenomenologists have said about phenomenology. And then you will read what your colleagues have said about what the key phenomenologists have said about phenomenology. And then you will read what your colleagues have said about what your colleagues have said about what the key phenomenologists have said about phenomenology. And then you will have to say something *new* that these colleagues will find interesting enough to consider publishing. In the midst of all of this, the meaning of what Husserl or Heidegger were actually *doing* gets lost, i.e. their intent to lead us into a direct encounter with the meaning of phenomenology itself. Instead the scholar's task becomes one of analysis, of making an object of what has been said, so we can bring to bear all the tools of our habituality in 'understanding' it. This typically means extracting the propositional content and subjecting it to rational criticism. Or it means finding the historical sources and influences that lie hidden in those propositions. Or it means comparing those propositions with the propositions of other philosophers, so we can know where these propositions fit in the historical structure of all propositional philosophical knowledge. Or it means finding a new reading of an original text and showing how someone else's reading is less convincing than yours. All this work, all this *thinking about phenomenology*, leads us away from the immediacy of now, from the living meaning of what we are studying, into the habituality of a purely propositional knowing. Instead of knowing nothing, we come to know *everything* about phenomenology. And so no one can tell us anything we don't already know. As a result we enter into a kind of *sealed container* of knowledge.

Of course, such sealed containers are not the special province of academic philosophers. As we have seen, it is in the nature of our system of habituality that it will attempt to assimilate everything we experience into such a structure of predictability. Unless we come to consciously recognise what is occurring, we will almost irresistibly be drawn into such a structure. And yet, on the other side, it is almost as if nothing has changed, even though everything has changed. For our system habituality does not cease functioning just because we become disidentified with it. What has changed is that the system is *no longer in charge of running our lives.* There is a gap, a space of consciousness, which, instead of already knowing everything in advance, knows nothing, except what is immediately manifesting now. And yet, in that knowing nothing we can start to inquire into this system of habituality—not by the trickery of one thought process starting to think about another thought process, but by means of a simple looking that remains conscious of its being conscious.

What is interesting about this state of looking into the immediacy of now is that thoughts start emerging in response to our questioning that are no longer habitual. They are, if you like, *transcendental* thoughts, in that they too, along with our questioning, refer to the immediacy of now. These thoughts put into language what it is that we are seeing. It is not that I, the one who is looking, am *formulating* these thoughts. They are simply arising in response to my looking. The core of this book is a transcribing of such thoughts. It is important to recognise that there is such a source of transcendental understanding available to us. It is the same capacity that we use to think habitually. But once we become conscious, then our thinking also escapes the trap of its own habitual reference to the past. That does not make it infallible. But it means we start to encounter new meaning rather than simply recycling and recombining all the old meanings. And it also means we can start to inform the old habitual meaning structures about how it stands in the immediacy of now. We can start to destroy the false habitualities we have inherited from the past, not by replacing them with new ones, but by means of the direct insight of our not-knowing. For this is destruction by falsification. We simply see that what we thought was true is not true. What lies behind such destruction is the mystery and intelligence of our *not-knowing.*

iconic, holographic, Bayesian probabilistic

NOTES

1. George Ivanovich Gurdjieff (1866?-1949) was a renowned Armenian-Greek spiritual teacher. His monumental tome, *Beelzebub's Tales to His Grandson* (1950), provides a unique (and, we assume, allegorical) account of the development of human consciousness on Earth from the perspective of an extraterrestrial being, namely Beelzebub himself.

2. Pyotr Ouspensky (1878-1947) was a Russian journalist and philosopher who is best known for his work interpreting the teachings of George Gurdjieff. *The Psychology of Man's Possible Evolution* (1951) was published posthumously and intended as a condensed introduction to the key elements of Gurdjieff's system. Ouspensky himself broke with Gurdjieff in 1924 but continued to teach and write about the ideas that Gurdjieff had introduced to him.

3. *The Life of Brian* was a 1979 Monty Python movie that satirised the life of Jesus by telling the fictional story of Brian, a man born next door to Jesus, who was subsequently mistaken for the Jewish Messiah. If there is one event that demonstrates that the 'Death of God' prophesied by Nietzsche (see note on page 170) has finally been recognised and publicly accepted, it would be the widespread release and commercial success of this film.

4. *Human, All Too Human* is the title of a book by Friedrich Nietzsche. In aphorism 35 he pertinently observes "that reflection on the human, all too human—or, as the learned expression has it: psychological observation—is among the expedients by means of which one can alleviate the burden of living, that practice in this art lends presence of mind in difficult situations and entertainment in tedious circumstances, that one can, indeed, pluck useful maxims from the thorniest and most disagreeable stretches of one's own life and thereby feel a little better: that was believed, that was known—in former centuries" (1878/1996, p. 31).

5. In W. Somerset Maugham's novel *The Razor's Edge* (1944), the central character, Laurence Darrell, has a deep and transformative spiritual realisation in India. Later in the novel, 'Larry', now back in the USA, instead of setting up as a spiritual teacher, expresses an intention to give up his private income and make a living driving a taxi in New York. The novel leaves it open whether this actually happens. The story itself is supposedly related to Maugham's real life experience of meeting the Indian sage Ramana Maharshi in 1938 and of his knowing several people like Larry who were similarly transformed by their contact with the sage and his ashram.

6. The story of the sannyas movement in Oregon ended dramatically in 1985 when the city of Rajneeshpuram was abandoned after Rajneesh attempted to flee the country. Upon stopping to refuel his plane in North Carolina, he was apprehended by US authorities, put on trial for immigration offences, and finally deported back to India in November 1985. Before that a scandal had emerged involving his former secretary Ma Anand Sheela, who had conspired in various criminal acts, including illegal wiretapping, mass poisoning and attempted murder. This occurred during a period when Rajneesh had retired into silence and left Sheela in charge of the organisation. It remains unclear how much he knew about these illegal activities. Sheela herself fled to Europe where she was arrested and deported back to the USA, finally serving a prison sentence for her various crimes (to some of which she pleaded guilty). The entire drama was made into a 2018 Netflix documentary entitled *Wild, Wild Country*.

7. On his death, Barry Long left behind two autobiographical manuscripts, the first of which, *My Life of Love and Truth* (2013) was published ten years later. The second part remains unpublished.
8. The free energy principle was originally proposed by Karl Friston, James Kilner and Lee Harrison in their paper *A Free Energy Principle for the Brain* (2006).
9. Rajneesh was referring to Plato's *Apology* where Socrates (c. 470 - 399 BCE) is musing as to why the oracle at Delphi declared him the wisest man. He concludes it is because, when compared with another man who was widely considered to be wise: "it is likely that neither of us knows anything worthwhile; but he thinks he knows something when he does not, whereas when I do not know, neither do I think I know; so I am likely to be wiser than he to this small extent; that I do not think I know what I do not know" *Apology*, 21d.

CHAPTER 10

THE UNIVERSAL BACKGROUND

A NEW BEGINNING

UNTIL now we have been looking at how we have arrived in this situation in which we find ourselves. We have been concerned with our *existing* habitualities, our collective understandings of the world and of our human condition. To a large extent, this has been a work of *deconstruction*—of seeing how certain of these understandings fail to correspond with our immediate experience. The result of such deconstruction is to open up a space of not knowing in the midst of all we thought we knew. The question now is whether it is worthwhile to build something new in the ruins of these former conceptions.

I think the answer here is to see that this is going to happen anyway. The issue is only whether we do it *consciously*, or simply allow our habitualities to work things out for themselves. For whenever a gap appears in the network of our understanding, these habitualities immediately and automatically set about closing it up. Our being conscious in the midst of this process firstly means that we do not become attached to these repairing projections. What such repairing entails is the proposing of new forms within which the unthinkable source can manifest itself. In remaining conscious we remain in touch with the essential mystery of this source—with the knowledge that it can only ever be revealed to us by means of these forms we project, and that every such form only reveals an *aspect* of the totality, an aspect that necessarily *conceals* what it is unable to reveal. We can never come to experience the totality as the totality that it is. And neither can we think this totality. That is the error of our materialistic science: we think, in our scientific modelling, that we are understanding the totality of the source, or that 'one day' we can get to understand this totality. And similarly, in perceptual experience we think we are looking out into the totality and seeing it as it really is. But we are not. We are seeing how it *responds* to our idea of how it really is.

The second aspect of our remaining conscious in this process of regeneration, is that we can include what our science has *actually* discovered, without falling into the habituality of the materialism that lies behind these discoveries. For the double helix structure of the DNA molecule is an objectively verifiable fact that in no way depends on there being some ultimate ground of materiality that determines everything that happens in the universe. And in any case, the process of the development of new forms of scientific understanding is *already* underway. It is not as if we have to invent such understandings. For what we have discovered *already* exceeds the scientific materialism we have inherited from our Newtonian past.

Even so, our situation is in many ways similar to that of Western Europe during the last scientific revolution. Both then and now, it was clear that the current, collective world understanding did not correspond with the discoveries that were being made. For medieval Europe the issue was focussed on the earth being the centre of the universe, around which all the other heavenly bodies were turning. It took nearly 150 years of controversy from the publication of Copernicus' *On the Revolutions of Heavenly Spheres* to Newton's *Mathematical Principles of Natural Philosophy* before it was more or less accepted that the earth revolves around the sun. The reason for this delay was not just to do with the astronomical calculations. It was to do with the collapse of an entire Aristotelian understanding of the being of the universe. It was just not possible for this understanding to collapse until another complete and more convincing understanding was ready to take its place. It was only with Newton's development of the idea of gravity that it became comprehensible how the moon and the planets and the stars could remain suspended in the void of space without falling into the earth.

For us, it is not a question of the earth no longer being the centre of the universe, it is a question of the being of our own consciousness. This is the essential component that our collective scientific materialism leaves out of its mathematical calculations. And yet it remains unclear just how we are to account for the undeniable fact of consciousness without destroying the one thing that gives our culture some semblance of coherence: the habituality of our objective scientific rational understanding of the being of the world around us.

The problem here goes to the heart of what we understand reality to be. Just as Galileo and Descartes had to carve our experience in two in order to unleash the objectifying power of mathematical thinking onto an unsuspecting world of peasants and artisans and merchants and aristocrats, we now have to find a way to bring the two worlds of matter and consciousness back into some consistent harmony that still respects all we have discovered. To do this we must look again at the question of the being of the material world.

The Phenomenon of Substantiality

What lies at the foundation of our Western common sense notion of the materiality of the universe is the phenomenon of substantiality. This was the basis of Samuel Johnson's kicking a stone to refute Berkeley's idealism (page 158). For I could easily be mistaken about something I see—such as Macbeth's vision of the dagger.[1] But when I actually touch something by picking it up or kicking it, I know it is real, for I feel its resistance, its weight, the space it encloses in its sheer *being there*. In our practical lives it is such substantiality that confirms a thing's existence. It is our final test, our final demonstration that something *is really here.*

It is this experience of substantiality that underpins our idea that ultimately everything that exists exists as *physical stuff,* and that the physicality of physical stuff consists in its solidity and impenetrability, and that behind the power of physical stuff to have an effect on us there has to be some kind of physical contact between its physicality and ours. In Newton's day we thought this physicality existed in the void of physical space in the form of impenetrable atoms, and then, when these atoms were found to shatter, we thought instead of impenetrable sub-atomic particles. But now it seems that nothing is 'impenetrable'. For behind the physicality of the particle there is something called 'energy' which can be transformed into all the various forms of physicality that it is possible for us to encounter. And then there is the quantum level. Here it appears that all the particles that we measure with our recording apparatus actually exist in a state of pure potentiality, occupying multiple possible positions and states at the same time, and only pop out of this indeterminacy into a fixed state when we make an explicit observation. And that means we now know, on the basis of our objectively physical observations, that our original idea of the substantiality of physical stuff, as the pure enduring solidity of tiny bits of matter, is not ultimately true.

This is like our recognising that the earth couldn't be the centre around which the rest of the universe is turning. Only this time we are recognising that our common sense notion of physicality is just *our* way of understanding the being of the things around us. It is part of our anthropocentric view of the universe, a 'built-in' feature of our perceptual systems, that we automatically take 'reality' to exist in just the way these systems present it as existing. We have been able to adjust to such things as the sun not really rising over the horizon, and the objects around us not really being solid, because we can picture some other way things are configured using our perceptual imagination—such as the earth going round the sun and matter being made of tiny particles held together by forces of attraction and repulsion. These are still things that made experiential sense to us. But now, in the world of quantum physics, even the *idea* of the being of a thing appears to be an anthropocentric projection of our own devising.

Do you see that from here Samuel Johnson's refutation of Berkeley is meaningless? For his experience of kicking a stone is something that can only occur for beings like us, with perceptual systems like ours. Such experience has nothing to say, one way or another, about the being of any ultimate reality that lies beyond that experience. Or, to put it another way, our experience of the substantiality of the things around us, is an experience of the *phenomenon* of substantiality. It is only this phenomenality that we know to be real, because that is all we *actually* experience. Such experience is the way the source manifests itself in the perceptual systems of beings like us. But that doesn't mean there is anything like substantiality existing 'out there' in some other place beyond this immediacy of experience. Samuel Johnson would have found this almost impossible to understand. But now it is not just George Berkeley who is trying to persuade us that all we immediately know is our perceptual experience of the world and that such experience does not reveal the totality of the source. Our *science* is telling us this as well.

What is at stake is our entire idea of the being of physical substantiality that we have inherited from Galileo and Descartes—the idea that what has real physical being is what can be fixed in the precision of an objective mathematical measurement. What we have discovered is that if we keep on measuring at smaller and smaller scales, this idea of the permanent mathematically determinable being of a physical substance finally breaks down. Our science *could* have ground out in a realm of tiny bits of physical stuff—but it didn't. It turns out, after all, it was just a hypothesis, and we have shown the hypothesis to be *false*.

But the situation for our culture is that this news has not permeated up to the higher levels of our collective cognition. We still go about our daily lives thinking about the being of physical substance in essentially the same way as Galileo and Descartes proposed. Even our physicists, to a large extent, keep quiet about this, treating the quantum level as a means to predict and control what occurs 'up here' in the 'real' world of determined and determinable physical entities. But, as far as the science is concerned, there is no absolute division between the quantum level and the level where we live and perform our measurements, there is just the absence of a coherent explanation that can unify these two 'views' of the same underlying source. The reason for this is not that no such explanations are possible, it is that we are simply unwilling to modify our thinking, because that would mean giving up our deep-seated conviction that the universe is really and truly made of physical stuff—the very stuff we experience it to be made of when we pick up a cup of coffee or kick a stone.

If we return to the phenomenology then all these problems start to disappear. For our idea of physicality is just a certain kind of habituality that we project into the stream of our experience. It is a way we are able to manifest the source and make it intelligible in the conduct of our day-to-day living. If we are not attached to this habituality, if we do not think that it presents reality as it absolutely is and must be, then the news from quantum physics is something to be welcomed, even to be embraced. It says things are not as we thought they were, it says how *fascinating*, it says let us look again into this source and see if there is not a new understanding of being that will better serve us, that is truer both to our experience and to our scientific discoveries.

THE CRACK IN THE EGG

Of course, just as the old Earth-centred understanding of the universe can be saved by making adjustments to the Ptolemaic model of the planets being attached to spheres revolving within spheres, it is possible to reconcile quantum physics with the old atomic model of the being of matter. We can do this by thinking of the quantum level as a purely mathematical construct, something that only serves to predict what is going to happen up here in the 'real' physical world. That means we can continue to think of atoms and molecules as things that have a definite, determinable, measurable physical existence—just not as the kind of permanent impenetrable 'grains of matter' that we once imagined. For it now appears that atoms are only relatively stable configurations of energy that can, in turn, be decomposed into sub-atomic energy forms. These sub-atomic forms can be manifested 'up here' according to the effects they have on our physical measuring devices. And these effects—such as a photon impinging on a photographic plate—suggest that sub-atomic phenomena exist as particles in definite physical locations. But such particle existence *only* manifests when a measurement is made. Otherwise it appears that what underlies the particle's manifestation has no definite physical location whatsoever. All we have are mathematical formulae that tell us how probable it is that a particle-like event will manifest *if* we make an observation. This is not just a matter of uncertainty about where the particle might be, it is uncertainty about the manner of being of whatever it is that lies behind the particle's manifestation. For the evidence suggests that when the 'particle' is not being observed, it takes on a form of wave-like virtuality, where it interacts with the wave-like forms of all the other potential particles that could manifest within its particular localised system.

To make this clear, we are not talking about a strange phenomenon that only occurs in particle accelerators and physics labs. We are talking about the fundamental being of everything we take to be the physical universe—it all has its foundation in this state of quantum virtuality. When we take the quantum level to be a mathematical construct, that means we only grant 'real being' to what manifests in our observations, and we leave the question of what is happening between observations unanswered. And that means we give up on trying to 'make sense' of what the universe is 'made of'. For quantum physics definitely shows that it is *not* made of discrete packets of energy occupying definite physical locations and following definite physical trajectories. Instead we are dealing with entities that appear to pop in and out of existence in a way that cannot be determined in advance but only predicted according to probabilities generated by the Schrödinger wave equation.[2] This quantum world is so alien to the dimensionalities of our sensory experience that we cannot even imagine what it would look like. All we have are mathematical expressions and the attempts of various physicists to produce metaphorical diagrams and analogies. No wonder we attempt to 'draw a line' over the quantum level, and treat it as something we can more or less ignore, and certainly not as something that requires us to fundamentally change our understanding of the being of the things we see around us. For even the physicists have been unable to agree on a coherent picture of what quantum physics really has to say about the being of the physical world. And then there is a second kind of line: the line of the limit of our own understanding. For mathematical physics has become so specialised that only a select group of people, who have devoted their lives to its study, can be said to properly understand the mathematics of its conceptual content. We, who cannot directly read this language, must make do with the greatly simplified picture that is passed on to us via popular science books, TV documentaries and the related internet content. Here, rather than encountering the 'equations themselves', we encounter something that has already been interpreted for us by the very physicists who developed the theories in the first place—people who, within the mythology of our scientific materialism, are seen, if not as gods, at least as the finest intelligences our civilisation has ever produced. Think of Albert Einstein. Or Stephen Hawking.

So what is the point of our inquiring into quantum physics in a book like this? If *Einstein* could not work out the true meaning of these quantum phenomena, then what chance have we? But let us take heart. Let us consider that the so-called quantum paradoxes are less to do with the mathematical models, and more to do with understanding these models in the context of our scientific materialism. At the centre of these paradoxes lies something called the *measurement problem.*[3] This is where the interface lies between the quantum and classical levels of being. At the quantum level it appears that the whole universe exists in a state of pure potentiality, where all the definite particles and events that we perceive 'up here' on the 'classical' level, are mathematically represented by probability wave functions (i.e. by forms of the Schrödinger wave equation). The wave functions are used to calculate the probability of our observing a particular classical level event *should we choose to make a measurement.* If we do choose to make a measurement, something extraordinary happens: the whole system of probabilities resets itself. This is known as a collapse of the wave function.

Let's say we are going to observe whether a packet of photonic energy is going to appear at a particular location on a photographic plate. We know in advance, according to the wave function, that the probability of this happening is 1 in 4. Before the observation we cannot say that the quantum level photonic energy has any particular location—we cannot even say whether it exists as a wave or a particle. The best model we have is the wave function itself which specifies, for each point in space, how probable it is that photonic energy will manifest there *if* we were to measure it. But when the observation is made, and we actually see something appear on the photographic plate, then the probability of this occurrence at that exact location becomes one, i.e. certainty, and zero everywhere else. This is the event of the collapse of the wave function. What it means for us, at the classical level, is that the quantum potentiality has manifested itself as an actual photon, right here, and nowhere else. The quantum effect of this observation is to instantly update the entire system of probability wave functions that are connected to this event throughout the entire quantum universe of potentiality.

Now all this may not sound too remarkable. For events happen in very much the same way at the classical level: something is uncertain, like the 1 in 2 chance of a coin toss coming up heads, but once the coin has landed, the uncertainty is over. What makes the quantum level so mysterious, is that our act of observation has actually *forced* the photonic particle to manifest at a particular location. If, instead, we had chosen to place our photographic plate at a slightly greater distance from the light source, then it is probable that our photon would have turned up in a totally different location, and appear to have taken a totally different trajectory.

That means the choices we make in our act of observation actually change the state and the future trajectory of the entire system we observe.

This can be further clarified using the well-known *double-slit experiment*. Here we have a screen containing two (very narrow) vertical slits with a detector screen on one side and an emission source on the other.[4] The basic form of the experiment was devised by Thomas Young in 1801. His aim was to test whether light consisted of tiny particles or of energy waves. He thought that if light consists of particles moving in straight lines (as Newton believed), then the experiment should produce an image of the two slits on the detecting screen (with some vagueness around the edges). In contrast, if light consists of energy waves then we should expect to see an interference pattern on the screen (as in figure 10.1). This pattern is formed in the same way as waves interact on water, i.e. when two wave peaks meet they get even higher, and when two troughs meet they get even lower and when a peak meets a trough they cancel each other out (figures 10.1 to 10.3 are based on diagrams from (Freiberger, 2012, August)).

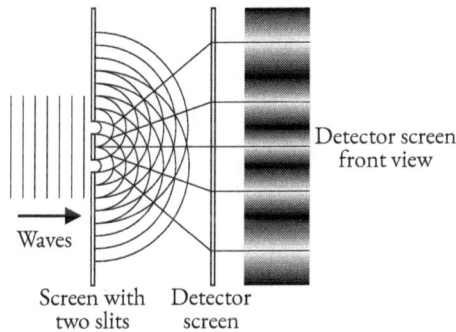

Figure 10.1: Double-slit experiment representing wave-like behaviour.

What Young found was the presence of an interference pattern, and so, for the next hundred years it was thought that light exists in the form of energy waves. Then, in 1905, Einstein showed that if light were to exist in the form of particles (photons), this could explain the so-called *photoelectric effect* where electrons are emitted from the surface of a metal in the presence of electromagnetic radiation—something that could not be explained by the energy-wave theory.[5] Since then it has been accepted that light can manifest either in a wave-like or a particle-like way according to how it is observed. For some time it was thought that this wave-particle duality only applied to light and other forms of electromagnetic radiation, and not to matter particles such as protons and electrons. But subsequent double-slit experiments have shown that matter particles also exhibit wave-like behaviour, with structures as large as atoms and molecules producing interference patterns. Experiments using electrons also show that if the electrons are fired one at a time toward the slits, so that they cannot interfere with each other, the usual interference pattern still emerges. This suggests that each electron, after leaving its electron gun, acts like a wave which passes through both slits at once, and then interferes with itself, before manifesting as a single particle on the detector screen (according to the probabilities determined by its previous wave-like interactions, as shown in figure 10.2).

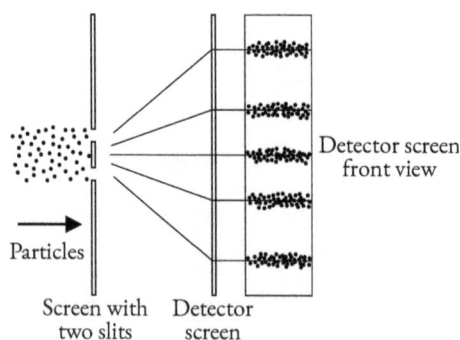

Figure 10.2: Double-slit experiment representing particle-like behaviour.

Finally, we can change the situation by detecting what is occurring at the slits themselves. There are various ways to do this, each involving a more sophisticated kind of experiment, such as measuring the state of particles that do not actually pass through the slits but are 'entangled' with particles that do. Measuring these entangled particles can tell us about the location of the particle that goes through the slit, without needing to measure that particle directly. These experiments have shown that, if there is any information in the wave functions of a system that specifies which slit a particle passes through, then the interference pattern disappears and is replaced by a pattern that corresponds to the slits themselves (as in figure 10.3).

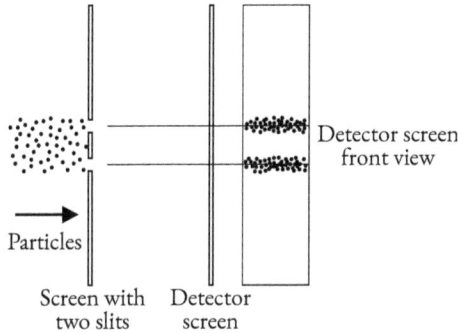

Figure 10.3: Double-slit experiment outcome when the path is detected.

This is a remarkable result. It means we can no longer think of particles or waves as things that enduringly exist independently of each other, separated in space and having to communicate their state by means of other particles or waves. Instead, an entangled quantum system forms a unified whole, where one part can be immediately 'aware' of what happens in another, such as our measuring the position of one entangled particle causing its partner to become localised, and to no longer behave in a wave-like way. It is only when we make a measurement that things appear (to us) to become separated out in space and exhibit independent being. Otherwise they fall back into a state of increasing interconnection and unity, a unity that can be actually demonstrated in the manifestations of our observations.

This is where the crack opens up in the edifice of the scientific project to explain every event entirely in terms of the interactions of physical particles. This is the crack where consciousness finally enters into the world of physics. A mild form of the crack appears when it is recognised that subjective human choices about when and where to make an observation actually change the behaviour of the system being observed. But scientific materialism can easily close this gap by saying that consciousness has nothing to do with such choices. It only *seems to us* that we freely choose. What really happens is that the physical particles that make up our brain and body set up the experiment. These particles do not 'choose' to do anything, they simply behave according to the same physical laws that govern every other particle. Whether or not the system that sets up the experiment is conscious is irrelevant to the actual outcome.

The more serious form of the crack concerns the act of measurement it-self. In our descriptions so far we have been speaking as if the event of the wave function reaching a detector screen is enough to make the wave col-lapse and manifest a particle. But experimental evidence shows that the screen and every other physical system that we encounter is also in a state of quantum superposition until they too are observed. You may have thought that when the wave functions of two potential particles meet then that will cause them to collapse. But what happens is that the par-ticles become *entangled* and *continue* in their superposed state until the system that they form is observed. As we have seen in the double-slit ex-periment, such entanglement means that the way one particle manifests can directly and instantaneously affect the way the other manifests. But it doesn't stop with just two particles. In the absence of an observation, the entanglement just keeps on growing, subsuming more and more of what we take to be the fixed and determinate material world. No one knows if this process has a limit. It could subsume the whole universe. All we can say is that as soon as a system is observed by human consciousness then the process of entanglement ceases and collapses back into a definite state.

This is the precise site of the crack. John von Neumann was the most famous and respected 'great intellect' to state the obvious conclusion: that it is the interaction between consciousness and the quantum dimen-sion that *causes* the potentiality of the quantum dimension to manifest as something definite.[6] And yet, of course, this conclusion is just too shock-ing to be embraced by the materialism of our age. For it *destroys* that ma-terialism. Consequently a great deal of effort has gone into developing theories that propose physical causes for the collapse of the Schrödinger wave function, and so eliminate consciousness as a causal agency. But, to date, there has been no convincing evidence that settles this question one way or the other (despite the usual claims to the contrary). In fact it is not clear, even in principle, how any physical experiment could ever bring about its resolution. For even if the collapse does happen before we ob-serve a system, how could we ever observe what happens *before* we make an observation? Given this apparent unfalsifiability, serious professional interest in the 'consciousness causes collapse' theory has declined, with only a handful of dedicated supporters (such as Henry Stapp[7]) keeping the issue alive.

But what about being the unknown?

The gist of the theory is quite simple. It says <u>the only way that we</u>, as human beings, can make an observation of a physical system, is to consciously and directly perceive that system. You may think this is not true, and that we can observe a system without being immediately present by using a recording device, such as a detector screen. For instance, we could set up the complete double-slit experiment from figure 10.2 in a sealed box so that no information can get in or out. We could also set up an automatic timer in the box so that the experiment starts and stops without any outside interference. Then, after the experiment is complete we could open the box and observe the detector screen, finding, as we would expect, the usual interference pattern. Now surely the detector screen *proves* the electron wave functions collapsed as they hit the screen while the box was sealed? Here we are imagining the electron waves are in a quantum state when they leave their source, and that everything else in the experiment is in the normal collapsed state we always observe things to be in. But we have no basis in quantum physics to think this way. For the Schrödinger wave equation applies as much to the matter on the surface of the detector screen as it does to the electron that reaches that surface. And as the wavefronts of the electrons meet the wavefronts of the screen, they form an entangled quantum system that encodes the superposed states of all the possible patterns that could appear on that screen *if* it were observed. But, of course, it is not observed (not yet). It is only when we open the box and actually look at the screen that a particular pattern is fixed. And even though, when we repeat the experiment, it always yields a similar interference pattern, each pattern is still a unique trace of where each electron manifested. It is not that our conscious observation caused the *general form* of an interference pattern to appear—that was determined by the structure of the experiment—it caused this *particular* interference pattern to appear, rather than some other.

This view says that *nothing* is fixed until it is *directly* observed by our consciousness. And then, as soon as we look away, what we no longer see or sense once more enters into a state of pure potentiality, with all the possible wave function states becoming superposed together again. Each time we look at something, we create a single fixed state by *destroying* all the other states the system could have been in. And each time we look away, all those other states remain destroyed—so the wavefronts must emerge once more, evolving out of our last observation and again turning it into something uncertain. At the same time there will be large sectors of what we think of as the material universe that we do not observe, such as (as far as we know) the interior of the earth. According to the theory, these areas remain in a superposed state, but are *constrained* by a principle of coherence which eliminates the superpositions that contradict what we have already observed. For everything is finally connected to everything else, and we cannot have a state in one part of the universe that will evolve into something appearing to us that cannot also have evolved from what we have already observed—otherwise our past would no longer uniquely connect with our present.

What opens up in this crack is the breakdown of the idea of the causal closure of the physical. Previously it was assumed that all the large scale events we observe are ultimately determined by the lower-level laws of physics. It was accepted that quantum physics added an element of indeterminism into this picture which meant we could not be certain how each particle would manifest. But we still knew, from the Schrödinger wave equation, the *probability* of that manifestation. And so we knew, if we repeated an experiment over and over, how things would manifest on average, over time. This still seemed like a form of causal closure, with everything in aggregate still being determined by physical law. And yet, *in principle*, there is no law and, it seems, no possibility of discovering such a law, that could determine exactly how each *particular* measurement is going to turn out. This is more than a crack, this is a *fissure* in the edifice of scientific materialism. For here there is injected into our understanding of the universe the presence of something that is completely undetermined by physical law. This is not something that just turns up occasionally in an experiment in a physics lab. We are talking of the manifestation of the entire physical universe, as it occurs, in each moment. All that the Schrödinger wave equation specifies are the *boundaries* of what is possible—the limits to what is not determined.

Physicists have tried to cover up this fissure with the word 'chance' and with the idea that 'nothing' determines the way the wave function collapses. But really this is only saying that we *do not know* what determines the form of this collapse. Even so this 'not knowing' has a unique and extraordinary character. It is the *end of the line* of our knowing. Everything else we think of as a chance event, like the throwing of dice, is to do with our inability to predict that event, because we cannot make sufficiently

[margin annotations: "Bishop Berkeley again!" and "Golden Ratio"]

accurate measurements. But quantum level chance is *pure* chance. It is not a matter of our not having fine enough measuring instruments. For we cannot even *envisage* how we could perform a physical measurement that would determine in advance the precise form of the collapse of the wave function.

The central pillar of evidence that holds up the idea of the causal closure of the physical is the statistical predictability of the wave function collapse over time. This is often thought of as a demonstration that consciousness cannot cause the collapse—for, if consciousness were involved, then surely we should find that the form of the collapse would be quite arbitrary, as if it were following someone's subjective whim—or that certain experimenters, by sheer force of concentration, could cause the outcomes of their quantum experiments to obey their intention rather than follow the form of the wave equation.

It is here we encounter a second difficulty. For, as we have seen, the general level of our philosophical understanding, including that of our physicists, has hardly developed since the seventeenth century. So, if one suggests that consciousness is involved in the collapse of the wave function, it is understood that we are falling back into the world of Descartes, where our individual immaterial minds are interacting with the substance of our physical bodies. It then follows that if consciousness is causally active, then it must be our personal minds that are influencing the collapse. And yet, the supposed influence of these minds is indistinguishable from that of mere 'chance'.

According to this dualistic understanding, we now have two theories: first, the 'chance theory' that says the physical universe has a capacity to make a probabilistically random selection each time a wave function collapses, and second, the 'consciousness theory' that says we each possess an immaterial mind that, in every split second, is making trillions and trillions of (completely unconscious) decisions concerning how each wave function of each particle that is involved in what we perceive is collapsed. It is here that scientists generally apply the tried and trusted maxim of Occam's razor. This says: "entities should not be multiplied without necessity". Clearly the 'consciousness theory' adds more than a new entity into the universe of the physicist, it adds a completely new domain of being, that of a mysterious immaterial causal agency that does nothing with its causal power but enact the probabilistic predictions of the Schrödinger wave equation. Behind this is the suspicion that all this talk of consciousness is yet another attempt to adulterate the purity of scientific objectivity by smuggling in a religious or metaphysical belief. In contrast, the 'chance theory' only asks us to accept that physical material can manifest itself with a certain degree of randomness. Seen in this context, we can understand why the 'consciousness theory' has had little effect on the underlying scientific materialism of our age.

A PHENOMENOLOGICAL COLLAPSE

The outcome of these breakthroughs in our understanding of the fundamental structure and behaviour of the ultimate physical constituents of the universe is that our common sense objective scientific view of the world has remained essentially unchanged. There may be cracks appearing in the bedrock of this materialism, but it remains the foundation of our understanding of the being of the universe. Of course there are quantum paradoxes that suggest our common sense notions of material being break down at the quantum level. But what this really means we cannot say. There are various theories, but no final answers. And so we 'soldier on' with a materialism we know cannot be true, partly out of habit, and partly because we have no alternative that we collectively recognise to be more plausible.

But what of the phenomenology? Isn't there something essentially phenomenological lying at the centre of the measurement problem—in the idea that nothing in this world takes on any definite form until it is directly and consciously perceived? Isn't this where phenomenology *begins*: with the recognition that the being and actuality of everything we perceive is something *we* bestow according to the feedback we receive from the unthinkable source of experience? And haven't we, through our phenomenological investigations, already seen that objective being is something that only emerges for *us* according to the dimensionalities and forms of *our* experience, and that the source upon which we project our expectations must necessarily transcend these dimensionalities and forms? The answer is: yes, we have seen this.

Now, would it not be reasonable to think that all we have discovered concerning our own direct and immediate experience of being conscious should be of some relevance to what quantum physics has discovered concerning the ultimate constituents of the objectified universe? For if we take the phenomenology seriously, then physics and phenomenology are investigating the *same* unthinkable source of experience. Phenomenology looks *directly* into that experience, whereas physics looks *indirectly* into the objectified forms that *appear* in that experience. Should we not expect that when physics gets to the very edge of what it is possible to objectively measure and observe that it will meet its own 'other side'—the very consciousness that, in its objective materialism, it excluded from its domain of inquiry?

Here we are not arguing from any kind of direct experience. I have never seen an electron gun or a screen capable of detecting electrons. I am taking all that on *trust*. It could be that, in another fifty years, quantum physics will have been replaced by a perfectly reasonable classically material explanation based on a deeper level of previously undetected physical particles. But what we are saying is that given the evidence we actually have, both phenomenological and physical, there lies before us the vague outline of a *much better* explanation. It is better because it can account for this evidence, rather than ignoring it and living in obvious paradox.

The explanation begins as follows: there is no independent physical material universe that consciousness somehow brings into being in acts of perception. As far as we can know, there is just consciousness and the source. We cannot say that the source is physical or mental or anything whatsoever. It is simply what lies behind or beyond the immediate field of our actual moment-to-moment conscious experience. What manifests in consciousness is the appearance of an objectively independent world of things and events and feelings and thoughts and so on. This appearance shows us an aspect of the source expressed in the forms of our experience. The source itself is not made up of a collection of aspects, it is that from out of which the aspects are generated, and into which they disappear, and within which they are also unified. For it is part of our experience that we experience the appearance of being separated from the source, whereas the source is just a name for something we can't describe, that includes everything within itself, including my experience of being separated from the source. *This* is what the phenomenology teaches us. We cannot even say that the source has being or that it does not have being. The only being that we know of is our own being and the being of the beings that appear in our experience.

So let's look again at the quantum physics from this place of the phenomenological reduction. What does the physics say about our experience at the very edge of what it is possible for us observe? It says that what we think of as the physical world only comes into being when we consciously observe it. Here, to come into being means to occupy a measurably unique spatiotemporal position and/or to possess some measurably unique energetic physical properties (such as a direction of motion). But, of course, we cannot help but conceive of material being in this way, because these are the very *forms* of our projective perceptual intelligence. What the physics tells us that is new, is that when we *stop* observing a system, then the particles that we previously observed go into a state of *non-being*, i.e. they no longer occupy a definite spatiotemporal position or possess a definite energy state. In other words, they return to a state of *potentiality*. In this state they even lose their separate particle status, becoming part of a unified system of potentiality along with all the other 'particles' with which they are 'entangled'.

Now, from a phenomenological perspective, we have learnt something quite fascinating. For, in ordinary perceptual experience, we can only learn about the source by projecting our expectations into it—and, in so doing, the source can only tell us how it stands in the moment of our projection. But now, quantum physics allows us to infer something about the source that we cannot directly perceive. It tells us that the things we perceive—not just the flashes on the detector screen but *everything* we perceive to have objective physical being—come and go, in and out of existence, according to their being perceived by us. And more than that, when they go out of existence, there is a certain mathematically definable *leeway* in how they will manifest again if we attempt another observation. This is entirely new information that without quantum physics we could not have known. It tells us that the evolution of the objectified universe is not ruled by mechanistic laws that define in advance what is going to happen here. Of course, it could be that, unknown to us, there are mechanistic laws operating *on the other side*—in the domain of potential being—that determine how things manifest on this side—in the domain of definite being. But here it seems that the physical universe is 'set up' in such a way that we cannot inquire any further. Our very act of observation destroys the potentiality we seek to understand—for as soon as we look it immediately *conceals* itself in an act of manifestation. And yet, for us, this is the question of questions: whether this potentiality that generates the quantum collapse is just another machine, another system of fixed mathematical regularity, or whether there is some intelligence, some freedom, some discrimination in the way the universe 'chooses' to evolve. And it is just here that our objectified scientific model of the universe, *refuses* to give us a definite answer.

It is for this reason that we must now begin a form of *construction*. For although we cannot directly perceive how it stands in the potentiality of the source, because we only perceive what emerges from this potentiality, perhaps we can intuit something on the basis of our immediate experience. For if the source has some intent in manifesting how the universe is to evolve in each instant, it is hardly the case that we have nothing to do with this intent. For it is occurring in us, in our consciousness, *right now*.

What we are proposing is that the power that manifests our experience, and, through this experience, manifests an objectively physical universe, is, finally, *what I am*. Obviously, here, I do not mean my personal self or my personal consciousness. For the place from out of which my experience manifests is also the place from out of which your experience manifests—otherwise we could not meet here, we could not touch each other, speak to each other, pass the jam between ourselves, or go out for a drive in the car. Clearly the world that manifests in our experience is emerging from the *same source* and we, in our perception, are projecting into that same source. What we are considering is the source itself. For each one of us, it is that invisible point from out of which our stream of consciousness is flowing into the past of our conscious retention.

This point, in itself, as an experience, is *nothing*. It is the pure nothingness from out of which everything that can be anything is manifesting. And, although each one of us experiences an entirely unique stream emerging from this point, the point itself is still singular. It is the utter unknown—that place that is always behind us—as I saw sitting on the trolley at Kuala Lumpur airport. It is an aspect of this nothingness of pure potentiality that quantum physics models using the Schrödinger wave equation. In this place, my stream of experience is no longer utterly distinct from yours—for both our streams are emerging out of the same source, a source that is here, present, within me, now, as it is within you. It is the point of impersonal universality from out of which the particularity of my life emerges. Do you see that this is how it stands, regardless of the being of any physical universe existing outside these streams of our experience?

Our perception of the objective being of a physical universe demonstrates the stream of our experience is part of a *greater unity*. Insofar as we *are* this stream, we are directly in touch with this unity, as are the other conscious beings with whom we live, although, instead of just being this unity, we are enabled, by means of our individuated consciousness, to appear 'out here' in a field of definite existence, and in this field, to look back onto the source from out of which we came. Of course we cannot experience the source in its totality, for the source has had to split itself up in order that it can experience itself in the first place. The only way that we can start to get a sense of the being of this totality is to withdraw from our individuated experiences and start to intuit that very point of nothingness from out of which this experience is emerging.

No ✗

So, to return to the measurement problem, what we are seeing is that the idea of a separate immaterial individual consciousness somehow causing actually independently existing material objects to manifest 'out here' in an objectively real physical universe, is a kind of absurdity. As far as we can tell there is only one process of collapse occurring. That is how we come to experience ourselves as living in the same objective material world. For if we each collapsed our experience in a different way then our experiential worlds would start to diverge. And yet they do not diverge. When I see the dot of an electron appear in one place on a detector screen, then that fixes it for everyone else. We do not end up with different opinions about this, or at least not for long, as we have clear ways of resolving such issues.

Yes!

And yet, all these acts of observation are still occurring in our *individual* streams of consciousness—streams that appear to remain utterly private to each one of us. This is where we have to trace back our stream of consciousness to its point of origin. This point is always now. What we are proposing is that it is the *same* point, the *same* source, for all of us. If we are to picture it, we could say that it is from out of this one source that all our different streams of experience are emerging. And, to be clear, this is not just pure speculation. There are many people who report having directly experienced or intuited this greater unity of consciousness. It is because of our prevailing scientific materialism that such reports are routinely dismissed, as if, one day, we could develop a scientific instrument

Why a "point".

No.

Conversational MYU

that could finally settle the question. But, of course, such an instrument is an impossibility, for consciousness is *exactly* what cannot be objectively measured by scientific instruments.

THE MOMENT OF CREATION

One of the first objections that is generally raised against this idea that consciousness is required to bring the universe into being, is to observe that consciousness was clearly not required throughout all the billions of years in which there were no life forms in the universe capable of making such conscious observations. During that time, stars and planets must have already come into existence, otherwise no life forms could have developed. Therefore it is absurd to think that conscious life forms are needed to bring the universe into existence, because the universe already needs to exist in order for conscious life forms to have developed in the first place. The answer to this objection is to think again of the double-slit experiment occurring inside the sealed box. If we accept that consciousness really does cause the quantum state of any system that we observe to collapse, and that otherwise the system remains in a state of superposition, then this condition should also apply to the universe as a whole—it too should remain in a state of quantum superposition inside its own universal 'sealed box' until some conscious observer brings it into the definite form of a conscious experience. There is nothing paradoxical here.

To see this, let us accept our cosmological theory that the universe began with a 'Big Bang' around 14 billion years ago *would* be correct if someone had been there to observe it. And let us also accept that in order for a quantum collapse to occur, there must be a life form present, like ourselves, capable of making a conscious observation. From this it follows that the Big Bang will only have been a *potential* Big Bang, as there were no life forms present in the moment of its potential occurrence capable of collapsing it into a definite event. From our perspective, it will take billions of years before such a collapse is even conceivable. Of course, we cannot be sure exactly what it takes for someone or something to be a 'conscious observer'. But we can be relatively sure that simple systems of atoms and molecules are not making any such observations, otherwise they would not exhibit the kind of wave-particle duality *we* observe in experimental situations. Beyond that it is not at all clear at what point a system becomes capable of conscious experience. At the moment we shall leave this question open and look instead at the state of universal potentiality that we reason must have preceded it.

Conjectures about this state of potentiality have already emerged in controversies about the significance of the so-called *anthropic principle*. This principle is a response to the discovery that the basic mathematical laws and constants we observe to hold in the physical universe have been set in just such a way as to ensure that life forms like ourselves would eventually emerge. If things had only been fractionally different, there would have been no galaxies with stars that could support planets where complex self-replicating organisms (like ourselves) could have emerged. Such extraordinary providence seems to imply the presence of a guiding intelligence capable of fine tuning all these laws and constants, just as we once thought an intelligence was needed to have created life on Earth.

The response of scientific materialism to this resurgence of deism has been twofold. Firstly there is the anthropic principle itself, which is based on the recognition that it *could not have been any other way*, for it is only in such a universe as ours that there could be beings like us capable of discovering these fundamental laws and constants, so of course we are going to find that they are set in just such a way as to enable us to be here. But there is still the fact of the extraordinary improbability of our being here at all. This is dispensed with in the concept of a *multiverse*, i.e. the idea that there are countless other parallel universes, from which we are completely cut off, each exhibiting different settings of the basic laws and constants, and where, consequently, the vast majority exist with no life forms whatsoever. In this way we are returned to an existence ruled by blind chance in which there is no need for any guiding intelligence.

This idea of a multiverse is an example of how scientific materialism maintains its boundaries. Ordinarily it would reject such a theory for being both unfalsifiable and for going against the principle of Occam's razor, i.e. of not multiplying entities without necessity. But here there is a necessity. For to accept the need of some guiding intelligence would be to destroy the ground of mathematical objectivity on which the project of scientific materialism stands, i.e. that everything we observe can be explained in terms of mathematically objectifiable entities and regularities. And so we come to entertain something that appears to be extremely implausible (i.e. the actual existence of countless unobservable parallel universes) but to which we can offer no irrefutable objection, just so that our prevailing orthodoxy can survive. That is not to say that scientists are personally invested in such theories. The idea of a multiverse is just one way we can make what we have discovered about the universe logically consistent with our materialist scientific presuppositions. Its implausibility is only a symptom that our basic assumptions no longer correspond very well with what we have actually discovered.

We meet with a similar implausibility in Everett's many-worlds interpretation of quantum mechanics.[8] There, in order to dissolve the measurement problem, Everett proposed that there is no wave function collapse at all, and that all the possible states represented in the Schrödinger wave equation *actually exist* as objective realities. That means, in each moment, countless different universes are branching off into separate existences according to the possible outcomes of all the quantum interactions between all the particles in all the currently existing universes. Again these parallel universes cannot communicate with each other, so there is no way to test the theory. And even more implausibly, it suggests that our consciousness branches off into all these universes as well, so there are also uncountable numbers of different 'me's, each one unaware of all the others, all leading, over time, more and more divergent lives.

As with the multiverse solution to the anthropic principle, Everett's solution is intended to close another crack in the egg where some intelligent agency may enter and guide the course of the evolution of the universe. If we put the two theories together we can once again save the idea of a pure determinism, where every event in the universe is determined by the mechanism of mindless mathematical regularity—but at what cost! For, to avoid the possibility of there being any intelligent (i.e. non-mechanistic) principle operating in the evolution of the universe, these theories assume that all possible universes exist together, branching off from each other in such a way that no universe can be in contact with any other. And what is more, this is assumed without any empirical evidence to back it up, and, at least at the moment, without any hope of finding such evidence. Do you begin to see how desperate the situation is for scientific materialism? For these are not fringe ideas held by members of an esoteric physics society, these are mainstream theories, logical consequences of paradoxes that cannot be resolved within the framework of scientific materialism, except by such extreme leaps of the imagination.

So what do we have to put in the place of these theories, when we take consciousness to be the primary reality and not the actuality of the objective universe? Let us return to the state of the potentiality of the universe in the moment of its first conscious realisation. We do not know when or how or even if this event occurred, but let us imagine it happened here on Earth, one day, when one human being suddenly woke up to an experience of being conscious[9] (bearing in mind that many different first moments of consciousness could also have occurred in other brain-like systems). If we take the quantum physics seriously then what we would expect to find at the quantum level are all the possible states the universe could have been in, as determined by the Schrödinger wave equation, from the initial state of the Big Bang, up until the moment of this conscious observation. What is more, and for good measure, we could include all the possible initial states of the universe—the various possible values of the constants and laws—in an immense quantum calculation, just as the multiverse theory proposes, only this time all these alternative universes do not *actually* exist. What occurs is a kind of inconceivably vast search through the space of all possible universes.

At this point we are not saying that any kind of intelligence is needed to perform such a search. To simplify matters, we could even allow that some form of survival of the fittest competition occurs, where potential universes that fail to develop as well others tend to become less probable—perhaps according to their degree of ordered complexity. Of course, these details are impossible to verify, at least from our current level of understanding. But what *is* necessary is the presence of a *criterion of consciousness* pre-built into this process that decrees when a configuration occurs that is capable of sustaining a state of consciousness, then there is an immediate collapse of the entire system of wave functions that led up to that moment, and only those wave functions that remain coherent with that configuration continue to evolve.

I would like you to check that what is being described here is only what the quantum physics suggests happens each time we make a conscious observation. The only difference is that this is the *first* such observation that has *ever* been made. It is in the moment of this observation that the entire universe moves out of its superposed state of potentiality and comes into objective being. *This* is the moment of creation. Before this *there was no time.*

THE GREAT REVERSAL

We are inquiring into how our understanding of the universe changes if, instead of putting the objective being of the material universe first, we put *consciousness* first, the consciousness in which that objective universe appears. This is not an arbitrary reversal. It is our direct experience that consciousness *does* come first. It is that *by means of which* the world appears. And it is that *of which* we are directly and immediately certain. The forms of time and space and substantiality are the dimensionalities which consciousness uses to articulate the being of an objective world for us to experience. To take these forms to be the ultimate reality and to take consciousness to be a phenomenon that appears within such a reality is to have got things the wrong way round. But to get things the wrong way round is our *natural attitude.* It is something we have inherited—it is our *habit of survival.* For the objective world that we perceive is not causally inert. It is not like a movie that we passively observe. Insofar as we are awake and alive, we are *in* this movie. Our being conscious, our capacity to perceive, to experience anything, is directly connected with the events we perceive to occur in this objectified world. Our error is to take this world as if it were the totality of all that exists, and as if that totality exists in just the way we perceive it to exist. We do not recognise that such an experience is an impossibility, because we do not recognise that we are having an experience, we think we are seeing things as they really are—that the forms and dimensionalities that belong to our experience belong to *reality itself.* And insofar as we remain in the natural attitude of our everyday perceptual experience, we don't just think this, we *experience* things to be that way as well. We continually need to remind ourselves of this situation because we habitually revert back to the natural attitude as soon as we become engaged in our day-to-day activities. The phenomenological reduction is our means of being conscious of being conscious—which means being conscious that our natural attitude is just that: an attitude that takes the being of the objective world that appears in our experience to be an ultimate reality, while forgetting that the ultimate reality is also what is *producing* this experience.

Contemporary quantum physics is an attempt to understand the phenomena that have been uncovered by empirical experiment within the framework of the natural attitude. It is this attitude that is the ground of the materialism we bring to bear when trying to picture what is occurring at the quantum level. And it is this attitude that cannot accept the idea that consciousness could be the cause of the collapse that manifests the objective universe in our experience. What stands in the way of making consciousness the foundation of our understanding of quantum physics is not the scientific evidence, it is our identification with the natural attitude itself. As soon as that attitude is transcended in a phenomenological reduction, it is immediately revealed that the objective universe is something that manifests within and for consciousness, and that the source from out of which that experience is emerging remains concealed to us. From this place, the findings of quantum physics, far from being paradoxical, start to provide a coherent empirically grounded account of how the manifestation of the objective universe is connected with our being conscious.

Placing consciousness at the centre of our ontological understanding of the being of the universe *reverses* the current scientific materialism. It says that everything, from now on, must be explained in terms of its appearing in consciousness and not in terms of interactions of matter and energy. For example, the old question as to why consciousness should have emerged in a material universe becomes the question as to why a material universe should have emerged in consciousness. This heralds the birth of a new kind of science.[10] And from the perspective of this new science, our previous manner of theory formation becomes completely inadequate. For scientific materialism considered its task to be the discovery of the mathematical structure of the material universe. In such inquiry it not only remained unconscious of consciousness, it remained unconscious of its own *meaning*. It thought it could dispense with questions of meaning by considering meaning to be a subjective phenomenon that could ultimately be explained in neutral objective mathematical terms. But, of course, the idea that you can dispense with questions of meaning in this way, is still a form of understanding, which is still a form of meaning. It is here that the unconsciousness of the objective stance becomes apparent. It thinks because it only deals with objective facts and mathematical formalisms that it has transcended human subjectivity and human meaning. But that is not true. The understanding that lies at the heart of scientific materialism is that the causal mechanism of the universe is *without* meaning. For once you have reduced and explained the phenomenon of meaning in objectively physical terms, then the meaning you were explaining has disappeared. It becomes an epiphenomenon of our being conscious, something appearing as a wave on the surface of our subjectivity, that can finally be explained away in the interactions of the ultimately real material substrate. And that means that our living here, and the being of the universe, is ultimately without meaning, not just without intrinsic purpose, but without any kind of intelligibility whatsoever, except the intelligibility of the mathematical expressions that describe the forms of its manifestation.

From the perspective of phenomenology, this represents a kind of extreme and irresponsible form of unconsciousness. For once we step outside of the objectively rational stance that people assume in their professional lives, it becomes quite clear that our scientific materialism is *not* neutral. It profoundly affects our way of life. It informs every level of our civilisation, from our government departments, to our educational establishments, our business organisations and our medical systems. What we serve now, what we value, are *efficiency* and *effectiveness* in the pursuit of greater wealth and power—all entirely objective mathematically determined quantities. And it is our scientific materialism that has underwritten the development of this overwhelmingly materialistic conception of what *matters*, of what *counts*. You may point out that it has also underwritten all the advances in the granting of greater freedom and equality to human individuals. But this is the freedom and equality of biological *machines*—the recognition that the race, religion or gender of such machines has little or nothing to do with their productive capacity—what matters is only the way in which they have been programmed. Of course, that is a very one-sided assessment of our way of life. But the other side, the humanity that we show each other, was not born out of our scientific materialism.

If we base our science on consciousness, then we can no longer remain unconscious of the role that science has in determining what has meaning for us. For we can see where the meaninglessness of such unconsciousness has led. Perhaps we can even start to see how *strange* it is that we should have allowed this to happen. For it is not a scientific fact that the universe has no meaning, and that it emerged one day as if by accident, and that our consciousness also accidentally emerged a little later on, and that the entire process is governed by laws and mathematical regularities that have no awareness of us whatsoever. Descartes did not think this. Neither did Newton. It is we who have allowed ourselves to fall under the sway of this mindless materialism, partly due to the failure of our original rationalism, and partly to our failure to care enough to develop a new and better understanding.

A first step in developing such an understanding is to recognise that there *is* irreducible meaning in the universe. I know this because my experience is an experience of meaning. For example, in looking at the red candle in front of me, I *directly experience* the meaning of the colour red. And I know that such meaning cannot be explained in materialistic terms because my experience of colour is self-evidently immaterial. In this domain of consciousness, what is self-evidently and immediately given to me are not atoms and molecules, but the sensory fields of my experience and the various qualities that manifest in those fields (i.e. colours, sounds, smells, tastes, touch feelings, bodily feelings, and so on). And, in looking at the red candle, I experience the *meaning* of the candle. I experience the roundness of its spatial form, the dull texture of its waxy surface, the way the wick is slightly twisted. Everything I perceive is a form of meaning. The colours in my visual field speak a language of shape and texture. They are not mathematically defined quantities of colour sensation appearing on a two dimensional screen. The way the colour varies expresses (means) the shape of the object I am seeing, and it also expresses (means) that the underlying colour is uniform, and it shows (means) how the light is coming in through the window. Each sensory field is like this. They are expressions of meaning that refer to what they are not, for the various colours that appear as the surface of the candle are not the surface of the candle, they are what reveal the surface of the candle to be the surface I perceive it to be. These colours are like the words of a sentence that describe the candle, only these are the lowest level of meaning we can reach to. For there is nothing 'behind' the colours I experience, no components or parts that colour can be broken down to, except other colours—whereas the words and letters of language *can* be broken down. They require the being of a certain shape or sound or thought image to express what they are, whereas, finally, only a colour can express what a colour is.

Meaning involves the being of a collection of one kind of thing (e.g. of colours or words or sounds) being able to manifest the being of another kind of thing (e.g. a candle or the meaning of this sentence or the blackbird's song), all within some form of unified dimensionality (e.g. of substantiality appearing within space and time, or of a thought appearing in the temporal space of the mind). Once we have the capacity to make something known to ourselves like this, that means we are *conscious*. In fact, to be conscious and to experience meaning are two ways of saying the same thing. And that means without consciousness, there is not just no experience of meaning, there is no meaning whatsoever. This was Brentano's insight: that *meaning* (what he called intentionality) is the defining characteristic of consciousness.[11] It is not that there is meaning over here and consciousness over there and that we can separate them out and have an experience that has no meaning. If we were to have an experience that we experienced as having no meaning, it would still have the meaning of having no meaning. For to have no meaning whatsoever, is to have no experience whatsoever, which is to be *unconscious*. Here we are understanding meaning *transcendentally*. And that means it has no opposite. What we usually think of as meaning is concerned with our making something *intelligible*. But now we are seeing that our entire field of consciousness is a field of meaning, that our experience of being in space is a meaning, our understanding of time is a meaning, our sense of being an independent consciousness is a meaning. From this place we never encounter an utter and complete meaninglessness, we only encounter phenomena that do not fit in with the already existing meaningful context of our understanding. And even if we feel that very background context is disintegrating into meaninglessness, that meaninglessness is still a phenomenon that is appearing in our field of consciousness, a phenomenon that we *experience* to be meaningless. For to experience anything *as* anything, even to experience it *as* meaningless, is still to have an experience of meaning.

Do you see that no account of what is happening in your physical brain can have anything to say about this transcendental dimension of meaning? Of course we can say that when *this* happens in the brain then I experience *that* kind of meaning. But the experience of meaning is occurring in an entirely different dimensionality to the firing of the neurons. Materialistic science is right to say that there is no intrinsic meaning in the behaviour of physical particles. But that does not mean there is no intrinsic meaning. Every moment of consciousness is an experience of intrinsic meaning. My idea that there is a physical universe is a meaning. My idea that there are physical particles is a meaning. To say that consciousness comes first is to say that meaning comes first, which is to say that our capacity to experience meaning is something that is foundational. It is not something that we are going to be able to explain in terms of something else. Just as materialistic science takes the being of energy as something primary, we are taking the being of consciousness as something primary. The difference is that the experience of being conscious is given to us directly in a way that we cannot doubt, whereas the being of energy fields is something we infer on the basis of our being conscious. And that means that our taking consciousness as primary is not something arbitrary, it is something that is *grounded* in the only certain being that we know of.

But what does quantum physics tell us about consciousness? If indeed it is the case that consciousness causes the collapse, and that the first collapse only occurred when the right kind of potential life form developed in the primordial quantum state, then we can infer two things: firstly, that some form of potential consciousness must have been associated with this primordial state, and, secondly, that there was some capacity in this potentiality that was able to recognise that the right conditions for the manifestation of consciousness had obtained. If we look into our own experience for something that connects with this primordial situation, we immediately come up against our own unconscious psyche. For it is in here, each night, that our perceptual consciousness—the consciousness that, through the quantum collapse, manifests the objective world of our waking lives—slips away into a subjective dream world and then, presumably, into a state of deep dreamless sleep. Here is the mysterious ground from out of which all our intentions and ideas emerge and disappear again. Here lies the potentiality of all our capacities for action, for speech, for remembrance, for thought. And yet this ground is more than just a repository of all we have experienced and learnt. For we did not learn the quality of red, the feeling of hunger, or the emptiness of space, we only learnt the names which define the boundaries of these qualities. And what of our desires, our sexuality, our instincts for self-preservation and for the protection of those we love. And what of love itself—its meaning—its beauty? We did not *create* these things, we only learnt how they are manifested. They are part of our collective inheritance: the space of the clearing in which the ready-made palette of experiential quality manifests the archetypal structures of our instinctual feelings and drives in the streaming temporality of the past flowing out of an eternal now. This is not the white sheet of paper that Locke envisaged. It is the ancient psyche in which the broad outline of our potential for experience is already pre-figured.

What we are proposing is that it is the same unconscious psyche and the same consciousness that is in me, and that is in you, and that manifests the entire objective universe through each one of us. If that is the case—and it seems that it is, otherwise we would each be manifesting our own separate universe—then we must also accept that the same psyche was manifesting the world in all those people we once knew but who have now died. For they were here with us once. And if that is true for the recent dead, then it is true for their dead as well, and for the dead before them, and so on for all the human beings who have ever lived and manifested this world in which we meet together. For if it is the same world that is manifesting to us outwardly, then it must be the same psyche that is performing and coordinating that manifestation—with the proviso that each one of us gets our own private corner of that psyche, at least for a little while, until we too join the dead.

This is not just a theory. We *are* this place from out of which the world is manifesting. As soon as we stop going along with the flow of our thinking and return to immediate consciousness, we directly experience the upsurging of the world in each moment. The question here concerns the manner of being of this world that appears in the clearing of consciousness. From the perspective of the natural attitude, it seems that we are saying that the external world does not *really* exist, that it is *all in our minds*. And, *in a way*, this is true. For our experience of the world *is* arising in our minds insofar as we equate mind with our being conscious. And what appears to us to be the physical world is actually an *experience* of physicality. We only *mean* the world to be a physical world. That does not mean it must also *exist* as a physical substance that is independent of the consciousness that manifests it. But, if the world does not exist as physical substance, surely that means it exists in our minds? And yet how can we know this with any certainty? After all, it is not *impossible* that Descartes was right, and there really are two substances, and a creator God who enables their interaction. Equally, it is not impossible that Berkeley was right, and there is only God's mind within which we move and have our being. Or perhaps the physicality of physical substance is at the same time a form of proto-consciousness that exists like the wave-particle duality of matter. Or perhaps there are *three* domains of being, the mental, the physical and the Platonic realm of ideas. From here opens out the entire labyrinth of all the various philosophical positions we could take on this question, each possible in itself, and therefore impossible to finally refute.

The way out of this labyrinth is to see again that we live in a field of meaning and that our idea of existence, of being, is itself a form of meaning. The ground of this meaning that we attribute to being is our direct experience of the enduring continuity of our being conscious. It is the direct experience of the enduring stasis of the eternal now, of my being here as the impersonal witness of the flowing passage of my conscious experience. Without this ground of being, of *my* being, there can be no other being for me. It is this *enduring of what does not change* that defines the being-for-me of what appears in the flowing changeability of my experience. Again we have to get things the right way round. Finally there is no *thing* 'out there' in the external world that does not change. All is in flux. And so our science proposes an ultimate unchangeable enduring substrate that lies behind the energetic flux, a kind of ground that supports the manifestation of the continual change that confronts us. And we regard this as the ultimate 'stuff' of existence—even though we can never encounter it directly. Once we thought of impenetrable atoms, then of sub-atomic particles and energy waves. But now we have the quantum level where we find no continuity of being whatsoever—only a potentiality to manifest something definite for a moment, before that definiteness slips back into a realm of pure potentiality. It is here that our search for the ultimate ground of the being of the physical universe has started to founder. For what we have discovered suggests that there is no such ground. Although who can say what may be discovered next?

Our situation is that we project an understanding of being into our experience. If we did not do that, then we would have no experience of there being enduring things within a world. Our entire perceptual apparatus is concerned with detecting regularities in the streams of our sensory input. We experience these regularities as the being of the things and events and qualities and attributes that appear before us. It is from *within* this framework of meaning that we start to speculate on the ultimate being of the universe. We *project* this idea of being, that is founded on the continuity of our being conscious, into our experience, and expect to find an answering being at the foundation of the world. And because of our scientific materialism we expect that being to be something objectively physical. But now something fascinating has happened. The scientific projection of objective physicality has discovered something—the quantum level—that cannot be assimilated into our understanding of being as the enduring of self-identity. We were expecting to find the ultimate causally effective component-things that comprise the beings we understand ourselves to be, whereas, instead, we found that the universe does not seem to have a fixed and determinate being at all. It rather appears to *come into being* when it manifests in consciousness.

To ask after the true being of the universe in a situation like this is like asking what the universe looks like when we are not looking at it. We cannot possibly answer such a question. For if no one is looking, the universe cannot 'look like' anything. In order for it to appear at all there has to be an understanding of being that determines how it is to appear, that, for instance, articulates it into a collection of enduring spatial forms. Even so, we still tend to think there must be a 'true' projection that could exactly capture how things 'really are'—like the eleven dimensional spacetime model of M-theory.[12] But this misses the crucial point: an understanding of being is a configuration of meaning that can only 'be' insofar as there is a consciousness that projects that understanding. So, whether or not the indeterminacy of quantum physics is true or false, there can 'be' no being without consciousness—because it is only 'in' consciousness that distinctions of continuity and sameness become possible. For something can only 'be' insofar as it retains its identity from one moment to the next. And that requires the 'glue' of consciousness—its spatiotemporal enduringness.

Of course you can *picture* a universe without consciousness, but what do you picture? Collections of stars and galaxies? But a star only 'is' insofar as it endures in time. What is it that holds the identity of a star together, that distinguishes its boundaries? Is the star aware of itself? If so then it is conscious and we have being. But if there is no consciousness whatsoever? In that case then nothing is distinguished from anything else. So there can be no 'things'. Do you see that we are trying to think of something that is unthinkable?

Finally this means that we are in no position to speculate about the ultimate being of the source of our experience. The best model physics can give us is that of a constrained field of potentiality that we make actual in our living experience of being conscious. It is this making actual that manifests the being of other beings within the clearing of our own being-here. What being might be, or what it might mean outside of this clearing can have no sense or intelligibility for us—for everything that we can and do experience, we experience in this clearing. The form of the clearing is the form of the world for us. The clearing itself is not a thing or a being, it is the place where being manifests. So we cannot think of there being my clearing and your clearing, as if they were objects existing in some kind of space. And yet we know our experiences to be connected and coordinated. There is something here of the one and the many that Henri Bortoft wrote of in his book *Taking Appearances Seriously*. For there is only one world, and it only comes into being insofar as it is experienced by someone, and yet it manifests differently for each one of us. Behind this manifestation there is a mysterious unity, and we are each in contact with this unity, and it is through this unity that we are able to meet each other, to affect each other, to love each other and to do harm to each other. It is an extraordinary, transcendental connection of the consciousness that we are. For, make no mistake, we are not the ants crawling on the surface of an obscure planet that our scientific materialism takes us to be. We are the transcendentally unified consciousness of life on Earth.

NOTES

1. This refers to the speech from Shakespeare's play where Macbeth contemplates Duncan's murder: "Is this a dagger which I see before me, the handle toward my hand? Come, let me clutch thee. I have thee not, and yet I see thee still. Art thou not, fatal vision, sensible to feeling as to sight? Or art thou but a dagger of the mind, a false creation, proceeding from the heat-oppressed brain?" (1606/2016, 2.1.33-39).

2. Erwin Schrödinger (1887-1961) first published details of the wave equation in his paper *An Undulatory Theory of the Mechanics of Atoms and Molecules* (1926).

3. For an introduction to the measurement problem and its relation to consciousness see Roger Penrose's book *Shadows of the Mind* (1994).

4. For a non-technical introduction to the double-slit experiment see *Through Two Doors at Once* (Ananthaswamy, 2018).

5. Einstein was awarded a Nobel prize for his work on the photoelectric effect which formed the basis for our modern notion of the photon (see (Einstein, 1905)).

6. John von Neumann (1903-1957) was one of the great mathematicians of the twentieth century. In his book *Mathematical Foundations of Quantum Mechanics* (1932/2018) he argued that as the entire physical universe is subject to the Schrödinger wave equation, its collapse must be determined by something "outside the calculation", i.e. the consciousness of an observer. This view was developed further by the physicist Eugene Wigner and became known as the *Von Neumann-Wigner interpretation* (see (Wigner, Eugene, 1961)).

7. Henry Stapp is perhaps the best known contemporary quantum physicist to support the idea that consciousness causes the collapse of the wave function. He has published a number of books intended for the general reader where he argues that the indeterminism of the quantum level underwrites the possibility of genuine human freedom, for example, see *Quantum Theory and Free Will* (2017).

8. Everett published the many-worlds interpretation in his paper *"Relative State" Formulation of Quantum Mechanics* (1957).

9. This idea of the first human being coming to self-consciousness is described in Barry Long's book *The Origins of Man and the Universe* (1984, p. 20) where the sun is mythically understood to have penetrated a 'veil of opaqueness' in the human psyche, one morning, when a particular human being first consciously 'sees' the sun rising.

10. I borrowed this phrase from the title of Stephen Wolfram's book *A New Kind of Science* (2002) While some of Wolfram's ideas are relevant to what follows, he definitely had a different 'new kind of science' in mind to the one we are presenting here.

11. Franz Brentano (1838-1917) was a German philosopher who reinterpreted the scholastic concept of intentionality to represent the defining characteristic of consciousness, i.e. that "every mental phenomenon includes something as object within itself" (Brentano, 1874/1995, p. 88). Husserl, who was a student of Brentano, extended this idea of intentionality in the development of his later phenomenology.

12. For an introduction to string theory in general and M-theory in particular, see *The Elegant Universe* (Greene, 2003).

CHAPTER 11

THE BEING OF FORM

THE SEARCH FOR MEANING

G IVEN that our being conscious is itself an experience of meaning, it may seem absurd that we should go in *search* of meaning. And yet, in our ordinary everyday lives, we do not explicitly recognise this continuous stream of meaning because we live *within it*—it is there in our understanding of where we are and what we are doing and why we are doing it. As long as we are immersed in the familiar routines of daily living, we have no problem with this embodied level of meaning as it simply and seamlessly assimilates our actions into the meaningful thread of the events in which we participate. It is only when we stop and think about the totality of this life, and its finitude, that we start to wonder about the ultimate meaning of our being here. Such wondering seeks after a reason that could explain why it is that we are born, why it is that we have this brief moment here on Earth, undergoing all these fleeting and contradictory experiences, before slipping back into that unknown and apparently unknowable place that we are calling the unthinkable source of experience. For most of human history we have lived in the certainty that this source is also the source of the ultimate meaning of our being here, and that in order to discover that meaning we only need to 'read the signs'—through prayer perhaps, or divination, or hearing the words of the prophets. It is here that we again encounter the symbolic contents of the human psyche—all the gods and the religions and the creation myths that are overflowing with reasons and explanations for our existence. But, in our embracing of modernity, we have rejected these symbolic forms. Our science has shown that such mythical accounts do not correspond with the world that is revealed in the spotlight of our strict objectification. It is this very objectification that has cut us off from the symbolic world of the psyche and all its intimations of a greater reality lying behind the materiality of our sensory experience. And so now, when we come to ask after the ultimate meaning of our being here, if we are honest and do not simply parrot an opinion we have picked up along the way, we are met with an inscrutable inner silence.

The question, of course, is how we are to proceed, now that we have come to doubt this god of scientific materialism? Clearly we cannot forget what we have learnt in the last four hundred years. For the sun that we photograph and study is not drawn across the sky in a chariot by a team of celestial horses. At the same time, phenomenology has taught us that the psyche is the source of *all* we experience, not just our subjective feelings and phantasies, but the objectivity of the world itself. Knowing this, as we turn within, we know that our reflective consciousness is facing an extraordinary potentiality . . . .

But the *meaning of life*—how are we even to approach such a question? For this is not something we can *calculate*, as the neo-Darwinists have attempted in their logic of survival. We cannot simply *measure* the trajectories of the events of our lives and somehow extrapolate a meaning by seeing where it is all tending. For it is all tending towards death, and, it seems, the disintegration of the universe itself. It is not a prediction of this final state that we are seeking, it is *transcendental* meaning, meaning that can make sense of our lives from the place of the inevitability of our dying. It is *this* that troubles us. All the effort, all the care we put into sustaining ourselves here, the cleaning of the teeth, the going to work, the falling in love, the bearing of children, the disappointments, the losses, the illnesses, the suffering and the eventual death—all the plans, the projects, the energy we focus on this little life, and then one day, perhaps much sooner than we think, it ends. We are seen no more. And when we realise this, we are almost forced to ask: well, what was it all for? For the suspicion is there, in the background perhaps, that there is no reason, that the very idea that there should be a reason is a human invention, and from the cold perspective of death, from the place where all the species evolved in a continual cycle of eating and being eaten, that reason, if there is one at all, is not concerned with the plans and projects of you and I and our idea that our little lives have some kind of meaning.

In the background of such thinking there lies the 'hard-nosed' view of our scientific materialism that says we simply live and breathe and eat and drink in order to survive long enough to mate and reproduce the species. There is no reason for our being here beyond that, for the universe is a giant mindless mechanism and only those who behave in such a way as to survive actually do survive. That's the end of the story. This is the *meaningless* explanation of our scientific materialism. It is meaningless in that it simply describes the operation of a certain machine. It does not say what the machine is for, or why it came into being, it simply says: this is the way the machine works.

And yet we cannot just dismiss our science. For it has taught us that the kind of meaning and purpose we once thought lay behind our appearance here on Earth does not correspond with the objective facts of our experience. Our little planet is not the centre of the universe. There is no celestial sphere upon which the stars are placed in order that they can revolve around us in perfect circles. Life on Earth was not created according to some pre-ordained plan devised by an all-seeing intelligence. It *evolved.* And the history of that evolution shows all the characteristics of a blind experiment involving the formation and destruction of countless trillions of life forms without any clear or obvious design in mind. What science says is that if there were gods and intelligences behind the creation of the universe, then they have certainly gone to great lengths to disguise their presence. For surely it would be easy for them to have left a sign, a message? Why should it be that wherever we look we find the inexorable regularities of the laws of physics in operation? If there is a transcendental intelligence, why does it not declare itself to everyone. It would be so easy, wouldn't it? To just appear one day and say: "Yes, it is I, your creator God, look, I shall raise a mountain right here before your eyes, I shall cause the seas to part, I shall bring back the dead!" But this does not happen. We are once again confronted with a profound and inscrutable silence. We stumble on, as if abandoned, from world war to world war, from Auschwitz to the ecological destruction of our planet. And in the light of all this, we are almost *forced* to ask: what kind of intelligence is it that should do all this, that should think that this is *necessary*?

What we have experienced, what our culture has experienced, is the death of the idea of there being a rational, benevolent creator God who consciously and deliberately created the human race and placed us the centre of existence in order to fulfil some intelligible pre-ordained divine purpose. The evidence, the scientific evidence, the historical evidence, the evidence of our own behaviour, simply does not bear this out. We thought there was a great father in the sky who knew what he was doing, who was guiding our lives, who held in mind on our behalf the greater meaning and purpose of our being here. But now it seems that this idea of God was something of our own devising. Something we projected into a transcendental heaven.

We stand in the ruins of this former meaning structure. Collectively we have given up on the task of making clear to ourselves the reasons for our being here. We may still have some feeling in moments of intense crisis that there is a deeper significance in the events we experience, but we no longer have a definite idea of what this significance amounts to. We are adrift, and really, we are in *anguish*. Our faith now lies in our science, and our science, our *hard* science, pretends to have nothing to say about the meaning of our being here. Like Pontius Pilate, it washes its hands of the responsibility of articulating the meaning that lies behind what it has discovered. Either that or it denies there is any such meaning. But this notion that objective science can somehow abstract itself away from questions of meaning and value is mistaken. For the idea that the material substrate of the universe is 'really real' and that this material substrate finally determines all that happens here, is not some kind of neutral, objective statement of fact. It is an ontological *theory* that contains a very definite conception of the meaning of our existence, i.e. the conception that it is ultimately without the kind of meaning that would grant any special significance to our being here. Such negative meaning is still a form of meaning. And our science and our scientists are still responsible for this meaning. For we are all responsible for the consequences of the meanings we live by, even if we live these meanings unconsciously. This is the simple justice of cause and effect. Our science has tried to escape this responsibility, just as we try and escape, by ignoring the question of the meanings that are embodied in the economic and technological systems that we serve. But we are not absolved or protected by such unconsciousness. For we *are* the meanings that we live and the meanings that we live determine *how* we live. And if we live according to the idea that our lives are without ultimate meaning—even if only by being unconscious of such meaning—that is still a meaning that we enact, and that enacting changes the trajectory of our lives.

That is not to say that being scientific automatically leads us to think that our lives have no meaning. All science has shown is the inadequacy of our former conception of a creator god. Our current materialism is only what has managed to survive from the wreckage of that former way of thinking. Such materialism, in itself, is neither scientific nor unscientific. It is no more than a habit of thought we cling to in the absence of a better alternative. And yet it is no accident that materialism leads us to think that our lives are without meaning. For materialism is a denial of the independent reality of consciousness, and it is only insofar as there is consciousness that there is meaning. So materialism, by definition, is unable to see what meaning is, in itself. And that means it is necessarily going to conclude that our lives have no intrinsic meaning. For, if it turns out that our lives do have intrinsic meaning, then that is the end of materialism. And, of course, our lives do have intrinsic meaning. Our whole experience of being alive is an experience of intrinsic meaning. This is not a matter of refuting materialism, it is a matter of seeing the meaning of materialism. For it is our materialism that cuts us off from any kind of direct knowledge or insight into the meaning that our lives are actually manifesting. It is not that we have to invent such a meaning, we have to *discover* it.

To consciously live in meaning is to live in conscious resonance with the unthinkable source of experience. Ordinarily we live in an unconscious resonance with the source, only seeing the fragmentary being of the things and events and thoughts and feelings that are the stock-in-trade of our daily experience. What we miss is the meaningfulness and unity of this continuous upsurging of life into the space of our being conscious now. We can intuit the presence of this greater unity that stands 'behind' the appearance of the world in the immediate consciousness of love or beauty. In such moments we no longer ask "What does it all mean?" Instead we find our being alive, our being conscious, is *already* intrinsically meaningful. Such meaning seems to suffuse our qualitative sensory experience and speak of the being of a deeper timeless reality. And yet it is all connected. For when we see the pure quality of the colours that manifest a sunset, we are also experiencing the transcendental beauty of the essence of colour itself—an essence that does not evolve or arise or pass away, but instead shows the form of what evolves and arises and passes away. It is here that we can understand what inspired the great impressionist painters. And it is here that we can understand how the source is able to communicate a meaning that transcends anything we can think or speak of directly.

In such experience we become *attuned* to the source itself rather than to the form of the world it is projecting. This is not something we can imagine, or treat as some kind of thought experiment. It is a matter of attuning ourselves in the immediacy of consciousness to the manifestation of the source in each moment. It is here that we start to encounter the methodology of a science that is grounded in consciousness rather than in the objectivity of the world. For we cannot measure the source with a ruler. We have to use the resonance of immediate perception, the same resonance that already reveals the actuality of the world (such as the actuality of the cat in the garden). This resonance of the natural attitude becomes the resonance of *transcendental* meaning just as soon as we become attuned to the source of the world's actuality (i.e. the consciousness of consciousness itself). The great problem here, of course, is that, in such transcending of the world, we lose our capacity for objective verification. For we are no longer looking at an objective something 'out there', we are looking at the meaning that lies behind our experience of there being something 'out there' in the first place. And, according to our usual way of thinking, once we lose our capacity to objectively verify our understandings, then we fall into the relativity of our various subjective viewpoints. From this place it seems that we all resonate differently, and so we fall into disagreement, and we find we have no standard by which we can resolve our disagreements. We form groups and schools, and we develop various philosophical theories of meaning, and instead of making progressive discoveries we become embroiled in continual conflict with people who do not seem to understand what we are seeing.

Again we enter the labyrinth, and again we remember that the way out is to disidentify with our self-reflective, personal thinking processes. Of course, different people will be disidentified to different degrees. So there will be disagreement. But that does not mean we cannot make progress. For there is still only the one source. And it is not that we are attempting

How do you know that?

to uncover, once and for all, the true and final objective meaning of our being here. We are setting up a line of resonant communication with the source of meaning that lies within us. And we are testing this line by looking into the meaning of what our science has already discovered. We started this process in our earlier consideration of the measurement problem. The idea was (and is) to show how we may go about creating a meaningful scientific account of the being of the universe that includes the incontrovertible reality of our being conscious. This is not a matter of everyone having to agree that consciousness causes the collapse of the wave function. That is only a working hypothesis. More fundamentally, what we are doing is pointing out the absurdity of a science that thinks it can get by without addressing the question of the meaning of what it discovers—that thinks it is not responsible for that meaning. What we are exploring is the development of a science that does take responsibility for the meanings it creates. A science that, instead of turning its back on the source, recognises the transcendental ground from out of which it is emerging, and recognises its responsibility to give a meaningful account of that ground.

That means we don't just test a theory by deducing its objective conse-quences, we also evaluate its *resonance*. The more resonant a theory, the more meaningful it is. We can think of this as a kind of parallel to Occam's Razor: that, all else being equal, we prefer theories that have a greater res-onance of meaning. In following this principle we actively look for theo-ries that correspond with the state of immediate consciousness that is our connection with the unthinkable source of experience. From the perspec-tive of materialistic science, this is a kind of heresy. For it seems as if we are choosing theories on the basis of what we like, or prefer, or what we wish and hope for. But, insofar as we are directly and immediately conscious of our being conscious, we are no longer ruled by subjective feelings of lik-ing and disliking. They too are phenomena that appear in consciousness, while consciousness itself is simply presenting itself to itself. The entire project of this new science hinges on the question of the impersonality of the consciousness of the one who is inquiring. And yet this capacity to remain impersonal is *already* a fundamental requirement of our *exist-ing* science. It is just that this impersonal objectivity of the conventional scientist is being shifted out of the natural attitude of the objectively mea-surable into the dimension of the transcendentally resonant.

To give a practical example, let us again consider the question of the quantum collapse. Here, all else being equal, we prefer the theory that proposes the quantum state of potentiality only collapses into a state of definite being insofar as it manifests in some form of conscious experience. Such a theory does not multiply entities unnecessarily. And neither does it understand consciousness to be an inexplicable, purposeless, epiphenomenon. That makes it more meaningful than a purely mechanistic account. It is more meaningful because it explains more, it covers a greater domain, and it resonates with our direct experience of the world manifesting in consciousness during each moment that we are able to tear ourselves away from the stream of our habitual thinking. If we follow this thread of meaning, it destroys our modern conception of the insignificance of humanity in the face of a vast and indifferent universe. It is almost as though we are returning to our medieval origins—with the earth at the centre of all creation—only now it is our consciousness that is the centre, for it is only in consciousness that the universe is able to come into being. Do you see the significance this bestows upon each one of us? For even though there appear to be billions of human individuals on the planet, all these individuals, and the planet itself, are only able to manifest insofar as there is a consciousness in which to manifest. And the only consciousness that we know of, that *I* know of, is this very consciousness that I am.

Here then is the beginning of a science that bestows a meaning on our existence: the meaning that we are here so that the source can manifest itself in a definite form. We are the very site of this manifestation. This is what ancient philosophy experienced as *wonder.* The wonder that this manifestation actually manifests. That there is *consciousness, being, meaning,* rather than the not-being of anything whatsoever—rather than the nothingness of a complete unimaginable absence. Could it be that the source is *pleased* by this manifestation? That through this it gains a sense of its own being? That it has *striven* to become manifest and that we are a living expression of this striving? Do you sense what is resonating here?

THE SCIENCE OF FORM

Let us look a little further into what science has revealed about the form of the potentiality of the source of our experience of the world. One thing is clear: modern Western science has been a *mathematical* science, a science that is concerned with the development and discovery of ideal mathematical forms that resonate with the source. In such resonance these forms acquire meaning and ideal being. And the resonance tells us something of the source—it tells us about the abstract mathematical structure of the world that we manifest in consciousness. We do not imagine this structure. But, in its pure mathematical ideality, it has very little to tell us about our being here, about our being conscious, about our being alive. It simply speaks an abstract ideal language of numbers and relations.

So how are we to make this mathematical science meaningful? How are we to bring it back to consciousness?

Firstly we should recognise that it has been exclusively a science of form. From the very outset Galileo abstracted away all idea of the objective being of sensory *quality*, considering it to be something subjective that could finally be explained in objectively mathematical terms (as our brain science attempts to this day). And we still think that if we can provide a mathematical description of the form of a certain process then we have somehow explained that process *away*.

No .
Relatedness
Geometry
Topology

But once we return to the undoubted and immediate reality of our being conscious, we find that the world, the actual world that we experience, is entirely manifested in the medium of sensory quality. This sensory quality is no more subjective than it is objective. It is that which manifests the world to us *before* we make any subject-object distinction. It is this fateful carving up of our experience into the quantitatively (measurably) objective and the qualitatively (perceived) subjective that has led us to develop a science that lacks intrinsic meaning—because, for there to be intrinsic meaning, there must be both form *and* quality. It is quality that gives form *being*, otherwise it is nothing. Even the ideal forms of pure mathematics require the qualitative consciousness of a mathematician in order to acquire their timeless and rarified ideal being. For all being is a form of meaning, and meaning cannot come to manifestation except and insofar as there is a qualitative field of conscious experience for it to manifest as and in.   *No - meaning = changing relational qualities .*

✱ So let us propose another principle of meaning for our new science: that nothing can exist, nothing can have being, without possessing both form *and* quality. But what does it mean to 'possess quality'? And what, for that matter, does it mean to 'have being'?

Does quality = change of relatedness among patterns ?

Here again, we must return to the phenomenology. The one ground level, indubitable experience I have of being, is this immediate being of consciousness that I am. My being is my enduring, my being conscious of the flowing temporality of the streaming forms of quality that make up the field of consciousness that I am. I know this unity of form and quality *directly*. It is not some kind of object from which I can separate myself. It is that by means of which objects are able to appear to me in the first place. For what is it to have vision? To see colour? When I see the world in front of me, what is the medium in which it appears?

Even if I withdraw from the streaming of these sensory forms, there is still a quality to my being here. I experience my endurance, my continuity, in my understanding of time, in my being poised here, in this immediacy of now, in this clearing that itself never moves, never flows, from out of which the past is continually manifesting and moving away from the horizon of now, even as the past remains present, while continually falling back and being replenished from out of an inexhaustible source. Being—meaning—quality—form: all are bound up in this unity of pre-reflective consciousness.

But now, when I look *into* my experience, I discover a whole other dimension of being. For I understand the qualitative forms that appear in the stream of experience to be the qualitative forms of *other beings*— beings in some sense like me, but at the same time *not* me. For I know myself from the *inside*, so to speak, whereas these qualitative being-forms only show themselves from the *outside*. How strange is this? It seems there is not just the one form of being (me-being), there is also you-being and it-being, and even thought-being and imaginary-being. And yet, in some way, I already understand that we and they are all still *beings*. It is as if we are back with Descartes. For how can I be sure there are any other *real* beings, beings like me, who have an *inside*, who are conscious, who are having actual experiences? The answer is, of course, that I cannot be sure, at least not with the kind of complete and indubitable certainty that Descartes was seeking. For this could still be a dream, and I the only dreamer. All I can say is that there is more here than this finite reflective consciousness, with its limited field of experience. There is also the source, the place from out of which all this is emerging. Perhaps, insofar as I am the source of the clearing *and* the being of the clearing then there is just the one inconceivable unity that I am. But that is not our concern at

! really?

the moment. Our concern is with our being here together, with the *meaning* of this being here, where we *do* appear to each other in these separated individualities, in the qualitative form of these bodies that manifest in our consciousness of each other. The question is, how are we to understand this? And what does science have to say? What has it discovered about these qualitative being-forms that we appear to be?

Perhaps the first, and most salient fact that objective science has discovered is that our consciousness of the world and of our being in the world is directly and immediately connected with certain objective events and processes that are occurring in our brains. Scientific materialism takes this fact as evidence that some form of materialism must be true. That is because it already assumes the universe is made of material, physical 'stuff'. And yet, as we have seen, if we take quantum physics seriously, the material being of the universe is really founded on an immaterial realm of potentiality that only manifests what we take to be material particles when there is an act of observation. At the quantum level there are only *patterns of probability*. Here what we take to be the material stability of an atom or a molecule becomes the stability of a highly probable pattern of manifestation. When we look for the 'real' substantial materiality of the atom, we find it disintegrates into something that keeps appearing and disappearing according to whatever it is that determines the collapse of the wave function that maps the superposition of its possible states. What we actually find is *the being of form* and not the being of matter. When we think of the being of form we usually think of something being carried on a material substrate, like the form of an ocean wave being carried on the movement of the material substrate of the sea. But if we look into that material substrate, we find molecules, made up of atoms, whose materiality is also a kind of wave-form carried on the immaterial potentiality of quantum probability. From this perspective, ocean waves are oceanic forms which are carried on molecular forms which are carried on atomic forms which are carried on sub-atomic forms which are carried on immaterial quantum forms of probable manifestation. Beyond that we cannot ask. Nowhere in this cascade of form do we find any materiality whatsoever, just form emerging out of form, emerging out of form . . . . At each level what we find is *the form of an event*, of a unified movement that can also be decomposed into the movements of the forms of the next level down.

No.
Participation

No
dissolves

why / not?

Resonance of brain / context, TOC = context neither.

And yet what kind of being can we attribute to this vast cascade of hierarchical movement-form? On what basis do we distinguish sub-atomic particles from atoms, and atoms from molecules, and molecules from ocean waves? Aren't these just distinctions that *we* have made, according to certain practices of objective measurement, distinctions that some other species or intelligence would simply regard as human inventions emerging out of the spatiotemporal dimensionalities of our perceptual systems?

If this is the case, then we cannot say that the pattern forms we distinguish have any intrinsic being of their own—we can only say that this is the way that we have found to divide up the otherwise unified streaming of the unthinkable source of experience. These forms have being because we invest them with being, just as we invest the sensory forms we perceive with actuality. For our human perceptual systems are just one way in which the unthinkable source is able to manifest.

Imagine if we could perceive time as a fourth dimension of space—if we could see the entire history of the universe as a single, complete form, from the Big Bang to whatever state it is that marks the end of objective time (if there is such a state). In this universal form there would no longer be atoms as we picture them now. Such discrete entities would become world lines in a vast interconnected system of paths that both join and split apart to manifest another kind of form—a *form of eternity*. Who can say what kind of beings such a perceptual system would distinguish, or whether there could be another dimension of time in which change of a different order could manifest?

And yet our mathematical science has already transcended the three dimensions of space and the one dimension of time with which we are perceptually familiar. It now represents a unified four-dimensional model of a curved spacetime within which the world lines of particles evolve in the mathematics of general relativity. It is this model that we now take to be the 'more real' representation of space and time, and not our naïve, perceptual notion that space somehow exists independently of time. That does not mean our idea of discrete atomic being is *false*. Our mathematical science is only transposing measurements of objective events from one system of dimensionality into another. Einstein's four-dimensional model of spacetime simply captures more information than we can directly represent in our three-dimensional perceptual imagination. The mathematical form of this perceptual idea still captures an *aspect* of what-ever it is that is generating our scientific observations of atomic phenom-ena. That aspect is a partial and approximate representation of the form of a deep invariance that lies behind the continually changing surface of our collective perceptual experience of the being of an objective world.

From this we can see that our scientific models of atomic being are not determined or limited by the dimensionalities of our perceptual systems. These dimensionalities only limit the extent to which we can picture or imagine what our scientific models 'look like'. But if we consider the meaning of these models more fundamentally, we can see that they are still based on a quite specific conception of being—namely an idea of *en-during mathematical form*. This too is a *human* idea that is 'built in' to our perceptual systems. For, according to our latest scientific theories, these systems have evolved to detect invariance in the streams of sensorimotor input and output, in order to form predictions of the future evolution of these information streams.[1] What emerges from this process are our ideas of a world populated with determinate and identifiable things and events. These are the entities we project to be the *causes* of the streams of sensorimotor information that continuously activate our nervous sys-tems. Here, in order for something in the world to 'be' or to 'exist' it must possess the kind of enduring mathematical form that enables us to predict the shape of our future. This is all to do with computational ef-ficiency. If you were to attempt to predict what you will experience in the next moment purely on the basis of the state of excitation of all your sensorimotor nerve impulses in this moment, it would turn out to be im-possible. There is just not enough information in the previous state to determine with any accuracy what will happen in the next state. For vir-tually nothing at this level is stable. It is the river in which you cannot step twice. What the brain is doing is forming connections between its inputs and outputs in such a way as to detect the presence of enduring or stable relationships. Once it has encoded the essential underlying relationships that exist in the sensorimotor streams that emerge from our interactions with a particular object—such as the table in front of me now—then in-stead of trying to predict how the visual field is going to change as I shift my gaze over the table by examining the immediate state of that visual field, it uses what it has stored from past experience with tables in general, and with this table in particular, to predict and project how the table will

TG - Form = delay + connection. 'Synchronous'

appear in the next moment. Of course the immediate state of my visual system informs this process, but it is <u>the mathematical form of the table</u> that is <u>encoded in the brain</u> that <u>determines</u> what <u>I am predicting</u>, both in terms of what I expect to perceive and how I expect to act.

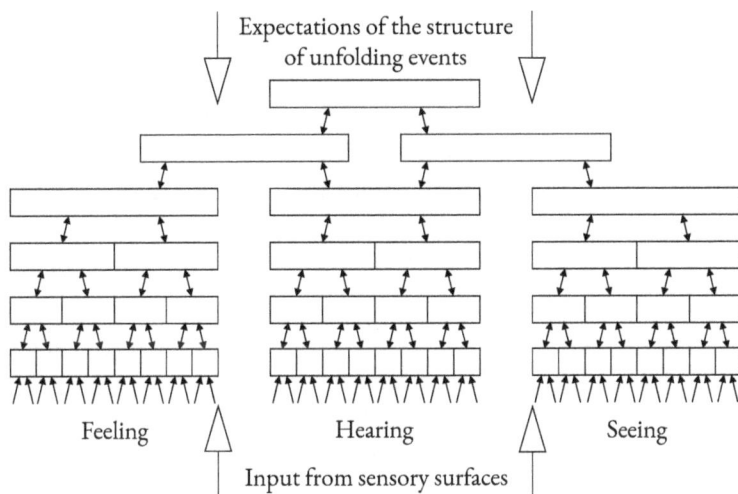

Figure 11.1: A simplified model of sensory input processing in the neocortex of a human brain. Motor output would be processed in the same way except that bottom layer signals would travel downwards ↓ rather than upwards.

This mathematical form is interesting in itself. It is not that the brain stores a perfect model of the table and every other thing we may encounter in our daily experience. It *constructs* its projection of what it expects to perceive 'on the fly'. Current theory[2] suggests there is a hierarchical structure to this projection (see Figure 11.1), where each level of the hierarchy corresponds to an area on the outer surface of the brain's neocortex. The hierarchy itself only emerges when we examine the pattern of longer range connections between these areas. The theory proposes that the top level of the hierarchy encodes our quite imprecise expectations of the general form of the context of the overall situation—such as my being at home, in the front room, on a Tuesday afternoon. As we go down a level or two, this context is brought into sharper focus by feedback from my more recent experience of sitting here in this chair, at this table, typing on this keyboard. This produces a more active and detailed expectation of the form of the things with which I am interacting, an expectation that will fade as soon as I leave the room and enter the kitchen. At each level, what is expected from higher-level projections meets with what has already been assimilated from the streams of information arriving from below. Ultimately we reach the lowest level where the raw information is being absorbed and emitted from the sensorimotor systems. But even here, in the retina for example, visual information is already being processed according to expectations 'built in' to the connections and sensitivities of the retinal neurons—neurons that are stimulated by photonic energy that is precisely focussed by a lens in an eye that has been guided to it current position by a higher-level expectation of what it will encounter.

In this model, our perceptual experience is produced by an extraordinary and seamless meeting of the general forms of our past experience (which are encoded in the connections between our neurons) with the moment to moment streams of information that are flowing in and out of our nervous systems. These streams bring our expectations into focus entirely on an 'as needed' basis. Right now we think the whole room is present before us in all its detail. But really, if we pay attention, we can see it is only the tiny area at the centre of our visual field that possesses this detail. As we look up, we think that the detail we see was already there. But it wasn't. The moment before we looked up, the part of the room we now see was just a vague premonition in the periphery of our visual field. It is the brain that has produced the expectation that both moved the eye and expected the form of what the eye will encounter, just in time for the visual input at the retina to fill in the detail of this expectation.

Here we meet the contemporary idea that the brain is a kind of extraordinarily complex prediction machine.[3] In order to make its predictions, it detects (relatively) invariant forms in the streams of sensorimotor information it processes, and, according to these forms, it distinguishes the things and events of the world around us. From this objectifying perspective, it follows that it is the structure of the brain that determines our idea of being. This idea is something that has been laid down by the process of the brain's own evolution—for the organism that predicts with greatest accuracy is the organism that has the greatest chance of survival (all else being equal). Such a predictive idea of being says that for something to 'be' it must endure by retaining an invariant form of identity. That means there must be *something* about the way it manifests that does not change. This 'something' is what defines it as the being that it is, so that, if it were

TG - 'forms' include personal values.

to lose this 'something', then it would cease to 'be'. It is this 'something' that the brain uses to predict how a being will endure and also how it will change over time. If there is no predictability then there is no being.

But, of course, this is only *half* the story. For the brain itself, considered objectively, does not distinguish any form of being whatsoever. It is only *No?* insofar as the brain is conscious that there can be a field of experience where one 'thing' is distinguished from another. Returning to our earlier principle, it is only when form and quality are unified in consciousness that being can manifest. It is consciousness that apprehends form in the first place—a mathematical expression that *describes* form does not *possess* form, for the expression is a configuration of meaning that again only has being insofar as it is apprehended by consciousness.

So where does this leave us? If we consider the brain objectively, we can start to explain the origins of the forms of our experience in the forms of the processes that occur in the brain. If we do not fall unconscious, and think that the brain, in and of itself, is causing our experience of the being of form, then we can start to explore the true significance of this remarkable correspondence. For it *is* remarkable to have recognised that brain processes occurring 'out here' in the objective world of things and events are also involved in determining the form of our experience of those events. Even though the manner of this involvement remains mysterious, there is no doubt that there is a detailed and definite correspondence. How else are we to explain the behaviour of people who have ✗ suffered severe brain injuries? Perhaps we are capable of sustaining states of consciousness that are independent of the activities occurring in our brains. But, if that is the case, these must be quite extraordinary and rare forms of experience. For there is little evidence, at least here in the world of our everyday interactions, that we can maintain any form of conscious experience without our also having a brain that functions in just the right way to allow such experience.

What challenges this common sense view are the many and varied reports of near death experiences, in which states of consciousness are described that appear to exist independently of anything occurring in a physical brain. For example, there is the case of the American neurosurgeon, Eben Alexander, who wrote an entire book documenting a series of psychic experiences that he believed were occurring while he was in a deep coma and his brain was known to be virtually inactive.[4] The problem with such reports is that they are based on the consciousness of a *memory* of a near death experience, not a consciousness of the experience in the moment of its occurrence. It could be there is no associated consciousness in the near death state itself, and that the memory is formed once the brain starts to function 'normally' again. This memory is then experienced so that it *appears* to have occurred during the near death state. In contrast, if we limit ourselves to experiences that are actually happening while the report is being made, we find, as far as we are able to measure, that such reports do correspond with the kinds of processes we would expect to occur in the brain of the person making the report. Of course, such an explanation still does not account for the *content* of a near death experience.

CONSCIOUS RESONANCE

The question we are considering is whether this division of perceptual experience into the various objective forms that we distinguish is something *arbitrary*, something *human*, something that is entirely determined by the kinds of brains that we possess—whether there are only 'plants and birds and rocks and things'[5] because our brains evolved to distinguish such forms. We are asking whether the distinction of form is also a distinction of *being*—whether the forms we distinguish also have their *own* being, independently of their being distinguished by us. This, of course, is a kind of impossible philosophical question. For how are we to see how things stand independently of our observing them? We can only speculate. Phenomenology avoids such speculation by remaining with experience itself and recognising objective being as a meaning that *we* bestow on certain aspects of our experience. But now we are stepping outside the domain of pure phenomenology. We are no longer 'bracketing' what objective science has discovered. We are building a bridge between the phenomena that we experience and the objective forms that science distinguishes from processes of purely ideal mathematical measurement. We are inquiring into the *meaning* of these objective forms.

The first indication of such a bridge is the correspondence of our conscious experience with certain processes occurring in our objectively physical brains. If we look at this correspondence we find it is the *immaterial form* of these processes that specifies whether or not we are conscious, and dualist

not the molecular material of the brain itself. For it is not the individual neurons that are the 'carriers' of consciousness. If that were the case then we would not be able to fall *unconscious*. And yet we do fall unconscious each night in deep dreamless sleep. In such states our neurons are still firing and interacting. All that changes is the *manner* of this interaction. It is the form of the manner of this interaction that defines whether or not we are having an experience of being conscious. At the moment, the precise specification of this form is a subject of ongoing research into what are called the neural correlates of consciousness.[6] So there is still uncertainty as to the manner in which our conscious states are correlated with our brain states. But there is little doubt that there is such a correlation and that its form will eventually be revealed by our scientific instruments.

What this means is that our consciousness is directly associated with an entirely abstract property of the collective behaviour of the neurons in our brains. This property is not something physical, it is a pattern of activity that the 'physical' neurons are enacting. For if the neurons enact a slightly different pattern, then consciousness disappears. And it is not that consciousness is associated with a *particular* collection of neurons. The pattern of activity that is our experience of being conscious is continually including and excluding *different* collections and populations of neurons. Current theory suggests that the crucial criterion is the manner in which these populations dynamically self-organise into emergent collectivities that pulse together in a coordinated and synchronous unity.[7] This coordinated pulsing is carried in the electro-chemical signals that the neurons transmit to each other through their axons and dendrites. Seeing this, we might think that consciousness is associated with the electrical fields that form as a result of this behaviour. But an electrical field is still not conscious in itself—for the brain is continuously generating electrical fields regardless of whether or not we are conscious. It is the abstract form of the process that the fields are enacting that finally objectively specifies the presence or absence of our being conscious.

Another remarkable aspect of consciousness is its demonstration of *nonlocality*. Nonlocality in quantum physics refers to the instantaneous connection between entangled particles, such that the measurement of one particle instantaneously affects the properties of another spatially separate particle (for a brief explanation and diagram see page 94). Einstein published a paper in 1935 (with Podolsky and Rosen), pointing out that

Very narrow concepts of physical into movement patterns.

Dissolve physical into movement patterns.

nonlocality is a consequence of Neils Bohr's Copenhagen interpretation of quantum mechanics.[8] He thought this argued against the plausibility of Bohr's entire interpretation, particularly the idea that the collapse of the Schrödinger wave function could be inherently random (which he famously opposed by saying that "God does not play dice"). But subsequent experiments have supported Bohr, and it is now widely accepted that nonlocal quantum effects do occur in practice. And yet, in all this scientific investigation of nonlocality, its most obvious manifestation in the brain has been overlooked. For, if we accept that our consciousness is connected with certain processes occurring in our brains, then we must also accept that our consciousness is *instantaneously* connected with these processes.   *Physiology of light.*

To understand this we must first be clear that consciousness is not some kind of physical field that could be laid out in the space where our neural processes occur, and where 'messages' could be passed to various 'points' in that field of consciousness. There is simply no objective evidence for such a field or for any particle or flow of energy that could transmit such messages. All our observations confirm that in the objective world of measurement, there is just a brain, made up of physical neurons that send signals to each other that move at a certain speed and take a certain time to arrive. That means there is no point or neuron in that physical space that instantaneously 'knows' what is happening in all the other neurons that contribute to the process that corresponds to my experience of being conscious now. And yet my consciousness does 'know' what is happening in this global process that stretches across my brain. It instantaneously unifies information that is being carried by millions of spatially separate neurons into the simultaneity of a single moment of experience. What I see now is being carried by the activity of neurons in my visual cortex, and what I hear now is being carried by the activity of neurons in my aural cortex, and what I perceive now is being projected by the activity of neurons in my frontal cortex. The activity of all these neurons, that are in physical contact only by means of sending single pulses to each other, where each pulse is separated by a few thousandths of a second, and each neuron only receives signals from around a thousand others, is instantaneously unified in my immediate conscious experience now. This immediacy means it has taken *no time* for consciousness to discover what is occurring in my brain—it is simply, directly and immediately conscious of all these spatially separate brain events. And, more than that, it *knows* what all these events *mean*. For I am not conscious of millions of little signals, I am conscious of the meaning of these signals, of their reference to the being of other forms that are distinct from the patterns of excitation occurring in my brain. Do you see, if we start by accepting the objective being of the physical brain, then consciousness emerges as something that transcends the limitations of ordinary physical spacetime, where—according to the theory of relativity—no information can be transmitted from one place to another faster than the speed of light? But consciousness *breaks this rule.* It instantaneously connects different regions of space in the manifestation of our experience, just as quantum states instantaneously connect spatially separate entangled particles. Both these phenomena point to the same conclusion: physical space is an *appearance* of something that is not, in itself, separated in space and limited by speed of light communication.

But what stands out, in the midst of our examination of consciousness and the processes in the brain to which it corresponds, is the *unique status* of these brain processes. For we are inquiring into the meaning of the being of the forms that science discovers. We have been asking whether our distinction of such forms is an entirely human-relative procedure, something that is determined by the structure of our physical brains, such that, if we had different brains, we would distinguish entirely different forms, or perhaps we would not even distinguish form at all. But what we have discovered is the being of at least one form that *distinguishes itself* independently of our acts of distinction—and that is the being of our own consciousness insofar as it manifests in the form of an objective process that occurs in the brain. It is not *we* as human individuals that have distinguished the form of this process, it is consciousness itself, or, if you like, it is the unthinkable source of experience that lies behind, and within, and is our experience of being conscious.

Now this may sound contradictory. For our experience of being conscious does appear to be an entirely human affair. But the *fact* that I am conscious is not something that has come to pass because of an act of human discrimination. My being conscious precedes all acts of discrimination, and all such acts depend on my being conscious. Of course, my recognising that I am conscious, my becoming conscious that I am conscious, does require an act of discrimination. But we are talking of consciousness itself. And I am conscious regardless of whether I consciously discriminate the fact that I am conscious. What science has discovered is that conscious experience also has an objective form in the world. It is the signature of the necessary and sufficient abstract form of the brain processes that correspond with my moment to moment experience. As we have already discussed, we do not yet know the exact form this signature takes. It is likely to be very abstract indeed. For it will say: whenever and wherever a form of this kind manifests, there will also be consciousness.

Of course, philosophers have already speculated on this. There are functionalists who say: whenever and wherever a process exists that takes the same inputs and produces the same outputs as an entity that we already recognise as being conscious, we can say that that process is conscious. But this is a *lazy* answer—an answer that steps around the problem of understanding exactly what it is about the processes in the brain that correspond with our being conscious. And it already ignores the most obvious fact: that, up until now, and as far as we can tell, consciousness has only manifested in *living* organisms. Functionalism ignores this because it is committed to a materialistic idea of being that already assumes there is no essential difference between a living organism and a machine.

Here we are going to take a different approach. We are going to begin by describing the form of a system that we know to be conscious, i.e. ourselves. We are starting here because we know this form has *intrinsic being*. Its intrinsic being is the being of our conscious experience. That means we are looking at something that discriminates itself by having an experience of itself. This not like an act of perceptual discrimination, where we discriminate the presence of enduring form in the streaming of our sensorimotor fields without reference to whether or not that form is having an experience of itself. Now we are seeking the form of our experience in the form of the processes occurring in our brains. It is not the underlying enduring stability of those processes that matters. Our guide is the *correspondence* between the brain process and our immediate experience. It is experience itself, its already given and immediately present form that provides our measure. This is not something arbitrary, something that evolved by chance on this planet. It is an extraordinary and primal feature of the universe that consciousness can and does emerge when the circumstances are favourable. And clearly, with us, the circumstances are favourable. But what are these circumstances?

Already we can say something of this correspondence on the basis of our existing science. For if we accept that the brain is a prediction machine, that means it is able to embody the forms it perceives. It is by means of this embodiment that it can set up a stream of predictions that then interact with the stream of information that impinges on its sensory surfaces. It is in the meeting of these two streams that our conscious experience of being in the world emerges. What occurs can be described *Yes* as a kind of *resonance*. We can say that the form the brain is projecting is resonating with the sensorimotor input it is encountering. But here, of course, we have again taken up the stance of the objective observer. Such an observer already pictures the human body in a world surrounded by objects and fields of electromagnetic energy, whereas, the objective situation is that this picture is formed by and based upon the very processes we are trying to describe. The fact is we only know that 'something' is resonating in us. This something is the unthinkable source of experience. We automatically think of this source in terms of the forms that our perceptual systems discriminate. But we are not entitled to do this. The resonance only tells us there is an agreement with the source—that *Cosmic* our picture corresponds. The source is not the picture. It is that which *epoch.* facilitates the very dimensionality of our being conscious, the space, the endurance of our being in time, the qualities of the colours, of the sounds, and then, within this dimensionality, the source grants an experience of resonance which tells us that the beings that are appearing in our sensory dimensionality actually correspond with an aspect of the potentiality of the source. *Prehension .*

What singles the brain out from all the other forms that we perceive around us is its capacity to embody the form of forms that are different from itself, forms that are able to *resonate* with the source. This embodiment of form is manifested in the brain's capacity to predict the form of its own input. Such prediction, when it succeeds (i.e. when it resonates) means the brain has *captured* an aspect of the form of something that lies outside of the domain of its own structure. If we examine this act within the context of our understanding of the being of form, we could say that the brain has literally captured an aspect of the *being* that manifests the form that is perceived. For we are no longer thinking in terms of a ma-

? mismatches of anticipations = process

terialism that identifies being with configurations of material substance manifesting at particular locations in an objective space and time. If it really is the case that there is no objective material substrate, that it is form supporting form, supporting form 'all the way down', then my perceiving of form is no longer a *representation* of something that exists separately from my act of perception. In capturing and experiencing the form of something that is 'not me', I am experiencing an aspect of the very source that manifests that form. At the same time, that form can only come into being as the form that it is insofar as it is able manifest (to resonate) in the perceptual consciousness of a being like you or I.

For example, consider an act of speech. Let us assume there is an impulse in you that manifests in your saying something to me. If we look into the world of objective form, this impulse will first be traced out in the activity of the neurons in your brain. The pattern of this behaviour will contain the essential form of what you are intending to say. This form will be *transformed* into a series of motor commands that will manifest in the movement of your lips and tongue and mouth and vocal chords (and so on). These motor commands, and the corresponding movements, will still embody the form of your speech act. And then these movements will set up a series of disturbances in the molecular structure of the air that will carry the form of what you are saying from your body to mine. The hair cells in the cochleas of my ears will then pick up the pattern of these disturbances and transform them into electro-chemical neural signals that are relayed to my brain. Again, another pattern of neural behaviour will be manifested as these signals meet other neural signals that, in each moment, project expectations into the incoming signals that result in an experience of my hearing and understanding what you are saying.

In this hearing and understanding, I have grasped the very same form—assuming I have understood you—that impelled your speech in the first place. Can we say that the 'true' form of what you intended to say only existed in your mind, or in the pattern of neural activity that originated in your brain? Or is it the *same* form that comes to manifest in my brain? Could we say that we are *both* hearing this one form manifest, that it is resonating in each of us at the same time—that you too, are first realising what it is you intend to say by hearing yourself say it? For where does this form really have its being? Surely only in the resonance of a consciousness that is able to manifest it as an experience of meaning? Even if I do not understand a word of what you are saying, my resonant consciousness will still manifest a true *aspect* of your speech act. I will perceive that these sounds were uttered by you and I will experience their meaning form just as I experience the meaning form of a bird song. Your resonant experience will be richer than mine, more meaningful, because you understand what you are saying, but there will still be an *intersection* in which we resonate to the same form.   *Why bother with material if direct transmission*

It is this meeting (in my brain) of the form of the utterance that orig-   *occurs?* inated in you (that has been transformed into the neural patterns prop-   *Purifying* agating up from the hair neurons in my ear) with the form that my neo-   *times* cortical neurons are predictively projecting in response to this input that   *heart.* is the objective event of the resonance that we are speaking of. Of course I do not expect the form of your entire utterance beforehand. But I do expect the general form and tone of the manner of your speech, and I expect it to belong to the context of our relationship and of this particular meeting and of what has just been uttered beforehand. And as your utterance progresses, I expect and project the form of each word on the basis of each previous word and on the way this current word has begun and on the emerging meaning of the entire speech act. And once we are past half-way through, I can usually predict, with reliable accuracy, how it will . . . .

This is the *signature of consciousness* in the brain, that unique site where the essential form of one event (your speech act) *meets itself* in the form of another event (my neocortical neural projections). This is not like the surface of a lake reflecting back an image of the sky. My brain is able to project the form of its input *before* and *as* it arrives, not simply copy and reflect it back. Do you see that this event of neural resonance is something extraordinary? That, in all the complexity and apparent randomness of the events occurring in the objective universe, how improbable it is that an event should occur that contains within itself the essential form of *another* event? And if we go further, and conjecture that each form has its own being (rather than thinking that each 'bit' of material substance has its own being), then we can begin to see that the event of a form meeting itself is an event of a *different order* from all the other events we observe in the universe. For all these other events are just forms transforming into other forms which transform into other forms. Forms—such as atomic forms—may repeatedly emerge out of processes that themselves repeat,

Predictive systems require relational comparison across time for change as difference.

but such processes never *meet themselves*, never configure themselves to embody the form of something that is coming towards them. To borrow Leibniz's term, such processes are *without windows*. Whereas we, who are conscious, *do* have windows—windows which objectively manifest in the form of our resonant brains.

If we look at the bigger picture of our place in the vast cosmos within which we appear to exist, what we are proposing is that consciousness like ours, when viewed *objectively*, only starts to emerge once corresponding brain structures have emerged that are capable of resonating with whatever it is that is the source of our experience. What this resonance signifies is that something 'out here', in the extraordinary process of the evolution of form, has at last come to correspond with the forms that (metaphorically) surround it. The presence of the potential of such resonance acts as a kind of trigger for the manifestation of consciousness. What we are considering is the evolution of 'something' that already had the capacity within itself for consciousness and self-experience but could not manifest that capacity until the right kind of resonant system was able to appear in the midst of the universal field of potential being.

Finally, what we are saying is that consciousness *is* an experience of resonance, so that the more *attuned* we are with the source, the greater is the resonance, and the greater is the consciousness. Our level of attunement grows the more we withdraw our identification with thought processes that are occurring in the orbit of our past experience—processes that recycle the material of that past by redirecting our attention away from the immediacy of now. Such redirection produces a derivative form of consciousness that resonates more or less habitually on material it has already encountered, even when it imaginatively projects that material into new configurations. This form of resonance is less conscious because it no longer directly resonates with the immediate *unpredictable* source of experience but instead becomes caught up with the secondary, *predictable* traces that this immediacy has deposited in the recording apparatus of the brain. That is not to say we cannot have new or surprising thoughts, it is to say that such thoughts only tend to arise insofar as we have disidentified with our habitual thinking processes and entered into an attunement with that place from out of which our thinking is emerging. For the immediacy of now is our only point of direct contact with the unpredictable source of both our thinking and our perceiving.

The first and most accessible way of getting in contact with this source is to become aware of what is happening in immediate sensory experience, to hear the wind in the trees, to become fully attentive to the quality of the sound, to enter into the silence from out of which all sound emerges. The more our own inner commentary recedes, the more we resonate with this extraordinary sensory manifestation—this meeting of our stream of sensory attention and expectation with the ineffable source that transforms our empty projection into a direct experience of the being of that which lies beyond the enclosure of our machinery of projection and expectation. This is the bridge between our scientific model of the brain as a prediction machine and our immediate experience of being conscious. And this is how we make such science meaningful.

The Pre-Being of the Source

Let us see where we stand. We are letting go of the old materialism—that pervasive understanding of being within which we collectively live and go about our daily business. We are taking a path out of a former world, partly to illuminate the hold that world still exercises upon us, and partly to see where this path leads. We are disidentifying with our belief in material being—our sense that the world is 'really' made of the permanent being of 'physical stuff', of bits of matter suspended in space connected by fields of physical force. We are seeing that the reality for us, when we come to examine how it actually stands in our experience, is the *qualitative being of form.*

The location of the crack in our prevailing system of thinking lies at the heart of our scientific materialism, in the quantum theories of physical science. It is here that our materialism has *already* broken down—and yet, as a culture, we do not have the consciousness, or the will, to follow through the consequences of this breakdown. It therefore falls to us, as individuals, to examine these consequences, and to lay down a path out of this situation in which we are collectively coming to accept and act as if our being were the being of a biological machine—a robot whose behaviour is predetermined by the behaviour of microphysical particles—when we already know that the behaviour of these particles is *not* predetermined. For their behaviour emerges out of a place we call the quantum level and is ruled by something we call 'chance'. In the discovery of quantum effects, our physics has returned to the *source*. Just as it has always been, behind our pretence of knowing, we stand on the brink of a supreme mystery— the place of the origin of our own being, and of the being of the universe that we perceive. This place is determined by no law. We can only say that there is a certain predictability within its unpredictability. The quantum level is an *objectification* of this source. It manifests as the being of this moment. It is the source from out of which our experience is emerging now, and now, and now . . . .

If we look into this objectification of the source, what we find is a sheer *potentiality* for being. It is only here, on this side, in the stream of our experience that we encounter *actual* beings—atoms, molecules, cars and phones, oceans and skies. The source itself is not a *being*. That is why it is unthinkable. For as soon as we think of it we turn it into a being, into something definite. Even to think of it as a potentiality is to try and encapsulate it into a familiar form. And so we can only refer to it in terms of a negation—as the *unthinkable* source. For if we look, we must accept that we cannot see the source, we can only see what emerges from the source—and that includes *this*—the stream of our own thinking about the source. The source always precedes whatever we can think or say or experience. That is because it is the source of any and every thing that can possibly *be* for us, including our idea of the source as the pre-being of a pure potentiality.

Our first step—our first *positive* step—is to propose that it is conscious-ness that effects the collapse of quantum potentiality into the definiteness of actual experience. Once we let go of our materialistic preconceptions, this becomes a simple description of how things actually stand—because, for us, there is no 'other world' of material being, there is just *this* world of experiential being—this dimensionality of space and time and feeling and sense, and, within that, the being of the forms that we think and imagine and perceive.

This is not a monism of consciousness that says everything exists as an idea in a cosmic mind. To speak like that is to pretend to know that we know the source—and we do not know the source—we only know the experience that streams out of the source. And yet, within the forms that we perceive in this streaming (remembering that our perceiving it-self is part of this streaming) we can distinguish *aspects* of the source. For these forms that we are and that we distinguish have emerged from the

I cosmic in pure consciousness,
symbolic in reflective as analytic states.

source along with our experiential conviction that these forms actually *exist*. This conviction of actuality, of being, is our warrant to say: yes, this form expresses *something* of the source. We could even say that our experience of form is *symbolic* of the source. It enables that which lies *beyond* being to have an experience of itself *as* being. When we think this way we can begin to see the significance of the objective form that is our brain. For it is the behaviour of this special form that objectively reflects to us the manner in which our consciousness of the world is being manifested. This reflection is only *partial*—it shows what has already been done. It does not show (it cannot show) that which is manifesting what is manifest, the power that lies behind the projection, that makes it into an experience of the being of form, that collapses the pure potentiality of the source into something definite, into an experience of being amongst other beings.

At the centre of our experience of the being of form, there lies this extraordinary resonance, that is the foundation of our conviction that we are encountering a world that exists beyond the immanent sphere of our immediate subjectivity. We are informed by this resonance that the forms we are intending and projecting are in correspondence with something 'other', that they are not just forms and figments of our imagination, but that they correspond with the transcendental source of experience itself. If we try and understand the *basis* of this correspondence we find it turns into something enigmatic. To begin with, we cannot say that the form we perceive 'exists' in the source—for the source, as far as we can tell, both phenomenologically and scientifically, does not itself 'exist', it is the source of a potentiality for existence. And yet we know, in the direct evidence of our living together, that the world has a common underlying form. When I use up all the milk in the fridge on my cereal, we do not find that this is only true for me, and that for you there is still some milk remaining. It is because of this unity of form that we know there is a common source. Collectively we believe this source exists in the form of the being of a material universe. Now we have let go of this idea of materiality, we are faced with the enigma of a source that we cannot experience objectively, because it is the source of experience itself. We can only think of it in terms of metaphors. And yet we know from our shared experience of resonance that the unity of the world, its having the same form for each of us, must have its common origin in this source.

dynamic.
+ topological
Not mathematical, geometric

If we pare down the possibility of this resonance to a minimum, then all that is left are the pure mathematical relations that define the forms that are resonating. Such ideal mathematical form shows itself in the speech act we considered earlier: it passes from one medium to another, from your immediate consciousness, into a pattern of motor commands, into the waveforms of air molecules, at which point it could equally be transformed via a microphone into the form of an audio recording where the entire utterance becomes a single static pattern of audio frequencies. Pure form, pure mathematical relation, can travel like this from one system of dimensionality to another, *and back again.* Seeing this enables us to *glimpse* an *aspect* of what is happening in our experience of resonating with the source. For ideal form does not 'exist' in any place or time. And in whatever way we think of such form, we can only ever picture a representation, never the form itself (as illustrated by the equilateral triangle in figure 5.1). For the form is a pure meaning that cannot be manifested as a spatiotemporal experience. But, as a pure meaning, it can still be *intended*—and in such intending we open up another dimension of non-sensible ideality. This is this same dimension we already intend in our perception of the objective forms of the everyday objects we see around us (as discussed in Chapter 4). What we are proposing is that it is our intending of ideal form (such as the ideal form of the table in front of me now) that reaches back from the domain of our unique and personal streams of consciousness into the common source of our experience—and that this intending can then invoke the resonance of a response from that source. All this is occurring in the immediacy of our being conscious, and yet, at the same time, it is appearing 'out here' in the parallel symbolic form of the objective behaviour of the processes that are occurring in our brains.

Here we should be clear that this description of the occurrence of resonance is still a *model* of something we cannot directly perceive, something that is pointing toward the ideal form of the process of resonance itself. In the immediacy of consciousness, there is no corresponding division of 'what is' into a stream of experience and its supposed source. We can only separate this out in the aftermath of a conscious reflection. And if we go far enough back into this immediacy, then, perhaps, we could say that I *am* this source, just as I *am* the experience of whatever is happening now. Here there is the greater unity of what we might call 'the whole'. This is the undivided, unreflected unity that embraces the potentiality of the source *and* the being of experience.

At the same time, the resonance of perceptual consciousness testifies that our experience is not the unity of a one-way stream. For we project back into the source and receive a *response.* This projecting back and this response are the very ground of our being conscious, the very site of the emergence of an experience of the being of the pre-being of the ideal potentiality of the source. Form itself only appears in this relation of projection and response. It does not 'exist' *in* the source, for existence, as far as we can tell, is the actuality of the appearance of form in consciousness, the very consciousness we presume is effecting the quantum collapse. And yet we cannot finally separate this experience we have of the being of form *from* the source. For the source encompasses *everything*—both potentiality *and* actuality, both being *and* non-being. And we certainly cannot think of the objective form that is our brain as the place where this resonance 'occurs'. For the objective form of the brain is yet another form that appears out here on the *basis* of such resonance. We can only say that the brain is a *symbol* of the form of the resonance we immediately experience in our apprehension of the being of the world.

Notes

1. One of the most influential theories in this area is Karl Friston's proposal of the *free energy principle* (2006), which understands the brain as a device that minimises the unpredictability of its input by anticipating the form of that input.

2. The idea that brain processing in mammals is hierarchically structured is originally founded on Felleman and Van Essen's famous paper *Distributed Hierarchical Processing in the Primate Cerebral Cortex* (1991, Jan/Feb). Rao and Ballard went on to connect the idea of hierarchical structure with predictive processing in their paper *Predictive Coding in the Neocortex* (1999). Jeff Hawkins then developed a more unified idea (related to Friston's *Free Energy Principle*) of the entire neocortex being engaged in hierarchically-structured predictive anticipation of its own input (2004), from which he developed the computational model of *Hierarchical Temporal Memory* (2016). Interestingly, Hawkins has recently distanced himself from this hierarchical model, and now thinks the neocortex is composed of many tens of thousands of relatively independent processing centres (called minicolumns) that interact via multiple pathways that go beyond the structure of a simple hierarchy (the final answer here is still a matter of ongoing research). For more detail see Hawkins' book *A Thousand Brains* (2021).

3. Two of the most influential and widely read books on the subject of predictive processing in the brain are Jakob Hohwy's *The Predictive Mind* (2013) and Andy Clark's *Surfing Uncertainty* (2016).
4. See Eben Alexander's book *Proof of Heaven* (2012).
5. 'Plants and birds and rocks and things' is a line from the song *A Horse with No Name* by the folk rock band America.
6. For a full explanation of this concept see David Chalmers' paper *What is a Neural Correlate of Consciousness?* (2000).
7. Recent findings on the relation between consciousness and the synchronous firing of neurons are reviewed in Christof Koch's multi-authored paper *Neural correlates of consciousness: progress and problems* (2016).
8. The Einstein, Podolsky and Rosen paper *Can quantum-mechanical description of physical reality be considered complete?* (1935, May) was the spur for the quantum physics research community to devise experiments to demonstrate whether or not nonlocal quantum effects actually occur. This became possible once John Bell (1964) developed his famous theorem which proposed testable empirical consequences of nonlocality. Subsequent experiments have consistently upheld that such nonlocal effects do occur between entangled particles, so that the majority of people working in the area now accept the reality of what Einstein once called "spooky action at a distance".

CHAPTER 12

THE SHAPE OF CHANGE AND ORDER

THREE CATEGORIES OF BEING

IN our consideration of the being of form, three broad categories have emerged. The first is the intrinsic being of consciousness. This is revealed in the indubitable knowledge that consciousness has of itself, that it is (that I am) *conscious*. This knowledge is not the knowledge of the being of a form, it is direct knowledge of being the being that *manifests* the being of form. This, for us, is not only the ground of being, it is the ground of there being any 'other' being whatsoever. For what could it mean for something to be, if there were no consciousness of that being? We could imagine such being, like we imagine the moment of the Big Bang. But such imagining only conjures up the being of an imagined experience. It does not make what is imagined into an actuality. And yet within the being of the unity of consciousness there lies the phenomenon of reflected consciousness, the experience of being an 'I' living in the midst of other 'I's. In this place I become one who sees another, who, at the same time sees *me*. In this reflection of seeing and being seen, I come to understand that I am a being amongst other beings, each of us having our own experience of the intrinsic being of consciousness. And in being seen, I see myself, and in seeing myself, I come to know that I am a separate being, and that there is 'being', and that there are *other* beings, and that my being is also a being-here-together with these other beings.

Actuality = shared patterning - relational change uniquely transforming interactive states.

It is on the basis of this recognition of the being of the consciousness of another (generally beginning with my *mother*) that we are able to perceive the being of any other form whatsoever. This possibility manifests in our secondary perceptual consciousness of the being of the objective forms we distinguish in the world around us. For here, instead of experiencing our constantly changing and never repeating stream of experience as a constantly changing and never repeating stream of experience, we distinguish the presence of enduring forms. And we experience these forms as enduring in the same way that we endure—as having an unchanging identity in the midst of a continual process of transformation. And we recognise the presence of such form by projecting an expectation of enduring form into whatever it is that lies at the source of the streaming of our sensory input. And if our expectation is 'on target' we experience the resonance of there being an actual being present before us.

Otherness

This is the *direct resonance of perceptual consciousness.* It presents to us a world that is fundamentally structured by the form of our human embodiment. For our capacity to perceive separable objects located in space ('plants and birds and rocks and things') has evolved on the basis that these are just the kind of things we can pick up, and manipulate, i.e. with our *hands.* As we discussed in Chapter 4, our primary mode of 'thing-experience' is the perception of *affordances*—we first of all see what we can *do* with the things around us, what they are *for*, not what they might be 'in themselves'. We perceive the forms that we are able to predict using the apparatus of our nervous systems and our nervous systems learn how to predict according to the manner in which our physical bodies are able to interact with the sea of sensorimotor signals in which they are immersed.

What emerges is the form of a *world-as-experienced-by-a-human-body.* Every 'thing' is experienced relative to the size of this body, according to what it can grasp, what it can change, and how far ahead it can predict. Our immediate sense of agency is the agency of this body, of how it moves, of how it thinks, of how far away things are from its immediate grasp. Time and space, far from being purely abstract dimensions, are primarily the way we experience this phenomenon of graspability—for the more graspable something is, the closer it is, until we reach the immediacy of the body itself, where the distance reaches zero, so that we experience actually *being* this body.

Thirdly there is our consciousness of those entities that science has discovered that lie beyond the immediate reach of our sensory perceptual systems. Here lies the world of atoms and molecules, of stars and supernovae, of the quantum level and the living cell. These are not 'things' that we can grasp with our hands. And yet neither are they pure mathematical ideas, like the number three. They are the forms that we have reasoned must underlie the forms we directly perceive. Instead of relying on our eyes, we have developed technologies that enable us to see further and in greater detail. This begins with magnifying devices, such as telescopes and microscopes, through which we see directly, but soon diversifies into electronic sensors that report on data streams we can only visualise indirectly. And then we develop models and hypotheses about the being and behaviour of the forms we infer to exist on the basis of these technological extensions of our senses—and we express these findings in the language of mathematics—for we no longer have the direct resonance of perceptual experience to manifest our expectations in the dimensionality of sensory quality. But, insofar as our science remains empirical, we have ways of connecting our theoretical knowledge of the mathematical form of the objective universe back to the immediacy of our perceptual consciousness. For instance, we can reenact experiments that provide perceptual evidence for the existence of our theoretical mathematical constructions—such as the evidence of quantum indeterminacy provided by the double-slit experiment.

Originally, in the mathematics of the scientific revolution, the forms we inferred to exist inherited the same kind of 'thing being' we grant to the things around us, like the tables and chairs and rocks and stones. And so we envisaged pebble-like 'atom things' made out of atomic 'matter' and we devised a mathematical physics that described the velocities and accelerations and momentums of these separable and movable bits of matter. But this Newtonian view of the universe broke down in the twentieth century. The idea that a perfected physics could start with the basic laws of motion and then mathematically predict the entire course of the evolution of the universe had to be abandoned. The differential equations of that earlier science are still useful for predicting the behaviour of certain relatively simple systems, but, as our scientific horizons have expanded, it has become clear that most of the events in the world that matter to us have a level of complexity that makes their long-term behaviour *inherently and irreversible*

unpredictable. This is not just a matter of quantum indeterminacy. It is a matter of the combination of *complexity* and *feedback.*

Behind this lies the deeper recognition that our old understanding of the being of the universe as something that can be explained in terms of the permanent being and interaction of 'bits of matter' is simply incorrect. It is not easy for us to accept this, because the idea of the underlying being of permanent form is built in to our perceptual systems. It is what we both project and expect, and consequently find confirmed in the forms of our sensory experience. Only, of course, if we are patient, and wait long enough, we find that the things around us are *not* permanent. Our tables and chairs were made in factories out of trees that grew from seeds, and finally they will disintegrate or be destroyed and recycled into some other form. The wood itself will finally decay and be returned to the soil from which it emerged, and its molecular form will disintegrate into its constituent atoms which will then recombine into other compounds. And even these atoms are not permanent—they too emerged in the process of the formation and disintegration of what we think of as the stars. Finally, if we look for permanence in the being of the sub-atomic particles that form the atoms, we find our notion of permanent objective being dissolves into the wave-particle duality of quantum potentiality.

Dissolve into fields

And what of ourselves, our bodies, our sojourn here on Earth? Can we possibly believe that our lives are going to continue indefinitely? And yet that is what most of us, most of the time, actually do believe, at least in the background, in the way we lead our lives and take our being here and the being here of those around us for granted. It is in this unconsciousness of our mortality that we encounter the foundations of our modern attachment to scientific materialism. For that materialism presents a mathematical understanding of being where there really do exist permanent (undying) mathematical forms that describe the being of the universe. In this objectified understanding that we inhabit whenever we reason and reflect 'logically', there is no personal death—for we step outside the time in which we actually live and die, and contemplate it in the aspect of its ideal form. This is what Thomas Nagel called "the view from nowhere".[1] From this place it really does seem as if we are immortal gods contemplating the universe and the being of our little temporal lives with supreme clarity and indifference. This too is the *lure* of philosophy: to so deeply inhabit the place of contemplation of abstract form that we lose touch with the reality of our being here on Earth. And yet, of course, we always return to this everyday life, where we must eat and breathe and maintain our bodily being. The only permanence we find in this stream of experience is the permanence that we intend in the ideas that we contemplate. We once projected this permanence into the being of physical matter manifesting as a material universe, but now we know differently. Our experience only tells us of temporary stabilities of form that manifest the temporary being of the things around us. We live and we die. We emerge from an unthinkable source and we return to that source.

Dynamical Systems Theory

If we look at this field of temporary stabilities both in the manner of its changing and the manner of its staying the same, the old mathematics of points of mass existing in fields of force no longer applies. That kind of science relied on us abstracting away all the detail and complexity of the world in order to uncover its 'ultimate' underlying forms and the predictable regularities that govern the behaviour of those forms, such as 'A causes B'. The guiding idea was that the underlying microphysical forms were the ultimate causal agencies that determined the underlying regularities of the larger scale events. This notion that the behaviour of the whole is completely determined by the behaviour of its parts still rules over our contemporary science. What has changed is the recognition that the behaviour of larger scale complex systems cannot *in practice* be predicted on the basis of the behaviour of their parts.

The standard example of such complexity is our inability to reliably predict the weather. This is not a matter of insufficient data. It is to do with processes of feedback that operate within the weather systems themselves. Feedback occurs when the consequences of a change in one part of a system return to have an effect on the behaviour of that same part. For example, if there is an increase in temperature in the earth's atmosphere (perhaps because of a change in solar radiation) this will increase the rate of water evaporation, which will increase the amount of water vapour in the atmosphere. Because of the greenhouse effect, this higher concentration of water vapour again causes the atmospheric temperature to rise, which in turn will cause more evaporation, releasing more water vapour, resulting in a further rise in temperature, and so on. Here it is atmospheric temperature that is caught in a *positive* feedback loop, so that if it increases, this increase, in itself, causes further increases, so that the temperature increasingly diverges from its initial state. Of course, this does not present us with the full picture, otherwise all the water on Earth would have already been transformed into vapour. What makes the weather complex is the presence of other feedback loops that interact with this temperature loop and stop it reaching a final stable state of complete evaporation. Here we immediately think of the loop whereby excess water vapour forms into clouds that then deposit water back onto the surface of the earth where it is once again turned into vapour that forms into clouds, and so on. In relation to atmospheric temperature this 'rain effect' is an example of *negative* feedback, because, if the temperature increases, the rain will act to remove water vapour from the atmosphere and so reduce and stabilise the tendency of the temperature to carry on rising.

When feedback loops start to interact with each other within the same system like this, the problem of predicting how the system will evolve over time becomes increasingly intractable. Here intractability means we are unable to find solutions to the mathematical equations that model the entire feedback situation. This has led to the development of a new branch of mathematics in the twentieth century that includes chaos theory, complexity theory, and dynamical systems theory.[2] These approaches, instead of trying to perfectly predict the behaviour of inherently complex systems, are primarily concerned with characterising the overall behaviour of such systems. For, just because a system exhibits complex non-linear dynamics, that does not mean it is completely unpredictable. It turns out that most complex systems have regions of reasonably predictable stability and regions of inherently unpredictable (or chaotic) instability, and that we can characterise these regions by observing how changes in different quantities (like atmospheric temperature) affect the systems's overall stability.

What is interesting for us, is that this new mathematics has developed a way of representing the *form* of the *behaviour* of an entire system. Until now, we have been using the general notion of a 'form' to refer both to the shape of a 'thing' such a table or a chair, and to refer to the form of a process, such as the form of the brain processes that are required for you or I to have an experience of being conscious. While it is easy for us to visualise the form of a table, even a 'table in general', it is not at all clear how we can conceive of the form of something as complex as a brain process. Of course we can give a verbal definition, by listing all the necessary conditions that the process fulfils, but it is still hard to see how such a formal specification could itself have a 'form' or a 'shape'. It is here that dynamical systems theory comes to our assistance.

Dynamical systems theory begins with the idea of a system of interest. From the perspective of mathematical objectivity it does not matter why, or on what grounds we have decided that something qualifies as a system, what matters is that we distinguish the boundaries of that system with unambiguous clarity. So, for example, we could define the weather, as a system, to encompass the behaviour of the earth's atmosphere. Then we would have to define where the atmosphere ceases and the surface of the earth and 'outer space' begins. In drawing such a line we also define what lies outside the weather system, like the oceans and rivers, and the forests and marine life. These become the environment with which the system interacts. The fact a system has such an unambiguous boundary already implies it possesses an underlying invariant form of its own. In each specific instance, this form will show itself in how the system is structured, and, if we consider all the possible ways it can be structured, we arrive at the idea of the general form of a system's *organisation*. This organisational form is what determines the being or identity of the system, such that, if a system no longer embodies its original organisational form, then it ceases to exist. Such inherent form is defined in terms of a system possessing parts or components that are interconnected in such a way that they form an interdependent whole, where each part or component can be considered to be a system in its own right. In this way, systems theory (which includes *dynamical* systems theory) can be seen as a general theory of the form of all objective forms.

Dynamical systems theory is a branch of systems theory that deals with how a system evolves or changes over time, by modelling the mathematical forms of temporal processes. It does this by representing the state of a given system using a set of real number variables. These variables measure aspects of the behaviour of the system that are considered relevant in determining its future behaviour. At any given moment in time, the values of these variables define a point in a mathematical manifold known as the system's *phase space*, in which each variable measures a dimension of that space (in the same way that we graph numerical information using the horizontal and vertical dimensions of the graph's axes). Each point in phase space represents a possible state of the overall system at a particular moment in time. As the system changes, so do the values of the variables that measure its state, and as these values change they define new points in the system's phase space. If we keep on observing and measuring the system at regular intervals, and transpose these measurements into phase space points, and then join together each successive point with a line, the line will trace out the *trajectory* of the system as it evolves over time.

This can be illustrated using the relatively simple example of the oscillation of an idealised pendulum—i.e. one that, after being set in motion, continues to swing back and forth, making oscillations of exactly the same magnitude, in exactly the same time period, in a single plane of motion. Of course, in the world of actuality, if you do set a pendulum in motion, it will lose energy to friction, and gradually slow down to a stop. But, as Galileo observed in 1602, even as it slows down, it still takes roughly the same time to complete a single swing. This led him to think of using this invariance to construct a device to measure time (i.e. a clock), although it was not until 1656 that Christiaan Huygens invented the first working example. Then, in 1657, Robert Hooke invented the anchor escapement in order to further stabilise the regularity of the pendulum's swing.[3]

In the example anchor escapement shown in Figure 12.1a, the axle of the toothed wheel would have a string wrapped round it, attached to a weight that causes the wheel to turn as the weight descends. This movement, combined with the oscillation of the pendulum, causes the wheel to give a small impulse to the pendulum at the end of each swing. In this way the pendulum is kept moving in a way that approximates to our ideal pendulum, i.e. it keeps executing identical swings, both in terms of the time it takes to complete a swing and the distance the pendulum moves. If you visualise the system in operation, you will see that the teeth in the wheel actually stop the pendulum from travelling too far, thereby creating the familiar 'tick-tock' sound at the end of each swing. To complete the invention it was found that the time period of a pendulum is most reliable when it is constrained to swing through an angle of between 4° to 6°, and at such angles it needs to be 994 mm long in order for a swing to last one second. This explains why a grandfather clock is so tall.

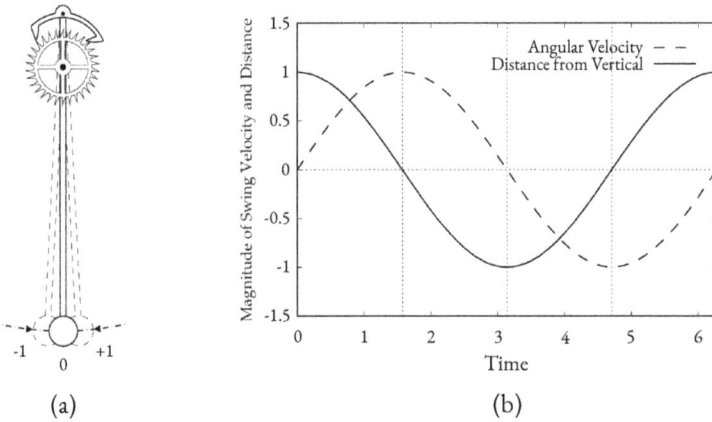

Figure 12.1: Left: A clock pendulum showing an anchor escapement. Right: Time series representing an idealised pendulum passing through a full swing and returning to its initial position.

We can represent the behaviour of an ideal swing (i.e. the one that a pendulum clock is attempting to recreate) as a dynamical system by measuring how *far* the end of the pendulum travels along its arc of movement (angular distance) and how *fast* it travels along that arc (angular velocity). In dynamical system terms, velocity and distance become the two variables that measure and determine the behaviour of the pendulum and so define the dimensions of the pendulum's phase space.

If we graph these quantities against time and normalise them to vary between -1 and +1 (where the vertical axis measures *both* distance *and* velocity and the horizontal measures time) we see (in Figure 12.1b) that they trace out two offset but otherwise identical wave forms. Here the velocity wave reaches its maximum (of either -1 or +1) when the distance wave is at a minimum (of 0, which represents the pendulum's vertical mid-point), and reaches its minimum (of 0, because the pendulum temporarily stops at the top of each swing) when the distance wave is at its maximum (of either -1 or +1). Note that here the + and - signs are used to indicate where the pendulum is within its cyclic movement, i.e. a positive velocity means the pendulum is swinging from right to left, and a negative means it is swinging from left to right, and a positive distance means the pendulum is to the right of the mid-point of its movement, and a negative means it is to the left.

If we transpose this information into phase space we obtain a *phase portrait* of the pendulum's behaviour. This portrait can be visualised by plotting the velocity and distance values for each time instant on a graph that only has velocity and distance axes (see Figure 12.2a). In this case time, instead of being represented on another axis, is represented by joining together each velocity, distance point in time order (as indicated by the arrows on the graph). This produces a line that describes the trajectory of the system through time, which, because we have normalised both quantities to vary between -1 and +1, turns out to be a perfect circle. This circular trajectory tells us that the pendulum is in a *stable state*, because it keeps repeating the same predictable movement. The circle is the *signature* of this repetitive stable movement.

In contrast, if we consider a more realistic pendulum, i.e. one that loses energy to friction, and so finally comes to a halt, we obtain a phase portrait in the general from of a spiral (as shown in Figure 12.2b). Here the 0,0 point at the centre of the graph represents the stable *fixed point* in which all possible movements finally come to rest. In this example we are only representing the single trajectory that occurs when we let the pendulum go from a point +1 away from its mid-point, but we can continue to plot all the possible trajectories, as we let the pendulum go from +0.9 away from its mid-point, then from +0.8 and so on. In that case our graph would become completely filled with trajectory lines and so lose its visual intelligibility. Nevertheless, all these trajectories would still end up in the same stable point state at 0,0.

The modelling of an ideal pendulum is an example of the kind of classical Galilean science that is interested in finding the deeper underlying mathematical regularities in the behaviour of events. It shows how a pendulum *would* behave once we remove the effects of friction, and once we force the pendulum to swing in the same vertical plane. This is the same kind of thinking that Newton used in formulating his basic laws of motion, where he imagined how a body in movement *would* behave if no forces were acting on it—concluding, in his first law, that it would carry on indefinitely travelling in a straight line at the same velocity. The grandfather clock is an example of a technology that has emerged from such science: its whole principle of operation is to exclude the complexity of our immediate experience in order to create the predictability of a clockwork, linear system.

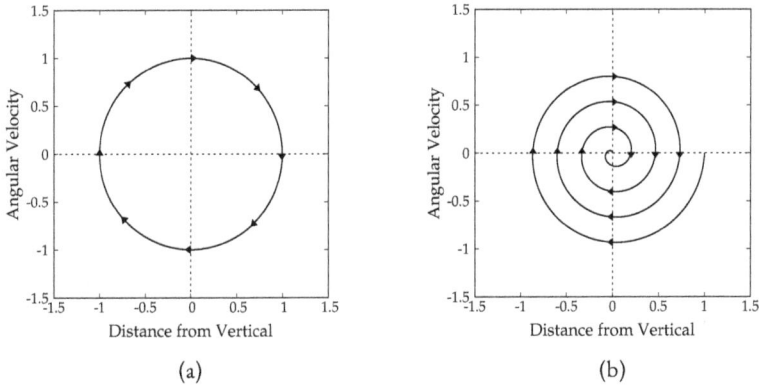

Figure 12.2: Left: Phase portrait of an idealised pendulum that swings continuously in a stable cyclic state. Right: Phase portrait of a pendulum that slows down due to friction and stops in a stable fixed point state.

It was only in the twentieth century that we started to reach the limits of such science, and to recognise that most of the processes we encounter in our daily lives cannot be broken down into underlying simple forms that can then be used to understand and calculate how a system will evolve. One of the significant examples of the failure of such thinking was the anticlimax of the human genome project. Here it was believed that human DNA was the underlying form that directly determines (i.e. *programs*) the structure and behaviour of the human body. And so it was thought that all we needed to do was to read this 'program code' in order to understand the causes and cures of most of the major human ailments. But things did not turn out that way. DNA is just one element in the system of a living cell. It, by itself, does not determine the behaviour of that cell, it rather engages in a complex *dance* with all the other cell components, a dance that does not follow a linear trajectory, like an idealised pendulum. It is a dance of feedback within immense molecular complexity. And yet the living cell *lives*. That means it exhibits an extraordinary stability in the midst of these continuous, unpredictable, non-linear processes of change.

It is here, in modelling the behaviour of such complex systems that dynamical systems theory is able to reveal the hidden structure that lies behind the surface unpredictability. A famous example of such structure is the *Lorenz attractor*. This mathematical form was discovered by the meteorologist Edward Lorenz in 1961.[4] He was modelling the phenomenon of rolling heat convection in the earth's atmosphere using a relatively simple system of three first-order differential equations. Rolling heat convection occurs when the sun heats air closer to the earth's surface faster than it heats air higher up in the atmosphere. As the hot air rises and the cooler air falls, the air can start 'rolling' and form long horizontal cylindrical cloud formations (similar effects can also be found in the heating of liquids). The equations that Lorenz used to model this phenomenon are as follows:

$$\frac{dx}{dt} = \sigma(y - x)$$

$$\frac{dy}{dt} = x(\rho - z) - y$$

$$\frac{dz}{dt} = xy - \beta z$$

Here dx/dt defines the rate at which the convective flow (x) changes over time (t), dy/dt defines the rate at which the horizontal temperature distribution (y) changes over time, and dz/dt defines the rate at which the vertical temperature distribution (z) changes over time. As with the distance and velocity variables for the pendulum, the x, y and z variables define the system's phase space, only now (because there are three variables) there is a *three*-dimensional space, where each point in the space represents a particular state of the weather system (i.e. in terms of the convective flow (x) and the horizontal (y) and vertical (z) temperature distributions). In addition, the system is described by three parameters: σ, the ratio of fluid viscosity to thermal conductivity, ρ, the difference in temperature between the top and bottom of the system, and β, the ratio of the width to height of the layer of air being measured. Once these parameter values are set, and an initial starting point is chosen, we can use a computer to calculate x, y, z solutions to the differential equations for increasing values of t. These solutions can then be sequentially plotted using a continuous line in a three-dimensional graph, providing a picture of the trajectory of the system as it evolves over time. Figure 12.3 shows one such plot projected back onto the two dimensions of the printed page. That means we need to visualise the third spatial dimension in which the two spiral shapes emerge like two similarly shaped butterfly wings, and in which no two trajectory lines ever actually cross.

If we were to see an animation of the way this system evolves we would see it start to cycle around one of the 'wings' of the attractor and then, seemingly randomly, jump and start cycling round the other 'wing'. This process continues indefinitely back and forth without ever exactly repeating itself. What is more, if we start the system off from slightly different initial x, y, z values, and compare its behaviour with the first trajectory, it will follow a similar but slightly different path for a while, and then the two paths will increasingly diverge. Even though the overall shape of this second trajectory will have the same recognisable two-winged form, when we compare the trajectories at the same time point of their evolution, they soon end up moving in different 'wings' and changing 'wings' after different periods of time, meaning their overall behaviour becomes increasingly dissimilar.

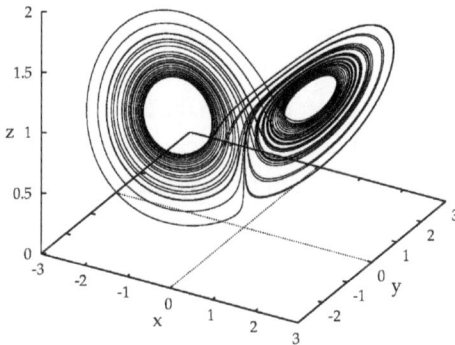

Figure 12.3: Phase portrait of a Lorenz attractor.

The phenomenon Lorenz uncovered came to be known as a *strange attractor*. This is where a system is *globally* stable but *locally* unstable. The global stability is represented in the characteristic shape of the attractor's two wings. If we consider the space of all the possible initial conditions for a given set of equations, there is a region around the attractor where all these initial starting points are 'pulled' into the attractor and exhibit the same basic pattern of behaviour without ever meeting or coinciding. These wing-shaped patterns mean that all these trajectories exhibit the same global form of movement, even though the precise location of a particular trajectory at a particular point in time cannot be predicted— except by solving the equations that generated the movement in the first place (hence their behaviour is *locally* unstable). In addition, however close and similar two trajectories are to begin with, that similarity will always, eventually, break down. It doesn't matter *how close* you make the starting positions, they will diverge at an *exponential rate* unless they are absolutely identical—in which case there are not two starting points, but only one.

It was from observing the behaviour of this attractor that Lorenz concluded actual atmospheric conditions exhibit the same kind of local sensitivity to initial conditions while exhibiting the same kind of global stability. That implies the weather is an *inherently* unpredictable system in terms of its local details, because even the smallest, most imperceptible differences in initial conditions will eventually result in entirely different trajectories for the system as a whole. This popularly came to be known as the 'butterfly effect'—partly because of the visual similarity between the Lorenz attractor and a pair of butterfly wings, and partly because Lorenz himself conjectured that the flapping (or not flapping) of a butterfly's wings in the Amazon rainforest could make the difference as to whether a tornado arrives in Texas a few weeks later (*all else being equal*).

The Lorenz attractor illustrates how an inherently unpredictable system can still exhibit an overall stability of form in its characteristic behaviours, so long as the system is not perturbed to such an extent that it leaves the region of phase space where it is pulled back into the attractor. This, of course, has direct implications for weather forecasting and ultimately for climate change. For we are facing the possibility that our human production of greenhouse gases will move us out of the phase space of the attractor that currently holds the earth's climate in relative stability. The problem here is that we cannot devise a system of equations that actually predict the evolution of the global weather system—to do that (because of sensitivity to initial conditions) we would have to model the system with infinite accuracy—meaning we cannot calculate exactly how far we can go in perturbing the system before it falls out of the region of its current attractor. All we can say, in a certain rough and ready way, is that the more unstable the weather becomes, the more likely it is that such a fall is going to occur.

ENTROPY AND SELF-ORGANISATION

We have already briefly considered the evolution of complexity in Chapter 7. But now we have the beginnings of an understanding of dynamical systems theory, we can look at this phenomenon in more detail. What stands before us is all the order and complexity and interconnectedness of the living and non-living systems that comprise what James Lovelock called *Gaia* (i.e. the *earth* system). And we are seeking to understand how all this complexity could have come to manifest from out of itself, so to speak, without the guidance of an overseeing conscious intelligence. Our first clue lies in the structure of the phase space of a complex system (like the weather). Such systems are generally composed of microphysical elements (such as air molecules) that, when considered individually, appear to be moved around quite randomly by chance interactions with their neighbours. And yet, when we consider the behaviour of the system as a whole, we recognise certain recurring patterns of stability—like the cyclic form of a storm, or the reliability of the trade winds, or the appearance of the tropical monsoons. Our question is, how are we to explain the emergence of such order and stability without invoking the presence of an overarching 'guiding hand'?

If we look again at the Lorenz attractor, and at the space of all the possible ways the system could have evolved, it turns out, relative to the size of the attractor, that there are much greater regions of chaotic behaviour, where the characteristic 'rolling' of the air currents does not occur. That means, if we were to start the system randomly, the most likely outcome is that it would begin in a chaotic place that does not immediately lead to the attractor. Instead the system would move from state to state without exhibiting any recognisable regularity—much as you would expect any disordered collection of tiny particles to behave as they bounce off each other.

So why is it that the weather (or the universe) appears to be so ordered? The answer, when viewed from a sufficient distance, is quite simple: even though there are relatively huge areas of chaotic randomness in the phase space of most complex systems, there are also attractors occasionally dotted about, that emerge, quite logically and predictably, as a consequence of the way the system is structured. So, if we consider all the possible ways a system could behave, there are going to be some rare situations, when the conditions combine in 'just the right way', where it is going to start behaving non-randomly. The great principle of the ordering of the universe comes down to this: once a system that is moving chaotically from one state to another 'bumps into' an attractor, it will become caught and start behaving repetitiously. For example, if a system has 999,999 unstable states and just one state that leads into a stable point attractor (like the spiral attractor of our pendulum), then eventually, perhaps after a thousand years, it will 'bump into' this one special state and then 'bingo!' that is the end of its random behaviour, not just for a thousand years, but for good, or at least until some other system comes and intervenes.

It is not that a system needs to be guided to a stable state. All that is needed is for both stability and instability to be possible within the same phase space. Once this is in place, the instability, in and of itself, will drive the system to find those rare and occasional states of stability, where it will then tend to remain. What we observe from the outside is really a dance between stability and instability. The remarkable phenomenon is not the occurrence of stability, it is that the universe is so delicately poised on the boundaries between these two regions, that it is neither caught in some completely static state, or propelled into such chaos that no stable structures are able to form.

Once again we meet the anthropic principle: the recognition that, even if there is no guiding hand behind the emergence of individual stabilities, it does seem as if the basic 'rules of the game' were set up very carefully so as to ensure that such stable complexity was capable of emerging. And again, our response is this: of course the universe is set up in this way, otherwise we would not be here (as the stable complex structures that we are) to observe it. For all those other universes where the balance between stability and instability was not quite right remain on the other side of the quantum divide as *possible* universes—universes that did not make it into the light of consciousness and being.

Nevertheless, what we have just portrayed is a highly simplified picture of the structure of the universe. To begin with, we have spoken as if the whole cosmos is continually engaged in this dance on the 'edge of chaos', whereas, when we look at the rest of the solar system, we find that the ordered complexity on Earth stands out as a rare and possibly unique exception to the general rule. What we actually observe are the presence of many stable, relatively unchanging systems existing in a state of near equilibrium with their environments. This begins with the stability of the atomic structures that underlie the material being of the physical phenomena we see all around us. *They* are not on the edge of chaos. Once we move away from the nuclear furnaces of the stars, only a tiny minority of these atoms ever actually disintegrate. The rest have already reached the stability of their respective attractors. But, as we move up the hierarchy of form and consider the behaviour of the molecular structures that these atoms comprise, things become more interesting. Even then, if we look at the surface of Mars, or the moon, despite a certain degree of complexity, and the enthusiasm of the documentary presenters, what we see is a fairly predictable and static equilibrium state, where unpredictable events certainly do occur, but also where the basic character of the situation remains the same for millions of years on end.

Figure 12.4: The surface of the planet Mars (courtesy of NASA).

What characterises genuine complexity is the interplay of *energy* and *entropy*. Entropy is a mathematical concept used in both physics and information theory. Informally it is a measure of the *disorder* or randomness present in a given system. In statistical mechanics this is measured by considering the number of possible *microstates* a system could be in that would give rise to a particular *macrostate*. For example, if we consider the macrostate of our just having dropped a sugar lump into a cup of hot tea, we create a molecular microstate where there is a very high concentration of sugar in one region of the tea and very little elsewhere. After a while a new macrostate emerges of the sugar having fully dissolved which is expressed in the microstates of the sugar molecules being more or less evenly distributed throughout the tea. In terms of entropy, we would now say that the first macrostate (before the sugar lump dissolves) had *fewer* possi-

ble microstates, which means it had *lower* entropy (i.e. *less* disorder) than
the later macrostate of it having fully dissolved.

The undissolved
sugar lump

One molecule
breaks free

All molecules
disconnected

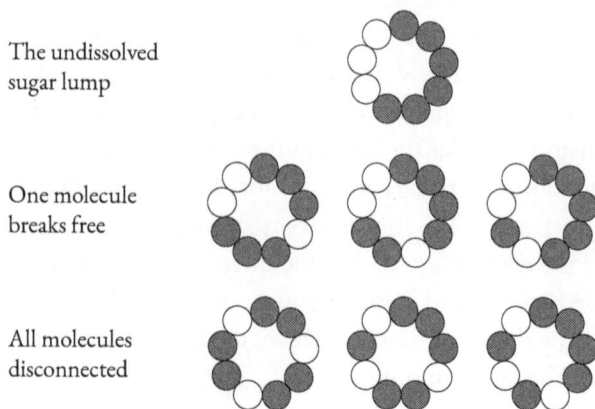

Figure 12.5: The seven possible ways of forming a ring out of three undifferentiated
white (sugar) molecules and six undifferentiated shaded ('tea') molecules.

To see how this works mathematically, consider an imaginary cup of
tea made up of three sugar molecules and six tea molecules that are connected together in a ring (forgetting for now that tea is composed of many
different kinds of molecules that do not form rings). Figure 12.5 shows
the seven different ways such a system can be ordered. The top ring represents the situation when the sugar lump is first dropped into the tea
and all the sugar stays connected together, and the remaining rings show
the various ways the sugar can mix with the rest of the ring. The seven
rings together represent all the possible microstates of the system, and, if
we assume each molecule can move freely in the ring, given enough time
and sufficient energy, each microstate will be as likely as any other (including the possibility of the sugar lump reforming again). From this we can
see that there is only one microstate associated with the sugar lump being whole, and six associated with its having fully or partially dissolved.
Therefore we can say (because 6 > 1) that the *macrostate* of the sugar being dissolved has greater entropy (or disorder) than the macrostate of the
sugar lump being whole.

So why should we care about entropy? The main reason is that it gives
us a quite definite and objective idea of what it means for one system to be
more ordered than another. In this way it provides a signature whereby

we can recognise the presence of ordered complexity in the universe. For there are many systems that exhibit great complexity in the details of their structure and in the unpredictability of their behaviour, without exhibiting a similar degree of order—such as the complexity of the relative positions of stones on a beach and how that alters as the tide comes in. What is of interest, in the context of our current inquiry, are those systems that manifest order *dynamically*, by using energy as a means of discovering and maintaining states that exhibit *greater* order. Such systems consume the disorder or entropy that is released from the available energy, to produce new lower entropy configurations. At the same time they export entropy into the surrounding environment, such that the overall entropy of the system and its environment actually increases. This property has been generalised in the *second law of thermodynamics* which says that the entropy of any closed system tends to increase over time until it reaches a state of maximum disorder, known as *thermodynamic equilibrium.* That implies order can only be produced in one region of the universe at the expense of creating greater disorder in another, and leads to the conclusion that the objective physical universe will eventually end up in a state of maximum entropy, or disorder, unless it is affected by some other system or entity that stands outside this law—such as, for example, the unthinkable source itself.

Ilya Prigogine[5] was a pioneer in the study of the relationship between energy and entropy. He was particularly interested in *dissipative structures* that produce and maintain low entropy structure in situations that are far from thermodynamic equilibrium. A canonical example is the formation of Bénard cells in liquids that are heated from below—such as a pan of oil heating on a stove. Initially the effect of this heating is distributed throughout the liquid in the form a continuous vertical temperature gradient that gets cooler nearer the surface. But as the molecules closer to the heat source gain more energy, they start to move more vigorously and so take up more space, which reduces their density, which reduces the effects of gravity, which makes them more buoyant. This buoyancy, relative to the cooler layers above, means the more energetic molecules start to rise, forming convection currents that displace the cooler liquid nearer the surface. At a certain point this convection behaviour 'spontaneously' *self-organises* into a collection of distinct hexagonally-shaped convection cells that maintain their identity and overall form by continuously cycling the hotter material from the bottom of the liquid to the top. This material emerges at the centre of a cell, where it flows across the cell's surface, radiating heat into the air, causing it to cool, until it reaches the cell's boundary where it flows back below the surface. Having cooled and become more dense it then continues to fall toward the bottom, where it again heats up, and repeats the cycle once more (see Figure 12.6).

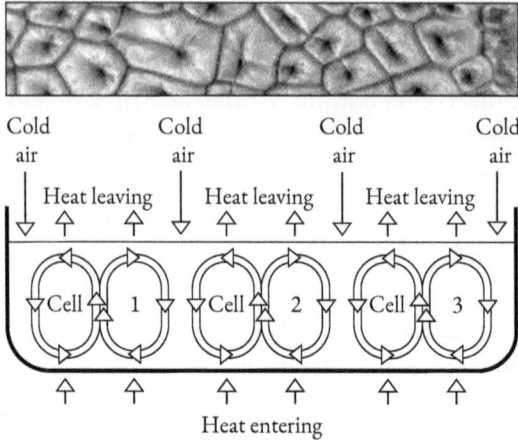

Figure 12.6: Above: Image of Bénard cells forming on the surface of heated silicon oil (University of Iowa instructional resource 4B20.50). Below: Diagram showing cross-sections of the convection currents.

As with our earlier discussion of Lorenz attractors and their relation to cloud formation in the earth's atmosphere, Bénard cells are an example of heat convection producing a complex phase space within which attractors are formed. In both cases, lower entropy ordered behaviour emerges from increasing the energy input into a previously disordered system. And in both cases the emergent ordered behaviour is sensitive to initial conditions and chaotically unpredictable in terms of the exact trajectory it will take in the phase space of its attractor. Such systems are classified as dissipative because the process whereby entropy is dissipated from the system (e.g. from the heat source into the excitation air molecules) is also the process that manifests their lower entropy form of organisation (e.g. the cellular rotation). Ordinarily, if we continuously feed energy into a system, the entropy or disorder reaches a maximum and stays there in a state of *thermodynamic equilibrium*. In contrast, a dissipative system manifests a lower entropy form of organisation by dissipating entropy into the surrounding environment, thereby maintaining itself in a *far from equilibrium* state. This phenomenon of the spontaneous (i.e. unplanned) emergence of low entropy order in a far from equilibrium system is known as *self-organisation*.

The exact moment in which the self-organising Bénard cell behaviour emerges can be represented as a *bifurcation* point in a bifurcation diagram (shown in Figure 12.7). Here the temperature difference between the bottom and the top of the liquid is fixed at various values and the subsequent final behaviour of the system is observed and plotted on a graph. What emerges is a bifurcation point at the *phase transition* where the liquid moves from one form of behaviour (random molecular excitation with no formation of cells) to another (cells forming, and starting to rotate). The bifurcation itself determines whether the cells rotate in a clockwise or an anticlockwise direction. After that the system remains in its 'chosen' state for as long as the temperature difference remains above the bifurcation point value. Finally, as the temperature difference increases, so does the speed of rotation of the cells, until they reach a maximum, which further increases in temperature difference cannot alter.

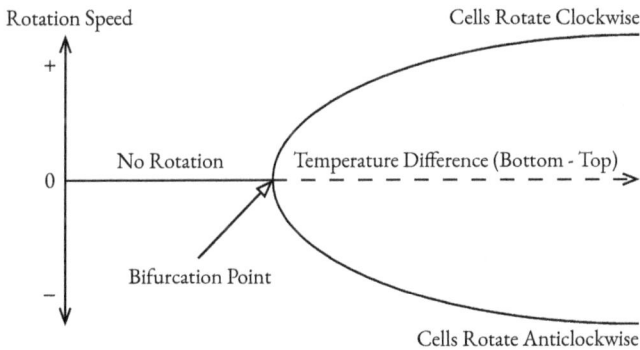

Figure 12.7: Graph showing the Bénard cell bifurcation point.

Bifurcation diagrams provide a simplified picture of the phase space, illustrating how the overall state of the system (e.g. its rotation behaviour) changes as the result of manipulating a parameter (e.g. the temperature difference). Detecting and measuring bifurcation points provides a map of the way the basic states of a system are connected and show how they may be controlled. In the case of Bénard cell convection, we can control whether the cells appear by controlling the temperature difference, but we cannot control the direction of rotation—this is determined by and is sensitive to the initial conditions in the system, and so, as with other non-linear dynamical systems, is inherently unpredictable.

THE INTENTIONAL UNIVERSE

If we put together all we have said so far in this chapter, then we have the beginnings of a *language of form*. It is by means of this language that we can speak more precisely about what distinguishes one form from another. And it is here that our science has opened a space of understanding that was not available before. For we can all see that something extraordinary has happened here on Earth, in the life forms we see around us, and in our own behaviour, our manner of living, in all we have created. But until now we have lacked a language with which to distinguish precisely what it is that *objectively* distinguishes the form of the living from the non-living, the plant from the animal, the animal from the human being, and so on. And despite the claims of modern materialistic biology, these questions have still to be definitively answered. In fact, we stand at a cross-roads where the meaning of our scientific discoveries stands in the balance. For the old scientific materialism is far from dead. It too has appropriated the language of form to extend its vision of a universal mechanism. And this is as it must be, for scientific materialism is a projection of being that determines, in advance, the meaning of whatever it discovers—the meaning of a kind of anti-meaning, or absence of meaning, that understands our existence as 'nothing but' the lawful mechanical interaction of microphysical entities that in themselves bear no meaning whatsoever.

And so it falls to us to *find* meaning in what our materialistic science has discovered. This is not a matter of invention. We are looking into the resonant source. We are placing the findings of science into the field of the immediacy of pure consciousness and asking after a response. It is from out of this science that our modern world has emerged, including our collective understanding of our place in that world. We cannot simply exchange this understanding for another one, and 'jump ship'. Such exchanges only affect the surface of consciousness. The deeper understanding of materiality remains. It is this understanding that rules the collective *phase space* of our manner of living together. For the mathematical concepts of dynamical systems theory do not just apply to physical systems like the weather, they also apply to the phase space of the forms of our human interactions. It is from within the phase space of our global civilisation that we form 'bubbles' of counter-culture and counter-thinking that oppose but do not fundamentally transform the attractor within which we are currently cycling. Such opposing phenomena are easily assimilated into trajectories that remain within this attractor, turning themselves into businesses, products, seminars, and books, slipping into the collective language, becoming political movements and therapies and 'spiritual practices' that are essentially *technological*, i.e. that embody objective procedures that are designed to produce objective results. Even the ancient practices of meditation and yoga are now marketed as ways to reduce stress and increase well-being. One only needs to live in Northern California to experience how the inspiration of the counter-culture of the 1960s has been drawn into the attractor of the system of corporate capitalism.

Our collective task is to find a way out of this current 'basin of attraction' that does not lead to the chaos of a complete disintegration. At the moment we are working on the understanding of being that structures the overall form of this basin. We are not setting out to destroy this understanding, but to see where it leads, in and of itself. For our science already contains the seeds of its own overcoming. These are the seeds we have been examining, firstly in the meaning of quantum physics, and now in the meaning of the mathematics of form itself—the very mathematics that originally transformed our medieval understanding of the being of a sentient universe into the mind and matter dualism of Descartes.

This mathematics of form is not just an abstraction in the minds of the individuals who have developed it. It tells us about aspects of the source that were entirely hidden to us before we starting thinking and inquiring in this objective mathematical way. What we are doing now is putting aside our assumption of materiality, our assumption that the forms science has discovered have some kind of independent, unchanging material existence that is devoid of meaning. We have begun with the recognition that the atomic and sub-atomic forms we once thought made up the foundation of this materiality, in fact only 'pop' into existence insofar as they manifest in consciousness—otherwise they remain in a state of quantum potentiality or pre-being that symbolically represents the unthinkability of the unthinkable source. And yet, if we look into the world of objective form that science has discovered, we find a remarkable fact: that the being of all the physical forms we objectively perceive to exist in the world around us, the tables and chairs, the birds and the trees, the clouds and the oceans, even the wind, is ultimately founded on the being of these atomic and sub-atomic forms that emerge from out of the quantum potentiality. Despite how it may seem to us now, the being of these atomic forms is not at all obvious. The Greeks conjectured that all may be atoms in the void, but they did not believe it collectively. It was only in the last two hundred years that we came to *know* of these forms and their specific structure. But what we did was to think of them as having a greater reality than the higher-level forms they comprise—such as the oceans, and the rocks, and the plants and the animals. That is because we thought of them as being made of an eternal unchanging material that is then merely combined to make all the other changing and impermanent forms we see around us. But now we know there is no such permanent stuff. And that means all the other forms are just as 'real' as the atomic forms. All is founded on something else, an unthinkable source that does not have a definite material being.

We need to absorb this quite deeply. What we are saying is that the form that is enacted when you run down the street, the form of your running, is just as 'real' as the form of your body, and the form of the atoms that make up your body. For an atom too is a kind of process, a configuration of vibrational energy that only manifests as something definite when it is observed and measured. And like your running, this energy traces out a relatively invariant pattern of activity, which manifests (for example) in the effects of its electrical charge. Of course there are important differences between the form of running and the form of an atom. For there is a *hierarchy of dependency* in which our ability to run depends on the form of our body which depends on the form of our atoms, whereas our atoms retain their form whether or not there is a body running. But what we have done, in our scientific materialism, is to think that this dependency means the *being* of our body and of our running can be *reduced* to the being of these atoms without remainder, so that everything about the being of that body and its behaviour can be explained entirely in terms of the behaviour of these atoms. For example, if we look only at the energy transfers that are involved in the activity of running, this *can* be reduced to, and explained in terms of fundamental forces that operate at the atomic level. But (of course) this does not explain my conscious *experience* of running. And when we reverse things in this way, and observe from the place of immediate consciousness, it seems as if the atomic forms have *less* reality, *less* being than our bodies and our behaviours, for these behaviours are directly manifest to us, both in our outward perception and in our inward experience, whereas we can only indirectly infer the being of atomic forms on the basis of experimental evidence.

For now we are going to remain neutral on this question of the relative reality of the being of the forms we perceive or infer to exist. Instead we shall *wonder* at this being of form. To do this we shall look again at the hierarchy of atomic dependency from the perspective of its cosmic origin. If we put aside our materialism and think purely in terms of the being of form, the great encompassing form for us, as the being of life on Earth, is the manifestation of the *solar system*. Of course, this system is embedded in the greater system of the galaxy, which in turn is embedded in the all encompassing system of the universe itself. But really we still hardly know what a galaxy is, let alone the universe in its entirety. We have a story about the Big Bang that was only devised a few decades ago, and now we have just realised that perhaps 90% of the universe is made up of dark matter and dark energy. The only thing we can be fully confident of here is that these theories of universal and galactic origin will be replaced by better ones just as soon as we can make some more accurate observations. Of course, everyone will be surprised that these current theories are wrong. And even though they won't be completely wrong, we can't be sure which bits are wrong and which bits are right. All we can say is that our models are making good predictions given the observations we have made up to now.

But back inside the solar system, we now know a lot more than Galileo. If there is life on Mars, it is well hidden. And the same goes for the other planets and moons. What we do know is that the forms of the atoms that we find on Earth are the same forms we find on the other planets. And we also have a good idea of the kinds of processes occurring in the sun that produce at least some of the atomic forms that we encounter. The further out we go from there, the hazier it gets. Certainly it appears that our solar system has formed out of clouds of interstellar gas that contained atomic forms that were created in extremely high energy events—such as the after-effects of the Big Bang and the formation and destruction of other stars. These were all cosmic events, i.e. galactic and universal events that occurred outside the domain of our solar system. And that makes the atoms into cosmic forms. They are what preceded the formation of the solar system, and they are the ground from out of which it has evolved. We can imagine all these atomic forms, largely consisting of hydrogen and helium, with scatterings of other elements thrown out in the explosions of ancient supernovae, distributed randomly in space, with the energy of gravity starting to work on pulling them together into the form of a kind of cosmic weather system. What starts off as high entropy disorder is then compressed by gravity into a massive centre which starts to radiate entropy in the form of electro-magnetism as it ignites into a process of nuclear fusion. Here the crushing energetic intensity of the gravity starts the process of transforming hydrogen into helium. It is the very extremity of the level of this energy that gives the atomic form its cosmically scaled invariance. For it is an effect of the stars, and can only be made and unmade in similar conditions—unless of course the atomic form itself has become unstable as a result of growing too large.

Figure 12.8: The formation of the solar system (courtesy of NASA / JPL-Caltech).

Already, in this cloud of solar origin, atoms have combined into molecular forms. And, as the sun continues to grow, it sucks in the lighter hydrogen and helium, leaving the remaining molecular 'impurities' to start cycling round the atomic furnace, creating a disk of rock and ice and gas, from out of which, through multiple collisions and accretions, gravitational centres begin to form: centres that are not massive enough to generate the energy of a star, but can still attract and crush together a body within which processes of molecular transformation can occur. These, of course, are the beginnings of the planets. And already, a certain division of responsibility has occurred: with the star being the centre of atomic transformations and the planets being the centres of molecular transformations. At the same time, the planets are utterly dependent on the solar centre as the source of energy by means of which they can piece together their molecular creations. If we stop thinking of this as a process of mindless and accidental collision between bits of insensate matter, and consider the system as a whole, we can begin to ask what this entire process might *mean*. For, as we know, it did not just stop with the formation of lifeless rocky and gaseous and frozen planets and moons, it produced *us*, the conscious intelligence of life on Earth.

When we shift our attention like this, and consider the form of the solar system as a whole, then something else emerges. For the solar system has an obvious unity: the unity of the sun's gravitational and energetic fields, and of everything that moves within those fields. From this perspective the sun is no longer defined by what we think of as its surface. Instead it stretches out and completely envelops us in solar gravitation and electro-magnetic radiation. If we could *see* these fields, they would extend and include all the forms that orbit around their centre. But we do not see this unity directly, instead we assemble a picture of the surfaces around us that have deflected a narrow portion of that energy. And so we see isolated things appearing in an apparent nothingness. That, of course, is as it should be. For we perceive what we need to perceive in order to maintain the identity of our form, i.e. to survive. And yet we still *feel* the enveloping warmth of this solar energy field, and we experience its gravitation in the regularity of our lives, and we know it to be the source of the light that illuminates our days. The shift we are making is to see through what we think of as the empty space that surrounds the earth and turns it into an isolated speck in the midst of a profound emptiness, and to recognise that we exist in unity with the sun, that the sun, in the form of the solar system, makes up a single energetic, interconnected unity, and that space is *not* empty, it is filled with pathways of influence—gravitational and electromagnetic influence—that connect us with the entire system.

⌐ and thermodynamic.

Here we could even think of the solar system as the 'body' of the sun, with the earth and the other planets as its 'members'.

At the heart of this system is the extraordinary phenomenon of *gravity. It* is the great attraction that pulls all this form together into such a unity that the original atomic structures break down and new form emerges. And in this breaking down and recreation, the attraction of gravity is turned into the light and heat, which irradiates the heavier elements that, in turn, are caught by this same attraction and compelled to orbit around its centre. Our scientific materialism cannot help but understand this process as something without meaning, without *intention.* But is that truly *logical?* For here we stand, as immediate demonstrations that there is meaning and intention within the solar system. On what basis do we deny the solar system a corresponding intentionality? Our science would respond that *our* experience of intentionality (of knowing meaning), is entirely an effect of our having brains that enact the kinds of processes that are associated with our being conscious—whereas the sun, having no brain that we can observe, cannot be conscious, and so cannot have an experience of meaning or intentionality. This indeed seems reasonable. The processes occurring in the sun, while generating vast amounts of energy and producing fundamental transformations of atomic form, nowhere near approach the complexity of the process that is a living cell, let alone of those that are occurring in my brain right now. But here again, by default, we are looking at the world through the lens of scientific materialism. The truth is we have no idea how brain processes are associated with an experience of intentional consciousness. We can only recognise the parallels between the forms of such processes and the forms of our experience. Intentionality itself, the experience of meaning, has its origin in the unthinkable source, as does our experience of observing and measuring human brains. It is not that the appearance of a brain *created* the intentionality of consciousness, it only *enabled* that intentionality to manifest as an experience of being in a world.

What we are saying is that the intentionality of consciousness as a *capacity* or *potentiality* is already latent in the pre-being of the source, just as the potentiality of the electron or the photon is latent at the quantum level. And if intentionality is potential in this way, then it does not *depend* on there being a brain in which it manifests. Of course a brain enables us to reflect and to realise that there is intentionality, but it does not create

TG A clearer triune process wholeness could have the feedback structure necessary for intentionality.

TIME ———→ Form as integration/comparison/reflection

Intention - 'aboutedness' = relatedness in feedback
The Shape of Change and Order   379.
sensitivity to self effect.

that intentionality, and neither is our reflection necessary for such intentionality to manifest. For who is it who means to do something—who means to write these words or read them? If we stop thinking and reflecting and actually look at what is happening in our experience, we find that each of our actions emerges directly from the source, without the interference of any personal decision maker. And if all our apparent personal intentionality is finally the intentionality of this source, the source that manifests the same world in each of us, then perhaps we can also say that *gravity* is something intended by the source, that it is not something physical and mechanical, but that it embodies a basic intention of atomic form, i.e. that it is *attracted to itself.*

The philosopher who first clearly distinguished this connection between our experience of intentionality and the universal intentionality of the source was Arthur Schopenhauer. He called this universal intentionality the *will*. His essential insight was to recognise that what Kant called the thing-in-itself—i.e. the true and essential being of the forms we perceive to exist—is not something from which we are cut off and that we can only experience indirectly in the representation of perceptual objects— for we have direct knowledge of the being-in-itself of one particular form, i.e. our *own form*. Schopenhauer thought that this direct knowledge is manifested in our experience of willing the actions of our body, in the very moment that an action is performed. From the perspective of phenomenology, we would say that Schopenhauer made a right distinction but drew his line in the wrong place. For it is not just in the act of moving our body that we experience intentionality. Our entire consciousness is intentional in its being an experience of meaning. So, for example, in perceiving a chair, I intend the chair to be an actual chair, something I can sit on, or pick up, that has a back that I cannot see at the moment, and so on. This intending is also a form of action, an action that manifests the chair—only I do not experience this as something I deliberately choose to do. In contrast, when I move my arm, or start to speak, I have the feeling that I am the author of these actions, that it lies 'in my power' to not move my arm or not speak, whereas I cannot help but perceive the chair (if I choose to look in that direction). But whether I experience my intentionality as voluntary or involuntary, it still possesses the form of a process that can be modified by changes in the structure of the brain, as in alien hand syndrome, when it feels as if your hand is moving by itself.[6]

TIME - FORM = feedback - active inference in enfolding/unfolding thermodynamics → spatial relatedness knowable.

Figure 12.9: Arthur Schopenhauer 1788-1860.

What Schopenhauer was saying is that in our immediate experience of being conscious we have direct knowledge of the source—of the universal intentionality from out of which the universe of form manifests. We *know* that intentionality, even though we do not know it as a being or form that manifests 'out here' in experience. We cannot describe it, or objectify it, we can only *be* it, and in being it we know it, just as we know what it is to be conscious. But this intentionality cannot be the 'me' that I experience myself to be in relation to 'you'—the 'me' who is the subject of my own particular stream of consciousness. For we know this source of intentionality is an essential unity, in that it manifests the same universal world form in you as it does in me. So 'I' can only 'be' this unity of the source insofar as I cease to be 'me' and fall back into the background from out of which everything emerges, *including* 'me'.

It is this 'sense of the source' that is pivotal to our overcoming the onto-logical pre-suppositions of modern science. Without such an intuition, we have only a mere abstraction, an idea of a mythical source without foundation in immediate experience. And if we are to hold to this in-tuition, we must understand that it disappears as soon as we attempt to subject it to the light of objective reason. For reason operates in a reflected world of ideality, breaking up our experience into distinct parts, separat-ing out one thing from another, whereas an intuition of the undivided unity of the source can only occur in the unreflected immediacy of 'now'.

It is an effect of the times, of our collective materialism, that we end up speaking this way, as if an intuition of our unity with the source were some kind of special experience, something rare and only available for the privileged few. But history teaches us that it is has taken several centuries of concerted effort to have more or less eradicated this natural human sense of being connected with a meaningful source. We look back now and think the people of the past were naïve and superstitious, victims of a corrupt priesthood, who used religion as a form of brainwashing in order to exercise their own form of tyranny. But such a picture is a poor caricature, something already distorted by our own materialistic presuppositions. For the power of the hold of religion on the minds of the people stretches back into the mists of prehistory and is not just the product of some clever act of persuasion. It rests in the direct sense of the earth and the universe and ourselves having 'come out of' a transcendental source. Our materialist science thinks it has disproved this idea through its discovery of the mathematical form of the material universe. But our materialism has not disproved our intuition of the source, it has simply denied it, by adhering to a procedure of mathematical objectification that renders the source *invisible*.

What has occurred is a *misreading of the story of form*. This misreading denies from the start that there could be any obverse side to the objective forms that mathematical science discovers. It is pre-assumed that what lies behind these forms is just matter, dead and lifeless and without intentionality. And this is assumed because the procedure of mathematical objectification is unable to find any evidence for the pre-being of intentionality in the universe.

Here we meet again with the same old absurdity. For intentionality is *exactly* what mathematical objectification cannot see, because, from the days of Galileo onwards, it has been deliberately and rigorously excluded in the very process of assuming an objective stance toward the universe. To be objective in this way *means* to eliminate the subjective, and the subjective (immediate consciousness) is *exactly* where the phenomenon of intentionality manifests itself directly.

But then, in the twentieth century, as we started to better understand the inner workings of the human brain, materialistic science finally recognised the 'problem' of consciousness. As a result, intentionality was superadded back into the universe, as a kind of special capacity or form of behaviour that emerged by chance in the unusual circumstances of the evolution of sophisticated nervous systems on the earth. Now, in the twenty-first century, it is finally dawning on a number of influential philosophers of mind that the prevailing materialistic orthodoxy is not rationally defensible.[7] Its absurdity, like the emperor's new clothes, is emerging from out of the complexity of the objective rationality that has sought so desperately to hide it. This has led to the re-discovery of *panpsychism*—the idea that there is a psychic component within all form and not just within the form of an animal's brain. But, like all reforms within the labyrinth, this one must make its way slowly and cautiously. And so it attempts to retain as much as it can of the original microphysical determinism. This is effected by adding a little psychic 'extra' to the old microphysical forms—so that they can combine together in an animal brain to produce a consciousness that is still determined by the interaction of microphysical-micropsychic entities. That way our materialism (despite getting a new name) remains essentially intact. For the little psychic 'extra' is no more than a kind of previously undiscovered basic force, one that was needed to explain the otherwise inexplicable presence of consciousness in you and I. With the addition of this psychic 'extra', the overarching meaninglessness of our materialistic outlook can survive—for it is still the case that we are ruled by chance and accident, and that our intelligence, our intentionality, is simply the result of the fortuitous (for us) combination of psychically 'charged' particles in our brains. To think the universe intended for this to happen would still be a very serious heresy indeed.

NOTES

1. *The View from Nowhere* (1986) was a book written by the American philosopher of mind, Thomas Nagel, where he examines what it means to take up an objective view of our being here in the world.
2. For an introduction see *Chaos and Dynamical Systems* by David Feldman (2019).
3. See *Origin and Evolution of the Anchor Clock Escapement* (Headrick, 2002).
4. Edward Lorenz first published the details of the Lorenz attractor in his paper *Deterministic Nonperiodic Flow* (1963). However, the original discovery of the phenomenon of sensitivity to initial conditions was made by Henri Poincaré in his 1892-1899 work *New Methods of Celestial Mechanics*.

5. Ilya Prigogine (1917-2003) was a Belgian-Russian chemist whose work on dissipative structures and irreversible processes had a significant influence on the development of complexity theory in the twentieth century. He wrote a number of books that made his work accessible to a general audience, including *Order Out of Chaos* (1984) with Isabelle Stengers.

6. Alien hand syndrome (Biran, Giovannetti, & Chatterjee, 2006) is a medical condition where people experience their limbs moving without their intending those movements. It is generally associated with some form of brain lesion. It is also known as Dr. Strangelove syndrome after the character, played by Peter Sellers, whose right hand and arm was out of control, executing Nazi salutes, in Stanley Kubrick's film of the same name.

7. Three contemporary philosophers of mind who have recently argued for panpsychism are Galen Strawson (2017), Thomas Nagel (2012) and David Chalmers (2017). Although this is still a minority view and largely ignored in the wider scientific community, the fact that such a position has gained serious attention and academic credibility is a sign of significant change in the mainstream debate on the relationship between mind and matter.

CHAPTER 13

THE PHENOMENON OF LIFE ON EARTH

FIRST PRINCIPLES

THE first principle for us is direct knowledge of consciousness. That means we are not concerned with explaining 'where' consciousness comes from, as if it were an object or a property belonging to some other thing (like our brain). It is what we are taking as given, as the ground from out of which we are to understand everything else. This is not something arbitrary. We start from the ground of our being conscious, because this actually *is* our ground, the only ground that we know directly and immediately. All else is something that *appears* in consciousness. But we need to be clear exactly what we mean by the word 'consciousness'. For, in our everyday language, we think of consciousness as if it were some kind of personal possession and consequently that there are many consciousnesses, and that they each get switched on and off according to whether we are awake or asleep, alive or dead. But this is only how consciousness appears when viewed objectively, from the outside, as if it were the property of an individual being appearing in a world of things (a world that itself only appears in consciousness). As should be clear by now, we are not speaking of consciousness in that sense. We are speaking of the very space by means of which everything appears, what Martin Heidegger called the *clearing of being*.[1] We are speaking of the consciousness that determines the collapse of the quantum level into the specificity of our collective world

?

experience. So this cannot be a *personal* consciousness. Insofar as we are human individuals, we too appear to ourselves within this clearing. But consciousness itself does not appear. It is the means of appearance itself, the same in you as it is in me. From the perspective of *what* appears, consciousness is nothing. And yet, from the perspective my being, consciousness is my source, it *is* my being, it is being itself. No - growth from foetus-baby

If we look at the stream of experience as it flows away in each moment, then consciousness is the means whereby this stream is illuminated, and it is the space in which it is illuminated, and it is the invisible, still point No of now, from out of which the stream emerges, from out of which it is Wrong 'thrown' into time, into the flowing past and into the expectation of a sequence future, into which I myself am thrown. And yet all this is always happening now—in the now that does not move, that does not even *exist*, for everything that appears to exist, exists within the stream, and the stream is already past. Even my projection of a future is already past, has already emerged from this source. Only 'I' remain, as the pure consciousness that knows this flow and knows itself as the 'still point of the turning world'.[2] No.

Despite our scientific materialism—i.e. our thinking of ourselves and Stative our consciousness as things emerging from and determined by the mate- in turnover rial configuration of the universe—and despite our thinking that such a and view is 'logical' and 'scientific', the *fact* is to think that way is *not* logical.[3] feedback We have direct knowledge of consciousness, and we only have indirect knowledge of the being of a material universe, and that only on the basis of being conscious. So it is *logical* that we start from consciousness itself.

When we do this, we do not *theorise* that there is an unthinkable source from out of which our experience of the being of the world manifests, we directly experience this to be the case. This is the second principle: that, in our being immediately conscious we also have direct knowledge of the source. It is the 'now' from out of which everything manifests, including my experience of being 'me'. The phenomenological word we use to express this 'action' of manifestation is *intentionality*. And at every level of experience, we know ourselves to be the source of such intentionality. This can be as simple as intending to switch on the kettle. But, as Schopenhauer saw, this intention is only fulfilled in the moment of actually pressing down the switch. Otherwise we are only *predicting* we will turn on the kettle, and our predicting is a quite different kind of intending to our actually acting. Here, *everything* we do or that happens to us is something we intend, including our perceiving and our thinking and our imagining and our feeling. Even the hammer dropping on my foot and the subsequent feeling of pain is something that is intended, for I *mean* that event to be a hammer dropping on my foot and I *mean* that feeling to be a feeling of pain. This is not a matter of choice. It is a matter of my entire living life being an experience of meaning, even when I experience it as being meaningless. This field of meaning, of intentionality, is the field in which we live in every moment, so we ordinarily do not recognise it for what it is. We cannot stop meaning, or stop intending the world to be the world that it is, but we can disidentify with our intending, and become the pure witnessing of unreflected consciousness, within which this process of intentionality is continuously occurring. Once again we are in a phenomenological reduction, where we can glimpse the truth that an essential characteristic of consciousness is that it is a consciousness-of that which it itself intends. Here the 'intender' of the world is no longer that egoic self we think intends the acts of our daily lives, for this *transcendental* intentionality is intending the very meaning whereby we experience ourselves to be that egoic self.

It is by means of this second principle that we can say the unthinkable source is the source of our intentionality—that it is the source itself that finally intends the meaning of everything we experience to be the world. And this too is something we can *experience*, insofar as we disidentify with our worldly egoic self and simply witness the impersonal movement of our perceiving and acting, as one thing follows another, which follows another, which follows another . . . . Here it is not a matter of ceasing to be a human individual, it is a matter of seeing what lies behind this experience of individuality. Once we have this 'feeling' of the source, of its being the origin of all our intentionality, then we can see it is the same source, the same intentionality in me as it is in you. We, of course, differ in our individual experiences. But the place from out of which these experiences emerge is not *something*, it is not a *being*, it is the unthinkable unity from out of which time and space and even the possibility of one thing being separate from another emerges.

So now, the question is, where can we go with this knowledge of the source that is entirely grounded in my immediate experience of being conscious? What has it to say about the origins of the solar system, for example, or the phenomenon of life on Earth? Here we must make a leap. The

[handwritten margin note at top: I prefer resonance as knowing with... Not a 'leap'. A relaxing into...]

[handwritten margin note at left: One I must not make!]

leap is to say that it is not just the same intentionality of the source in you and in me, but that it lies within all the beings that we perceive to exist around us, not only in the cats and the dogs and the other animals that appear to be conscious, but in the very fabric of what we take to be the material universe, in the atoms and molecules, in the sun and the earth and the solar system that unifies them. Here we must be clear that this is a leap. We have no direct experience of what it is to be a molecule or an atom or a planet or a star. All we have is the intuition of the unity of the intentionality of the source as it manifests our lives here on Earth. From this place we can only *reason by analogy*—like the analogy of the correspondence between the forms of our experience and the forms of the objective processes occurring in our human brains. If we take the analogy of this correspondence further, then we can start to infer the character of the intentionality that expresses itself in the being of the other forms that surround us. Here we cannot proceed in a purely scientific and objective fashion. For scientific objectivity cannot see the character of intentionality, as it is, within itself—it can only see the objective forms that intentionality intends. It is up to *us* to fill in the gaps that our mathematical objectivity has left open. It is not enough to simply say there is nothing there, no intentionality whatsoever, because our mathematical objectivity cannot distinguish it. We *know* by means of *empathetic* perception that the cat is conscious, for we literally experience its being conscious, in its looking at us and seeing us. Of course we do not know this with *objective* certainty, for we can never know the experience of anyone or anything but ourselves with complete certainty—but we have the objective evidence of the similarity of structure of the cat's brain and ours, and we have the direct evidence of our own experience. Between the two we can arrive at an understanding that remains true to both the science and our perceptual intuitions. But what of all the other animal species, and the plants, and the bacteria? And what of the form of the matter that comprises these organisms? *[handwritten: Know by responsive relational development... Conversational consciousness is non-local...]*

BRINGING THE SOURCE TO CONSCIOUSNESS

Our guide here is intentionality itself. For, if we are correct, then it is the same intentionality lying in the wellspring of our own being, that also lies behind the manifestation of the natural world. So you would expect there to be at least some form of correspondence. And given this wellspring is the source of intentionality itself (and not its reflection in an objectified world), the natural place for us to start is to inquire of ourselves, of our own experience of intending, and to ask what it is, *essentially*, that we intend in our being here on Earth. Here we are looking into the heart of what we have called the phenomenological reduction, at the very impulse within consciousness that drives it to manifest this world, this life, this being I am. Our methodology, in the overarching framework of this reduction, is to simply describe what we see. We are making the assumption that the intentionality of the source lies open to us, that it does not conceal itself, that what it intends *is* what it *manifests*—and what it manifests *is* this life, this world, this being I am. But what does it mean, to manifest? For us it means to *bring to consciousness*. It means to *knowingly apprehend* what lies before us. In Heideggerian language, we do not simply act and react, we 'comport' ourselves understandingly toward the beings that we encounter in a field of pre-existing meaningfulness. It is part of this comportment that we inquire, that we question, that we seek to make the beings we encounter more intelligible, more meaningful, more understandable—to reveal, to make manifest, to make conscious. But what is it that we seek to make conscious? It is what lies *outside* our field of immediate experience, but still remains *accessible*, in the sense of being *potentially* 'within reach'. And what is potentially within reach is the source itself—which is, for us, the very definition of potentiality. Accordingly, we can now say that our essential intentionality, the impersonal intentionality of the unthinkable source of our being conscious, is to bring that unthinkable source to consciousness.

Now, it is generally the case that whoever attempts to define the ultimate character of human intentionality, ends up characterising their own (personal) intentionality, or an aspect of it. Here we need only think of Nietzsche and his concept of the will to power or Freud and his concept of repressed infantile sexuality.[4] And so it is hardly surprising that someone like me, who is writing a book that is concerned with bringing the unthinkable source to consciousness, should conclude that the essential intentionality of humanity is a kind of philosophical 'will to consciousness'. At the same time, I can hear Nietzsche saying that this idea of a will to consciousness is really only a disguised will to power, and that the only reason we wish to bring what is potential to consciousness is so we can control and dominate it. The answer here is that we are not asking after what motivates our *egoic* human behaviour, we are asking after the intentionality of the source. That means we looking from the place of immediate consciousness at whatever it is that is manifesting in the moment of our looking. There is no thinking, and no reflecting. If something like a will to power emerges in this place, then it emerges as a phenomenon, an impulse passing through the stream of experience. It does not define that stream. What defines the stream is the intention that is manifesting the stream, not what appears as a being or an impulse within the stream. And so, to reiterate, our essential intentionality is to bring the unthinkable source to consciousness. The 'proof' is that this bringing of the unthinkable source to consciousness is what is actually occurring in any and every moment we care to check by looking again from this place of the immediate consciousness of a phenomenological reduction.

Our next step is to start reasoning by analogy. To do this we take the intention to bring the unthinkable source to consciousness and 'read it back' into the evolution of the objective forms we see around us. We are saying that it is the same source in us that is the source of the manifestation of the forms of the objective universe, and, consequently, that it is this intention of the source to manifest itself in consciousness that has driven the entire formation and development of the objective universe that science has discovered. This is our basic transposition: instead of thinking that the universe has no intrinsic intentionality and has formed entirely by accident, we are thinking that the universe is suffused with intentionality, and has formed entirely on the basis of a will to consciousness. And we are also recognising that we, the humanity of the earth, are the most developed expression of this intentionality that we know of, at least within the local region of the solar system. This is not a wish fulfilling projection of human subjectivity onto an inanimate collection of rocks and gas. It is a natural recasting of our place in the universe that emerges as soon as we relinquish our attachment to an irrational and unfounded materialism.

At the same time, this is still an act of faith: it is the same faith that moved me, as a young man at Warwick, to remain true, as far as I was able, to what had been revealed to me from this 'source within'. And again it is our *freedom* whether or not to 'go along' with this bestowal of meaning. For we are saying that everything that happens here, from the manifestation of a single photon, all the way up to the events that comprise our human history, is *intended* by this source that we are. It is perhaps easy enough to look at a sunset and see a transcendent meaning in the form and beauty we experience. But what of all the cruelty and suffering and deception that we have also manifested? What kind of source is it, we must ask, that could have intended all that? Of course these are old questions, questions upon which our former faith has foundered. And this is where materialism gains its converts. For it says that no one and nothing intended all the chaos and catastrophe that we see unfold in the universe—it is all the relentless blind action of a universal mathematics.

Behind this turn to materialism is the rejection of the idea that the intentionality of the source is 'watching over us' like a kind of benevolent parent. It was this idea that died in Europe as a result of the two world wars—for it could not survive the reality of those catastrophes. If we are to find meaning in our being here, it cannot be *that* kind of meaning. Instead we must look again, and pay attention to the way things actually manifest here, and see where this can lead us. ???

One striking fact is that we are *alone* in the objectivity of the solar system—that we have not found any other self-conscious, articulate life forms, capable of asking after the source of their own being. At the same time we know there is consciousness and intentionality in the universe, even though we only find it fully manifested here, within ourselves. We do not find a god who is manifest in the objective world, who can speak with us, and explain the 'meaning of it all'. Instead we have to inquire backwards, into the source, into the meaning that we ourselves experience and manifest. What begins to emerge, if we start to reflect on this situation, is that the source is simply *unable* to manifest directly, out here, as an objectively manifest being, like God appearing to Moses on Mount Sinai. That is not to say we cannot have visions of such events, that express something of the source *symbolically*. And it is not to say that the source cannot manifest *through* a human being. For that is exactly what is happening anyway, in each one of us. It is only to say that the source cannot manifest here *directly, as* the source. It must always show itself indirectly, in the forms it intends. For the source is the source. And that means, logically, if it becomes something particular, then it is no longer itself, i.e. it is no longer the undivided unity from out of which all particularity emerges.

Personal experience illumines the numinous

\ omits the core concept of relationality.

/ 3 persons thinking about each other,

So we cannot make the source into a being. That is why it is unthinkable. We can only know of it indirectly, symbolically, in terms of what is manifest here, and in terms of the intentionality that we directly experience ourselves to be, the intentionality that brings what manifests to consciousness (to being). Our task then, is to *read* this intentionality, both from within and from without, in a way that is analogous to the way we read a work of art—only now the symbol that speaks of the source is the objective mathematical form of the universe itself.

READING THE SYMBOLIC FORMS

We have already begun this reading of the source in our consideration of the meaning of quantum physics. Here the source becomes the symbolic potentiality of the quantum level, from out of which it manifests itself in the actuality of the consciousness of definite form. And we have also started to read intentionality into the manifestation of gravitational attraction between atomic forms. Here, in gravity, we find something universal, as we do in the manifestation of electromagnetic energy. There is no 'edge' or 'boundary' to these phenomena, unless it is the boundary between the 'absolute something' of the universe and the 'absolute nothing' into which we must imagine it is manifesting. For it did not manifest into a pre-existing time and space of some imagined past—time and space emerged along with the objective universe as the dimensionalities of its being-for-us, i.e. for consciousness. And quantum physics also teaches us that this atomic realm remains in instantaneous contact with itself through quantum entanglement, so that each event of observation that fixes the evolution of atomic form for a particular observer instantaneously determines its evolution everywhere else. If we are to think of this atomic realm in terms of our experience of intentionality, then what appears is something that has yet to distinguish itself as a being that is separate from any other being. It remains itself, within itself, and only gains the distinction of possessing atomic 'parts' insofar as it is observed by some higher-level form, such as ourselves. Otherwise we must picture a kind of energetic gravitational unity, that subsists 'quantumly', i.e. continuously on the boundary of potentiality and actuality.

Could we go further and imagine that each atom has some kind of experience of being an atom, that it not only has its being-for-us but also a being-for-itself—that it *feels* the forces we have observed to operate be-tween its subatomic components and the components of other atomic forms? Here we can only compare the form of an atom with the form of something we do know has an experience of being a being amongst other beings, i.e. our own form. In such comparison we soon discover that our ability to distinguish ourselves from other beings is founded on our pos-sessing a functioning nervous system, a system that is capable of *predicting* the forms of these other beings. The system of an atom, as far as we can tell, has no such capacity to register the form of anything else within it-self. It simply 'undergoes' energetic exchanges or transformations. For these exchanges and transformations to be 'felt' there would need to be some centre of feeling, some sense of self, of separateness. Of course, it could be that the entire universe of form, in all its particularity, simply possesses a sense of self at every level. But if we observe the extraordinary complexity of form that is required for us to experience our own separate-ness, and our complete dependency on the integrity of this form—such that when it is damaged, our world too is damaged—then the idea that an atom has an experience of selfhood and otherness starts to look im-plausible. Instead, it looks as if these atomic forms are only the ground of the possibility of such experience, a kind of unified field of potential-ity within which no particular is distinguished, but from out of which more complex forms can manifest, forms that will eventually make the distinctions within which separate atomic forms can appear.

At the same time, if the life and consciousness we know now is to emerge from this atomic potentiality, then it cannot be the entirely inert material we have imagined it to be over the last four hundred years. The psyche that emerges in the evolution of form, also lies within this atomic potentiality. So, although it may lack any conscious *sense* of being, it is still charged with *potential* for such being. It is in this charge that the in-tentionality of the source is driving the evolution of form. The source *is* this potential, but there is no form as yet that can express it, that can make it actual, that can make it known to itself. And so, although, in a certain sense, all that is to come is pre-figured in this intention, it remains *blind*. It does not know its own articulated form from the 'outside', only from the 'inside'—and then only as an unthinkable unity. In order to know itself outwardly it must manifest a suitable form, and all it 'knows' is that this atomic form, as it stands, is not suitable, is not articulated enough, not sensitive enough, not complex enough. And so the intentionality of the source expresses itself in the gravitational attraction that crushes together the atomic proto-forms so they can explode apart into new configurations of complexity—all guided by the overarching galactic gravitational fields, that form clusters of stellar transformation, that mix together the elements that can then form ever more complex compounds in the lower gravity of the orbiting planets. All the time there is the striving, the intention, to create and climb a hierarchy of form.

Here we are simply describing what our science tells us actually happened, with one small but significant change: we are saying that the source *intended* this to happen. But this intention is not that of an all-seeing God. The source is not a super-human intellect, making a plan, or a blueprint for the construction of a universe. It is a pure potentiality that intends to manifest itself without knowing in advance how this is to be done. It simply acts, it explores, it 'tries things out'. Isn't *this* the kind of universe we actually live in? It is this blindness, this apparent lack of intelligent foresight, that has persuaded so many that there is no intelligence whatsoever behind the evolution of the universe. But that, of course, is nonsense. For it is obvious to each one of us that there is intelligence in the universe: it is right here, right now, in you and I.

What has disappointed us is that this intelligence has not created a universe according to our expectations. We thought it should see everything, know everything, that it should be a kind of perfected version of ourselves. But that is not how it is. The source had to *discover* how to manifest intelligent foresight through a process of exploration that we see expressed before us in the historical past of the objective universe. We *are* this manifestation of intelligent foresight. Why should we have expected some other manifestation of intelligent foresight to have manifested us? There would have been no point, would there? For there would have already have been intelligent foresight, and no need for the whole drama of climbing the ladder of form to discover such foresight.

What we see in the evolution of the universe is not intelligent foresight, but intelligent *hindsight*. For, as we have already observed, there are an uncountable number of ways the universe *could* have evolved, that would *not* have produced any kind of interesting complexity. Materialistic science has tried to explain the extreme improbability of the existence of the universe in which we actually find ourselves by imagining there are (in ac-

intelligent love now?

tuality) an uncountable number of other parallel universes where there is no life and no consciousness whatsoever. Of course, if you start out by denying the universe could have any intrinsic intentionality, then you are almost forced to think of these trillions and trillions of lifeless, meaningless universes that no consciousness can ever know or visit. But now, from where we stand, we have a more plausible account of the fact that we live in this improbable universe. It is because the source from out of which the universe emerged, *intended* for there to be life and consciousness, and so *selected* just that path of evolution that manifests the forms that are able to manifest life and consciousness. It did not do this by looking ahead or by design—if that were possible then the whole process would have been unnecessary, for the source could have created a perfect universe in an instant. Instead, what we are proposing is that the universal intentionality, rather than manifesting all the countless trillions of lifeless universes, simply 'envisages' the possibility of these universes so as to 'discover' a universe that manifests life and consciousness.

Here we are using the notions of envisaging and discovering metaphorically, for we are speaking of a manifestation of pure creativity where there can be no actual overseeing entity that envisages or discovers, there is just the potential-actuality of the creativity itself. Perhaps we can say that within this pure creativity, one of these possible universes sufficiently *resonated* with the potentiality for consciousness that it collapsed from possibility into the actuality of a conscious experience. At this moment—the moment of the advent of time—the particular path of possibility from the 'Big Bang' that led up to this collapse also becomes fixed as the history of what 'must' have happened in order for the universe to have become the universe that it is. But, of course, that history never 'actually' happened, for there was no consciousness to experience it—it is only in *hindsight* that it becomes that particular path of possibility, from amongst all the uncountable others, that led to the actuality of this first moment of consciousness.

What we are doing is looking at the story that science has uncovered in its observations of the cosmos from the perspective of an altered understanding of being. Those observations and reasonings still stand, only now they have a new significance. And yet there is still an air of arbitrariness about this shift from materialism to 'intentionalism'. For none of this pre-history of the universe actually occurred in our direct experience.

We are only recognising that this is one way it *could* have happened. If we are to shift the very ground of our understanding then everything depends on whether this view of an intentional universe can be justified according to the available evidence—justified in the sense of offering itself as a *better* explanation than our current materialism. To which end, let us consider the following points:

1. Occam's razor says we should not multiply entities beyond necessity. In creating the need for an uncountable number of parallel lifeless universes, materialistic science is multiplying entities beyond necessity. In response, materialistic science will say that in granting the universe an intrinsic intentionality, it is we who are multiplying entities beyond necessity. But materialistic science forgets that we know already, according to our direct and immediate experience, that there is intentionality in the universe, i.e. *our* intentionality. So we are not multiplying entities beyond necessity, we are *extending* the capacity of intentionality to produce certain effects.

2. Materialistic science would go on to say we have no evidence to suggest our human intentionality can have any effect on the evolution of material processes, so that even if there were a universal intentionality, it could still not have any material effect on what happens here. Our response is to note again that our conscious intentionality does have an effect on the material world insofar as we are capable of realising that we are conscious and can subsequently speak and write about such realisations. For the supposed entirely materialistic processes in our brains are incapable of forming a correct idea of our being conscious on the basis of their entirely localised physical interactions. In fact, if they were to reason correctly, according to entirely materialistic assumptions, they ought to conclude that there is no such thing as consciousness, and that anyone who says there is, is under an illusion. And, as if to prove this idea correct, there are eminent philosophers of mind who are so reasonable, that they do say this.[5] But anyone who has followed what we have said about direct knowledge of consciousness will know, from direct experience, that our brains are responsive to realisations that they could not have formulated on the grounds of a pure mechanism.

3. Quantum physics demonstrates there is an inherent indeterminism at the foundation of all the events that occur in the objective universe. Whether this indeterminism can itself be determined by a universal intentionality is not something we can easily verify—for what would it look like if a photon appeared at a certain location on a photographic plate because of universal intentionality as opposed to a universal randomness? Outside the physics lab, there is also the evidence of parapsychology, which strongly suggests that human intentionality can exert a non-physical influence over objective events.[6] But such influence is not reliably repeatable—and scientific materialism demands that an effect can only be considered real if it can be reliably repeated— whereas it is the very hallmark of psychic experiences that they generally only happen in unusual and highly charged circumstances. This puts us in the strange situation that if we were to collectively surrender our scientific materialism, then we would suddenly be in possession of evidence that showed it was false all along. In fact, I think we would be overwhelmed with such evidence. But, as it stands, the phenomenon of non-physical causation does not correspond with our collective understanding of reality, and so is generally 'kept quiet' for fear of ridicule and loss of professional standing.

4. Finally, there is the direct intuition that it is the same intentionality in me as outside of me. We are dealing here with something that is *pre*-conscious, something that 'lives on the other side', in the unthinkable source, beyond the reach of our explicit reflection. And yet we can still look into the human psyche to see if there is a resonance of recognition for this idea that our universe is animated by the same intentionality that animates us. We can look again at the wind blowing through the trees—really look, with silent attention, at the extraordinary harmony of the fluttering of the leaves, of the bending of the branches. Is it really so unreasonable to think that the wind intends to blow? Not that it has a choice, but that it is the expression of the intention of something greater, the entire weather system perhaps, or the earth itself, something within which we are all included, just as our living cells are included in the system that we are. We have lived so long in a civilisation that thinks such intuitions of unity are only groundless, subjective, wish fulfillments, that we push them aside. But they

happen, don't they? As when the sun rises on a beautiful day and we forget our personal concerns and feel ourselves to be included in something vast, something profoundly significant. Is it really so reasonable to reject such experiences? Are they not, in fact, the most real, the most significant experiences of our lives? On what basis do we turn away? Because science has 'proved' such intuitions are illusory? We may think this, but if we actually look, our science has not proven anything of the sort. It has just measured and classified and predicted the evolution of objective form. It has nothing to say of our immediate experience of emerging from, and returning to, the unity of the unthinkable source.

Need to replace 'intentionality' with 'the relational between'.

MORPHIC RESONANCE

Locus of being. Me or MYU ?
Self-advancement! Ambition... *Trinity*

And yet, despite this evidence of our direct experience of being conscious, we must also admit, when we look out into the universe our telescopes and probes are revealing, that we are confronted with a deep sense of isolation. If there is life on another planet in another solar system, it is so distant from us in space and time that we have little or no chance of ever making contact with it. And when we look around our own solar system, we again meet with an extraordinary emptiness—not just the emptiness of the vast stretches of space that separate the tiny planetary spheres that circle the sun, but the emptiness of the absence of life, the absence of anything like the organised complexity that we find before us on the earth. This is what was revealed to the astronauts who first landed on the moon: all the danger, all the expense, all the technological brilliance, only to realise that the extraordinary experience they expected to find was not there in the dust and rock of the moon, but in the sight of the living jewel of the earth rising in the sky of an otherwise lifeless desolation.

And if we look at what science has discovered of the formation of the stars and the planets, it certainly appears as if a lifeless universal mechanism were in operation. We see no signs of sentience in the gravitational processes that condense the stars from clouds of interstellar gas and dust, or in the formation of the orbital trajectories of the planets and asteroids. Instead we see the functioning of inexorable mathematical regularity. The same goes for the microphysical manifestation of atomic and molecular form. Each hydrogen atom behaves just like every other hydro-

molecules break/How does it work? Is it good to eat?

gen atom. If there is an intelligence operating at this level, then it appears simply as an intention to repeat the same thing over and over.

But there is also no doubt that this universe of galaxies and stars and planets is the ground from out of which life has emerged. So it cannot be that everything just repeats itself fixedly and mechanically through all eternity. It is this very fixity that forms the stability that is needed to manifest new and higher levels of order. For the way the universe manifests a higher level of form is to combine together the elements of a lower level. And so we see atoms forming out of the combination of sub-atomic forms, and molecules forming out of the combination of atomic forms. And for this to happen, the lower-level forms must first become *stabilised*.

If we reconsider the scientific account of the formation of the universe, as if time 'began' with the Big Bang, we can see that these lower-level forms were not always caught in an eternal fixity, but themselves went through a process of formation from out of an earlier state. Again we encounter a universal origin where all possible universes have their possible beginning. It is from here that the atomic forms emerged. And if we think of them as 'chosen' by the source, then they seem to have been chosen specifically for their stability and their capacity to combine into new forms of molecular complexity—a capacity which itself depends upon the gravitational and electromagnetic and quantum unity that emerges from and encompasses this atomic form of organisation. It is this universal form of organisation that maintains the balance between atomic stability and the capacity of these atomic forms to break down and recombine into new forms (new elements and new compounds). It is here in the extraordinary improbability of this balance that we start to feel the presence of an underlying intentionality—for our universe exists on a knife edge between order and chaos. It is only in this region that we can expect *both* stability *and* complexity to manifest, for if there is too much stability then all form will become fixed at a low level of complexity, and if there is too little stability then no lower-level form will last long enough to enable a higher-level form to emerge.

Rupert Sheldrake developed the hypothesis of *formative causation* to explain this manifestation of order and stability in the universe.[7] His proposal is that whenever a certain form or pattern of behaviour repeats itself, then it becomes slightly more probable it will repeat itself again. Behind this lies the idea of there being immaterial *morphic fields* that, in a sense,

Figure 13.1: Rupert Sheldrake (Photo by permission from www.sheldrake.org).

record and remember the forms of the behaviours that occur 'out here' in the objectified universe. These fields get stronger the more the pattern they embody is repeated, and the stronger they get, the more influence they exert to make future events conform to that pattern. Sheldrake conceptualises this influence as a kind of *resonance*, i.e. a two-way interaction whereby the 'potential energy' of a morphic field tends to 'pull' on the forms of events that are sufficiently similar to itself, causing them to fall into a correspondence (i.e. by means of a resonant attunement where each adjusts to the other). At the same time, when an event does fall into correspondence with a particular field, it incrementally adds to the 'pull' of the field's potential energy.

We can represent this potential energy as a basin of attraction in the phase space of form within which each universal process follows a trajectory of change (like the phase space of a Lorenz attractor). Using such a model, we can understand the Big Bang as the primordial emergence of a universal phase space of potential form-trajectories. Within this initial chaos of transformations, as soon as one process 'happens' to re-enact the behaviour of another, a resonance occurs which creates a tiny basin of attraction around the common average of their repeating behaviour. Immediately, all the other form-trajectories in the local area of this attractor will be slightly inclined toward it—but only those few that pass close by will actually enter into it. As soon as this occurs, the basin again grows slightly deeper, and its sphere of influence extends, increasing the number of potential processes it can attract. But while this basin slowly grows, new attractors are forming elsewhere in the phase space, and all are growing at different rates, with some growing so deep that they engulf and dominate the smaller attractors in their local area. As this process of amalgamation continues, the entire phase space starts to become dominated by a collection of large, deep attractors. Now nearly every potential trajectory will be pulled into one attractor or another, thereby eliminating most of the original chaotic behaviour. Here, in this evolving phase space, we can see how the initial chaotic randomness of the beginning could have settled into the stable forms of the hundred or so natural atomic elements we find today (see Figure 13.2 and (Seaborg, 1964, p. 14))—all without the need for any deliberate selection or forethought of design.

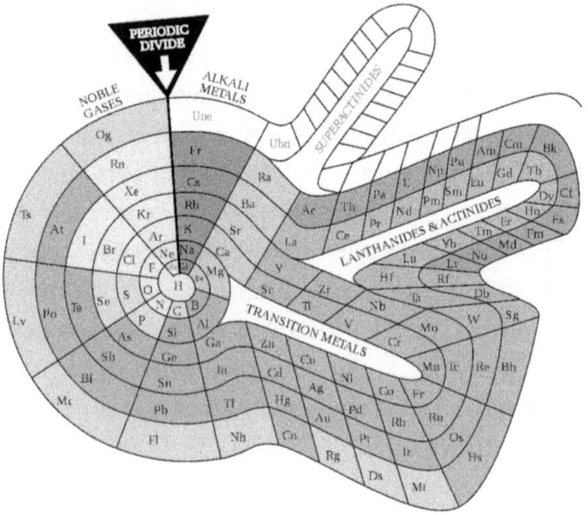

Figure 13.2: Theodore Benfey's spiral periodic table of elements. Image created by DePiep under a CC-SA-3.0 license.

Why not?

Creating dualism

At the same time we must recognise the 'potential energy' that forms these basins of attraction is not *physical,* i.e. it does not show up in the inventory of the basic forces we can measure using our scientific instruments. For the underlying morphic fields represent an immaterial tendency for a pattern to repeat itself. If they 'exist' at all, then they are pure psychic expressions of the intentionality of the source, i.e. the intention

No!

to manifest an ordered stability of form. Here Sheldrake's hypothesis provides an alternative to our story of an intentionality that 'envisages' all possible universes and 'waits' until one of them is able to manifest a consciousness-bearing form. Instead, from the outset, the universe, in

Yes.

and of itself, would have a tendency to evolve stable forms. We would still have to think of this occurring on the 'other side' of the quantum divide, in a realm of pure possibility—for at this early stage, there could be no 'outer' consciousness to make these forms explicitly manifest. But the development of form would certainly be much more focussed, as the possibility of those countless universes that remain in a state of chaotic instability would be progressively ruled out. Only *this* kind of universe, with its gravitational atomic stabilities, would resonate sufficiently with *OK,* the intentionality of Sheldrake's morphic fields.

At the same time, we have to avoid the fixity of an utter stability. For morphic resonance only pulls in one direction: toward the repetition of *Ah! A* the same, which, instead of producing the ordered complexity we need *problem-* to manifest consciousness, produces order by *eliminating* disorder. That *not* means we must assume another principle is operating in this primordial *diversity* intentionality—an opposing will to *instability* that balances out the con- *in unity* servatism of the morphic fields. And, in the form of our mathematical science, we find just such a will expressed in the second law of thermo- dynamics, where low entropy order must be *purchased* at the cost of a greater disorder that is *dissipated* by the very process that is manifesting the order. That means there can be no permanent form of low entropy universal order—for such order can only exist in the context of a greater disorder, a disorder to which it is continually contributing. But this is not some kind of doom, for these are the very conditions that are necessary for complex ordered forms to emerge in the first place.

Mainstream science has rejected Sheldrake's theory because the idea of immaterial morphic fields exerting influence over material events simply does not and cannot fit with the prevailing materialism. But from our perspective, morphic fields represent just the kind of organising princi- ples you would expect to find in an intentional universe. Their immate- riality is no obstacle, for we already dwell in the immateriality of imme- diate consciousness. If we embrace Sheldrake's vision, we can begin to see the outline of an intermediary world of morphic form that stands be- tween the objectivity of the world we experience and the unthinkability of the source itself. For these forms are not unthinkable—they are the very forms we came across in Chapter 4 in our consideration of immedi- ate perceptual experience. It was there we realised that when we perceive an everyday object, what we intend is not what we *actually* see, but the *ideal form* of what we see—and that what we see is only an *aspect* of that form. What we are suggesting now is that the ideal forms we intend and with which we resonate in each moment of perceptual experience, are the very same forms that Sheldrake has conceptualised in his idea of morphic fields—only now, instead of theorising that such fields are organising our experience, we have direct phenomenological confirmation that this is actually occurring.

Of course we cannot conclude from this that the ideal forms we perceive have the same causative power that Sheldrake proposes. We only know that they guide our perceptual experiences into common pathways, so that we come to experience the common form of a common world. But we at least bring the idea of a morphic field out of the domain of sheer hypothesis, and start to ground it in direct experience. For our entire consciousness of the objective world is mediated by these ideal forms, forms that we take to be the 'truth' that lies behind the aspects of the objects that present themselves in our sensory fields. These forms are neither subjective nor objective, they are the 'intentional objects'[8] that enable our subjectivity to manifest the objectivity of the world (that connect subjectivity and objectivity into the unity of our experience of being in the world). They are not *actually* out here in our sensory manifestation of the world, and neither are they some inner 'idea' that we assemble and project. Instead, they express the underlying objective structure of the things and events we encounter—they are that with which our projective, predictive intentionality *resonates*. It is in this resonance that we know our protending projection corresponds with something 'other', something transcendent to our immediate subjective 'innerness'. And it is here that our particular and individual perceptual intentionality meets with the universal, impersonal intentionality of the source. For where does the form with which we resonate reside? It resides on the 'other side of now', concealed within the source—and yet not *wholly* concealed, for the resonance of our perception tells us of a *transcendental correspondence of form*.

Do you begin to see this realm of transcendental form that structures our experience? This must have been what Plato saw in what was later called his Theory of Forms.[9] And yet now we think Darwin has refuted Plato, and that there cannot be any realm of eternal Platonic Forms, because the forms we see before us have *evolved in time*. But our ordinary concept of time does not apply here. It lies on *this* side of the source—it is the means by which *we* experience our being in the world. That does not mean the source 'exists' in time. And neither does it mean that the forms of the things we experience 'exist' in time. Think again of our example of an equilateral triangle. That does not come into being and pass away on a certain date. If it exists at all, it exists in a domain of ideal perfectibility, where its infinitesimally narrow lines always and forever meet at *exactly* 60°. When we speak of the objective form of some worldly thing, like a book, we are speaking of something that lies within the pure potentiality of the source, just like the equilateral triangle. Of course, the form of the book can change—for we can always cut it in half. But we can also cut an equilateral triangle in half. That still does not bring it into the time of our world experience. What Plato was suggesting is that our experience of forms coming into being and passing away is only how it *must* appear to us here, in our perceptual consciousness, because *everything* appears to us here in the aspect of its being in time. It is only in the stream of experience that 'time flows'—whereas the ideal forms with which our experience resonates do not flow—for even as they appear to change they are only tracing out the stable form of their own transformation. They lie 'on the other side', 'behind' the ever changing stream—and yet we 'see through' the stream, into what endures, into what does not change, into what is the same for all of us, even though we are still 'in' time, and in the midst of change.

The Form of Life

One of the most profound discoveries of modern science is the form of the living cell. The true significance of this discovery has yet to permeate our collective understanding because our mainstream biological sciences are still looking at the living cell as if its form were a simple and secondary consequence of the mechanical interactions of its molecular components—a kind of epiphenomenon that is completely determined and explained and grounded in the being of these components. And that means we do not see the *life* of the living cell—all we see are the mechanical interactions of lifeless components enacting an essentially lifeless process. To see the life we have (as ever) to let go of our materialistic understanding of being and look again at the *form* of what is occurring. For nature is quite transparent here. It is not trying to *hide* what it means for something to be alive, it is displayed straightforwardly in the very molecular interactions that our mathematical objectification has uncovered.

After Henri Bergson,[10] the great modern pioneer in this area was the German philosopher Hans Jonas (1966/2001). It was he who recognised the distinguishing characteristic of life to be the *form* of the metabolic process. To understand this we must take a few steps back and examine the possible pre-history of the development of life on Earth. Of course we cannot know with certainty how the first living cells came to manifest. But we can piece together a plausible account on the basis of studying the *offspring* of those first cells, i.e. according to life as we find it today and according to the traces such life has left in the rocks and soil and oceans and atmosphere of the earth. And while it could be that the basic chemical constituents have been 'seeded' via the intervention of extraterrestrial intelligences, or through the accidental impact of interstellar material, we shall dismiss such conjectures on the basis that we would face the same question of origins on whatever other cosmic body that material had itself originated.

Essentially, when we look at the form of life we find a new form of molecular organisation. We already touched on this idea in our discussion of circular causation and feedback in the evolution of self-replicating systems (see Chapter 7). There we sketched a picture of a primordial ocean containing an inconceivable number of molecular components, all in a state of incessant interaction, within which new forms of molecular structure were continually emerging via a simple process of collision and chemical combination. Then, using the conceptual tools of modern complexity theory, we started to explain how this initial state of disorder could have manifested the first living cells. The basic explanatory step is to recognise that any complex system that traverses a phase space containing large regions of chaotic behaviour, and relatively small regions of stable behaviour, will eventually discover and settle into the regions of stable behaviour, because a process that is caught by a stable attractor will remain in that attractor and continuously repeat its predetermined behaviours. Processes that have not been caught simply wander chaotically across the phase space, eventually 'bumping into' another such attractor, at which point they too will start exhibiting its repetitive structure. Of course, much depends on the distribution and accessibility of these attractors, and the form of stability they embody. For not all attractors have a simple fixed point or limit cycle structure that permanently capture a process by bringing it to a halt or by making it repeat the exact same series of steps (e.g. see Figure 12.2). There are also strange attractors that never exactly repeat themselves (such as the Lorenz attractor of Figure 12.3) and other *unstable* attractors that only temporarily alter the trajectory of processes and then return them to their chaotic surroundings.

The reasoning behind the mechanistic account of the formation of living cells is that, because we can directly observe the emergence of order in the behaviour of many initially chaotic complex systems (such as the formation of Bénard cells shown in Figure 12.6), we can also assume an analogous process of self-organisation was responsible for producing the stable complexity of the living cells in the earth's primordial ocean. This further assumes, given the right initial conditions, that such complexity is a simple consequence of the unguided interaction of the low-level microphysical properties of the various atomic constituents of that ocean.

Here it is as if complexity theory has thrown a lifeline to mechanistic science so it can overcome the absurdity of thinking that a living cell could have assembled itself from a process of purely accidental molecular collisions. But this only replaces one improbability for another, i.e. the idea that the universe has selected, by complete accident, exactly those atomic forms that are capable of producing a phase space within which a living cell can emerge. The problem here is that we cannot calculate how improbable this is—for that would involve simulating all the possible molecular interactions in the primordial ocean, starting with the Schrödinger wave equations, and working on upwards—a calculation that is just too complex to perform in actuality. Instead our scientific materialism assumes that this is how it *must* have happened, because it can think of no other plausible way it *could* have happened.

But, just because we observe many stable attractors in the behaviour of the universe, does not mean their existence is a logical consequence of the low-level 'laws' of physics. The truth is we do not know how these stable attractors have formed, we simply know they have formed. It is an act of faith on the part of scientific materialism to think that the forms of these attractors are 'written in' to the low-level physics. For the laws of physics themselves are a mathematical description of *another* unexplained set of stable attractors. We only think that these attractors stand outside of time and determine the form of every other higher-level attractor because these ideas are *already* assumed in the mathematical reductionism of our materialistic science. It turns out the universe looks just the same if we allow that the source intended things to be the way we find them— only now we do not have to assume the existence of countless parallel universes where life did *not* evolve, in order to explain the extraordinary improbability of the being of *this* universe. For we know that life is not just a set of stable attractors. We know this because we know what it is to be alive, and we know life to be the ground of our being here, our being conscious. Perhaps we cannot say precisely what it is to be alive, but we can *feel* it.

If we put together the idea of morphic fields with our immediate experience of form perception and our knowledge of the molecular form of the living cell, we can begin to provide a meaningful account of the manifestation of life on Earth. Firstly there is the question of the phase space in which the extraordinary complexity of the first living cell was able to manifest. What we are proposing is that the intentionality of the source is able to *operate* on this phase space. That means it is in some sense (that we cannot imagine) *aware* of this space, i.e. it is aware of the forms that are forming, but only *as* forms, in their ideality and generality, not in the actuality and particularity of their manifestation in space and time. For, once again, if the source were aware (i.e. conscious) of these forms in their particularity and actuality, then there would be no need of *us*, the consciousness of particularity and actuality of life on Earth. And consequently there would be no need to have manifested all the trillions and trillions of life forms that led up to us, that have come and gone, transforming the earth, eating and being eaten, forming species that then go extinct, all without apparent foresight, and yet still manifesting forms that themselves possess a finer and finer ability to distinguish and perceive the particularity and actuality of the forms of the earth surrounding them.

What we see in the evolution of life, once we relinquish our materialism, is the manifestation of an *inner* intention of the source to become *externally* conscious of itself, in time and space, in its very action of manifesting such consciousness. And within this innerness of the source, all that is initially 'known' of the possible externality, are the morphic fields, i.e. the ideal forms of those processes that are repeating themselves. It is from this place that the intentionality of the source is guiding the process of evolution by 'selecting' those forms (i.e. those attractors) that maintain

KEY

But the rest is no!

Yes, except not external to God — panentheism

the balance between stability and chaos that is necessary for higher-level form to develop. This intentionality does not 'foresee' where this process will lead externally, but it still 'knows' what it intends, i.e. the formation of a form that is able to resonate with the forms that 'surround' it. Once such a form emerges, then the source also comes to know itself externally as well as internally. And that, of course, is what is happening right now, in you and I.

It is important to understand that this account is, and can only be, a *myth*. For we have translated something we cannot know objectively (i.e. the transcendental reality of the source) into the dimensionality and particularity of our experience, i.e. into the domain of spatiotemporal being. In 'reality' the source did not 'do' anything, because it was not 'in' time, making decisions about the phase space of universal form. From our spatiotemporal perspective we could equally say that the source ran through the entire possibility of our universe and everything that could possibly happen in it, in the instant of an eternal now. It is only we who have to live through this linear time, where one thing has to happen before another. And it is only by seeing this that we can gain a correct perspective on anything we may say about the source. It is simply not possible that there could be some literal experiential, spatiotemporal truth about that from out of which our experience emerges. We can *only* approach the source symbolically and mythically, and in doing so it is important. *ironically* we understand this, otherwise we make a 'thing' of the source, and the source is not a 'thing' or a 'being', it is that which we can only approach by means of symbol and myth and intuition. ?? *resonance!*

Having said that, we still have the intermediary world of ideal form. ! *No.* This world is not the source itself, and neither is it our spatiotemporal world of actuality. It stands between the two, allowing the timeless forms of the source to resonate in immediate experience. It is here that we are 'in touch' with the source, where it reveals something of what lies behind experience, from within experience. And it is here that the mathematical *symbol only* thinking of our scientific materialism also has its home. But we must be clear that this is not some epiphenomenal shadow world that simply reflects the ideal forms of what already exists 'out here'. These ideal forms are 'real', in that they actually form our common experience, not just in the resonance of perception, but also in observable, empirical effects that can be measured using the existing apparatus of our empirical science.

Bishop Berkeley – Idealism.
Does he limit resonance to perception?

Rupert Sheldrake has long been advocating the pursuit of serious research into these effects, so that the being of the underlying ideal forms can be incorporated into our collective world understanding. But here we run into the problem that no single experiment can demonstrate that one's entire understanding of the being of the universe is incorrect. It took two hundred years from Copernicus to Newton for the Aristotelean world view to be relinquished in Europe, and it seems to be taking a similar amount of time for us to let go of our scientific materialism. The experimental results Sheldrake has published that indicate morphic fields exert causative effects in animal and human behaviour have been fiercely rejected by the scientific establishment—not on the basis of their invalidity, but on the basis that it is assumed in advance that such effects are impossible.[11] Therefore, according to our mainstream science, it must be the case there is another explanation that is compatible with the current materialism, however extreme or defamatory.

But let us return to those primordial oceans. And let us accept that what we now experience to be the planet Earth was once in a state of unobserved quantum superposition. From this place of pre-being, all the possible quantum states of all those possible primordial oceans would have been potential together at once. In this potential lies the inconceivably vast space of all the possible chemical reactions that could have occurred, and the possibility of distinguishing just those reaction forms that are able to repeat themselves. If we allow for a corresponding potential of morphic resonance, then once one such repetition occurs in one possible ocean, it will become slightly more likely to occur in all the other possible oceans. We can even envisage that morphic resonance is not just attuned to repetition, as Sheldrake proposes, but favours just those forms of repetition that are most likely to lead to a new level of complexity. Once we adopt this perspective, then the problem of the extraordinary improbability of the development of life on Earth starts to dissolve. Instead of asking exactly how such a miracle could have occurred, and attempting to recreate the event in a laboratory, we can perhaps come to accept that the evolution of complexity is something that was intended from the very beginning.

What is of more interest is to understand the *meaning* of the form of that new level of complexity that came to manifest itself as life on Earth. The first step is to recognise that life begins as the manifestation of a new

[handwritten margin note, left: YES, LIGHT]

[handwritten note, bottom: Why does this sound so strange? Godly intentionality described in p406 - he is missing conversational trinity in the unspeakable source - harmonically resonant in us]

form of *molecular* organisation. Before life, there were already countless ways that molecules could bind themselves together, and countless ways they could move and be configured in space and time. We had storms and waterfalls and volcanoes, the whole system of the ocean currents, the interactions of the tectonic plates, and the atmosphere, the tides, the gravitational equilibrium with the sun and moon, and so on. In the oceans we already had the formation of complex organic molecules, and self-replicating molecules, and autocatalytic networks of chemical reactions whose final output was also an input into the same network. And yet in all this complexity there was still nothing that we would recognise as being *alive*.

A basic missing element was the being of an *intrinsic* separation between one process or event and another. Of course, from our human perspective, we cannot help but distinguish a wave on the ocean or a storm that is passing over us, or the volcanic eruption that is ejecting molten rock onto the surface of the earth. But from a *molecular* perspective, all we find are bonds being broken and remade, and transfers of energy that result in some kind of molecular movement or transformation of atomic structure. There is nothing at this level that says: yes, this molecular aggregate has become a 'thing', something more than its individual molecular parts, something that *distinguishes itself* from its surroundings. All sinks back into the anonymity of molecular being, where each particular molecule of H_2O is just like another, and each molecular movement is just like another, and each molecular transformation is just like another. We have no distinction of *individuality*. We are still somehow 'within' the source, dealing with no more than *universal* forms, fixed ideas, caught and determined by those universal fixed ideas we call the laws of physics.

What occurs with the first living cell is the emergence of a new form of being that transcends the molecular form on which it stands and depends. Now there is *birth* and *death*. There is the emergence of a process that has an unambiguous beginning and end in time, and an unambiguous boundary in space—a process that is no longer associated with any particular set of molecular constituents, but is still defined by the forms of movement and transformation of the molecules that continuously pass across its boundary. This temporary taking up and dropping off of molecular material is analogous to the way a wave passes across the surface of the ocean, where the oceanic molecules move up as the wave reaches them and then move down again as it passes them by. And yet clearly this manifestation of form in the medium of molecular displacements does not represent a new order of complexity, for wave forms are ubiquitous throughout the universe at every level of complexity. So what exactly is it that defines the form of life as we understand it?

Figure 13.3: Humberto Maturana (Photo by Rodrigo Férnandez CC-SA-4.0).

After Hans Jonas, the next great pioneers in this area were Humberto Maturana and Francisco Varela.[12] It was they who first recognised that behind the metabolic processes that both create and define the form of a living cell on Earth, there is a more general form of organisation that transcends and encompasses the particularity of these processes. They called this form of organisation *autopoiesis*. Autopoiesis literally means *self-creation*. What distinguishes life as a process is that it is continually *producing itself*. That means, as a living cell, it forms a *circular* network of molecular transformations that produce components that produce components that eventually again produce the first component in the cycle. In this way the cell enters into a process of a continual self-renewal. But such renewal, in itself, is still not enough. For there are many chemical reaction chains that can be experimentally induced to repeat themselves without showing any recognisable signs of life. What is also needed is that the self-producing living system, as part of its process of transformation, manifests a *boundary* that *contains* the process by *separating it off* from the rest of its environment. It is this boundary that defines what belongs to the living process and what does not. For no process of transformation can be completely enclosed within itself—it will always need access to a form of external energy to power the transformations. And that means allowing a continual influx of energetic form into the living system, that then releases some of its energy, and in so doing is itself transformed. But this influx itself causes a further problem. For if the living system is to maintain its own form of stable order, it cannot indefinitely keep ingesting such energetic 'food', otherwise it will keep growing larger and larger and finally dissolve into the general disorder of the surrounding environment. It must also *eject* what is not assimilated into its own form as 'waste'. And that again is why a boundary is *necessary*: to allow ordered energetic forms to enter into the process and to allow the disordered by-products to leave, while at the same time 'keeping hold' of the forms of the components that constitute the process, which includes the form of the boundary itself. For if the process is to be genuinely self-producing it must also produce (and maintain) its own boundary (see Figure 13.4).

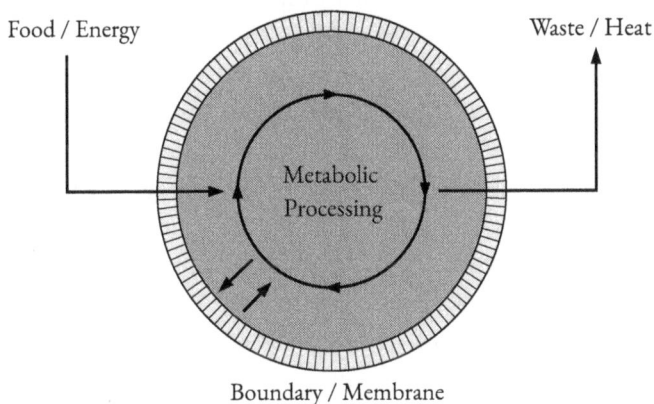

Figure 13.4: A schematic representation of a living cell.

It is this boundary that is essential to the form of organisation that is life—without it every self-producing process remains dissolved in the general chaos of the universal transformation of molecular form. Once there is a boundary, there is the first manifestation of a self-sustaining, self-separating process of stable molecular transformation. The important thing to see is that this form of organisation represents a *different level of order* to anything the universe had manifested beforehand. It is like the jump from the energetic sub-atomic forms to the form of the first atom, or the jump from the atomic form of organisation to the molecular form of organisation. At each step we find the manifestation of a new kind of unity. Before the living cell there was only the unity of molecular forms. Admittedly these forms had to climb a ladder of self-organisation, which included the formation of self-replicating molecules, and of autocatalytic repeating networks of molecular transformations. But only when these forms are contained within a boundary of their own making is there the emergence of a new and distinct form of unity—not the unity of a particular repeating process, but the unity of a self-producing process that has separated itself and distinguished itself from all the other processes that surround it.

We should pause and make sure we are not making unsubstantiated claims here. For there are many other apparently individuated 'things' in the universe, like the stars and the planets and the moons and the rocks and stones we find on the beach. Each of these have their 'boundary' that is clearly defined by their movability in space and time. For the rock and the stone can be washed up on the shore of the ocean while still maintaining their general spatial form. And the moon too goes around the earth, while maintaining its own form relative to that movement. So what is so special about a cell having a boundary and an enduring form?

The difference is that this enduring form is not that of a fixed arrangement of enduring molecules. The form of a cell is traced out in the continual formation and replacement of its molecular components. It literally (with the exception of the DNA in its nucleus) continuously constructs the forms of the molecules that are involved in its metabolic process—building them up from out of the molecular food it harvests from its environment, using the templates of its RNA and the genes in its DNA to assemble the proteins and RNA it needs to perform its 'work' of self-manifestation. The 'shape' of the cell is the shape that this process of transformation traces out over time. So we cannot say that the form of the cell *consists* of the molecular forms that happen to constitute the cell in any particular moment. The form of the cell is the form of an aggregate *process* of molecular transformation. Of course we can equally say that the form of a molecule is the form of a process (i.e. of the vibrations of the energy that manifests the molecule). But now, in the living cell, we have the form of a self-contained higher-level process that is traced out in the movements and transformations of these lower-level molecular processes—so we cannot simply treat the boundary of a cell to be like the boundary of a stone. The cell's boundary is a boundary of a different order.

Equally we cannot consider an ocean wave to be a process of a higher level of order, even though it too is manifested in the aggregated movement of molecular forms—for an ocean wave does not produce itself by transforming its intrinsic molecular structure and it does not distinguish itself by producing its own boundary. It is simply 'nothing but' an aggregated movement of molecular forms. Life, as an *individuated* process, requires more than a repetition of aggregated movement. It has to be combined with the phenomenon of bounded self-production.

Even so, you may ask why we are making such an absolute distinction at this level of the living cell. There are two main reasons: Firstly, the living cell is clearly the unit out of which higher-level living systems, such as ourselves, are composed. And that is not just to say that our bodies are composed of living cells at this particular moment. It is also to recognise that our entire body grew out of the initial division of a *single* living cell. And that makes the cell into a biological 'molecule' of a quite new and different level of order to that of the chemical molecules that comprise it.

Secondly, we know there is a clear distinction between living forms and non-living forms, because we directly encounter this distinction in our immediate experience of embodiment. For we know there is an *edge* to such experience that is defined by the reach of our embodied nervous systems. This is demonstrated in the absolute privacy and uniqueness of that immediate experience. For only I am seeing the room in front of me from this particular perspective at this particular moment in time, feeling these particular feelings and thinking these particular thoughts. Of course I can *infer* what you are seeing and feeling, but I can never know it in the direct and immediate way that I know my own experience. And in knowing this I know there is such a thing as individual being, and individual consciousness. The question is *where* this distinction of individual being can first be made. And it is here again that the living cell distinguishes itself. For it is here we first recognise the processes of birth and death and the continuous process of self-maintenance in between—processes that we too experience in our time here on Earth. And it is here we find the fundamental connection between our being and the being of a living cell, in that we ourselves, insofar as we are this collection of cells that comprise our living body, grew out of the manifestation of a *single living cell*.

To think that a molecule can distinguish itself as a molecule, that it could have some sense of its own separation from the rest of the universe, seems highly improbable. For it has not produced itself or separated itself as a process from the surrounding energetic unity in which it manifests. That means it is essentially without the possibility of experience as we know it, because, once formed, as far as we can observe, it ceases to change. And if we rule out the molecule as the first unit of individual being, and the ocean wave, and the repetitive but dispersed molecular transformations we find occurring in the oceans and the atmosphere and in the scientific laboratories, we are left with the living cell as the first example that we know of truly individuated, *intrinsic* being. Its being is intrinsic, because its entire activity is concerned with one outcome, its own self-continuation.

As soon as we start to consider the life of an individual cell, we almost cannot help but feel sympathy for our shared fate: to be born and to die and to continuously be engaged in avoiding the inevitability of that death. For the being of the cell *is* the being of the process of its own self-production. It *must* maintain itself in an environment where there is the right kind of molecular 'food' and avoid the molecular contaminants that would cause that process to breakdown. And as soon as a cell gains some limited power of locomotion, it behaves just as we behave: it moves toward situations that are congenial to its self-production and moves away from situations that are uncongenial. Do we see intentionality in these actions, or the functioning of a fixed mechanism? Perhaps both. For the being of the cell is not the being of the molecules that comprise it in any moment. Its life is not situated in any particular molecular component or process. The cell is only alive insofar as it is able to sustain its autopoietic form of organisation. And this autopoietic form is no more than an idea, an ideality that the molecular forms are reproducing in the basin of their cellular attractor. Do you see that this attractor, this form of self-creation, does not *exist* here as some material entity? It is a pure form, a morphic field, i.e. a form of the intentionality of the source. And the cell, being a self-enclosing individuality of form, *is* this intentionality, only now it is no longer a universal intentionality, like the intentionality of gravitational attraction. For the first time, we have an *individuated* intention, and intention that some part of the unity of the source should *survive*, should *endure*, that there is a being here that is not the being of the entire universe, but the being of a 'this-ness' that is dependent on an 'otherness' to provide it with forms to ingest into itself. And the form of this intention *is* the cell's life—for it is alive only as long as it persists within the form of its intention.

Of course, this is not just about an individual living cell. We too are living beings, and we too only live for as long as we are embraced in the intention of our own autopoietic living form. The point is to see that life, the very life that enables us to experience our being here on Earth, is the intentionality of the idea of life. What we see outside of us are life forms. But we also experience being a life form from within, and this experience in-forms us that to be alive is to be an intentional being. In our being alive we literally *are* the intentionality of the source that is manifesting this life. But we only know ourselves to be this intentionality for as long as this form persists, for as long as it can maintain the form of its own self-production.

Reproduction and Inheritance

In considering the form of a single living cell, we have only touched on the surface of the question of the form of life. As yet, we have not even mentioned the question of reproduction. This is partly because Maturana and Varela left reproduction out of their autopoietic conception of life. Their thinking was, given we still consider a sterile organism to be alive, it follows that the capacity to reproduce cannot be something intrinsic to life (otherwise a sterile organism would be considered to be dead, which would be absurd). But this reasoning only applies if we think of life as something individual, something that becomes present with birth and goes away when the autopoietic form of an organism disintegrates. What this misses is the obvious fact that our being alive is *inherited*. We may not get to pass on this inheritance in the form of offspring, but it was certainly passed on to us. And all the life forms we know of *came out of* such reproductive inheritance, whether or not they go on to reproduce themselves. Of course, there is the notable exception of the *first* life forms, that, by definition, could not have inherited life from another life form. But these forms are the special case of a unique beginning, of a transition about which we can only speculate. Our current science thinks this beginning must have been an incremental process, where some form of inheritance was already occurring in the self-replication of complex hydrocarbon molecules that eventually came to be enclosed in semipermeable membranes. Perhaps life 'flickered' on, at first appearing in several shallow morphic fields, each partially manifesting autopoietic tendencies, and then, as these basins of attraction became deeper, and transformed more and more of these complex self-replicating molecular processes, the ideal form that they came to embody, the form of life, started to shine forth, and to manifest a correspondingly new and higher form of intentionality.

Behind this objectified view of the origin of life lies the deeper question as to whether or not life itself is an objectifiable property of the life forms we distinguish, a property that comes in discrete packages which appear and disappear along with the forms to which they are supposedly conjoined. Here again, if we consider the evidence impartially, and we include the obvious role that reproduction plays, then it no longer appears that the 'standard scientific view' is correct. For if we look at the process of reproduction, from the first living cells up to and including all the living cells that comprise your body at this very moment, what we find is the manifestation of a *single system of life,* of which your body is just one more of its temporarily manifesting extremities. Within this system lies an unbroken chain of inheritance leading from those first living cells, that connects together all the life forms on Earth into the unity of a single organism.

To see this we only have to think backwards: for each and every cell that we find living now on Earth was produced from out of a parent cell. This could have been via a process of sexual reproduction, where two parent cells join together, or through asexual reproduction, where a parent cell divides itself in two. In either case, each living cell has emerged from out of another living cell. And here *there is no death.* The life of the parent cell is not extinguished, it does not cease to exhibit its autopoietic form of organisation, it just undergoes a process whereby it divides itself in two. If we put aside, for the moment, the larger scale forms of the bodies that these cells comprise, we can discern the continuity of an unbroken process of autopoiesis. Clearly an almost uncountable number of cells have also 'died' without reproducing. But those cells are not the parents of the cells that we find today. They are the 'outer surface' of life that falls away and disintegrates. Within the living core there is this one unified autopoietic process that has been continuously functioning since the first cells produced and reproduced themselves. If you think of one of these first cells, it becomes clear that it is *still* alive, it is *still* here. The life that it was (that it is) was certainly divided into two cells—but it then *became* these two cells, it did not 'go away', it just underwent a change of form, a division. If we identity life with this process that manifests autopoietic order, we find it is the same process manifesting in the two cells after division that was manifesting in the one cell before division. And, then, of course, the same transmission occurs when these two cells divide, and when their offspring divide, and so on, up to this very moment in which we find ourselves now.

It is *we*, in our objectification of existence into the separating dimensions of space and time, who think the life of each cell is somehow absolutely distinct from its parent. We are thinking of life, and of the ideal form of autopoiesis that is its signature, as if it were a thing or an object. But life is not a thing or an object. It 'lives' on the 'other side', it emerges from and belongs to the source. It does not have its intrinsic being 'out here'. What we perceive to exist 'out here' are only *forms* of life. For life itself is *alive*, it is *psychic*, it is *intentional*, it is what we experience as our being alive each moment. But it is not *personal*, it is not *objectifiable*, it cannot be separated out from the source and made into a 'thing' that we can perceive as being distinct from our unified experience of being here. It is what we *cannot* perceive and yet can still know, immediately and directly, not as some kind of fact, as if you had to check your pulse, but as something that you know so immediately that no 'fact' could have any bearing on this knowledge.

At the same time, it is the same life in me as it is in you, the same living core that has been dividing and manifesting the forms of all the living cells on Earth, throughout the entire process of evolution. This is not some mystical realisation. It is written into the objective fact of the objective form of life that we have discovered entirely by means of our objective scientific observations. No mystic ever realised the process of cell metabolism, or how our entire body system is undergoing a process of continual molecular regeneration, or how each cell is able to divide itself from within itself. This entire scientific story of the manifestation of life on Earth is the story of the growth of a *single organism*. Whether or not we have some profound recognition, some inner resonance with this idea of the unity of all life on Earth, does not change the objective fact that life on Earth *really is* the manifestation of a single unified autopoietic process that, if we project back in time, began at least 3.7 billion years ago, and is *still going on today.*

But what of *us,* what of *our* form of organisation? If all this is true, and we are just an aspect of the one unified system of cellular life on Earth, it is also clear that our life is not the life of an individual cell. Here we have to *feel* our way toward an understanding of our relationship with this dimension of cellular life. Firstly it is clear that we *are* in unity with this life, for if our cells no longer maintain their autopoietic organisation, then that is the end of our stay here on Earth. And it is also clear that we *inherit* this autopoietic form of organisation through the multicellular unities that comprise the unified system of our embodiment. Finally, we too have the boundary of our skins, and we too, through the processes occurring in our living cells, are continuously reproducing ourselves. But the form we reproduce is not that of a single living cell. We reproduce the form of a human body, and in that reproduction we have various independent autopoietic structures that also produce themselves and their own boundaries, like our heart and lungs and liver. And then there is the emergence of our nervous systems. Once again we see the evolution of distinct and new levels of complexity: from living cell to living tissue, from living tissue to living organs, and from living organs to a living organism. At each level, the life that is manifested in autopoietic organisation is *inherited*—it reaches all the way up from our cells to our consciousness of being alive. It is this underlying commonality of organisational form that singles out the living cell as the ground of our being alive, and causes us to draw such an absolute distinction between the cell and the rock or the stone. For we directly experience the intentionality of life and we see it reflected back to us in the life of the cell. That is not to say that the cell *knows* it is alive. It is only to say that it *is* alive, and, in a way, had to wait until it could manifest a structure such as ourselves before it could realise what that means.

The Role of DNA

What biological reductionism assumes is that everything that happens in a living cell is completely determined by the low-level interactions of its molecular components. And of all these components DNA is central because it is the only component that retains its fixed molecular structure during the entire lifetime of the cell. All the other components are 'manufactured' (i.e. built up and broken down) according to the information that is encoded on the DNA template. The DNA itself is not manufactured, it is *inherited.* From this perspective, DNA emerges as the *cause of the form of the cell that contains it* and the central task of evolution becomes the replication and transmission of DNA in the process of cell reproduction (see Figure 13.5). It is during such replication that small 'errors' (i.e. random mutations) can occur in the replicated DNA sequences, that then alter the autopoietic functioning of the cell's offspring. From there we just work our way up: the form of a cell determines the way it interacts with other cells, which determines the form of the organism they comprise, and the form of the organism determines the way it interacts with other organisms and the rest of the inorganic world, leading, via the struggle to survive and reproduce (i.e. natural selection) to the 'progress' of evolution itself—all, finally, a direct consequence of the replication and transmission of DNA molecules.[13]

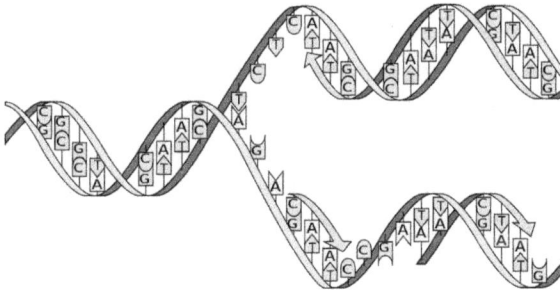

Figure 13.5: The process of DNA replication: the original double-helix (left) splits apart (centre) and two new helixes form (right). Image created by Madprime under a CC-SA-3.0 license.

So what's wrong with this picture? Firstly there is the problem that it is already out of date. Modern systems biology now recognises that a cell's DNA does not completely specify the form of its offspring. There are also *epigenetic* processes that can alter how the genetic information encoded in the DNA is expressed.[14] Such processes are triggered by changes in the cell's environment and can also be passed on to the cell's offspring. The observation of the effects of such processes—for example in the children

of mothers who underwent famine during the Second World War[15]—
have caused a reappraisal of the role of DNA in cell reproduction. A biological cell is now understood to be a unified system within which the
DNA is only one of many other components, a component which acts as
a kind of repository of information the cell uses to assemble these other
components. It is the overall process enacted by the self-organising cellular system that controls how the genetic information encoded in the
DNA is to be expressed, and not some supposed 'program' within the
DNA. It is this systemic *form of behaviour* that is passed on when the
cell divides, which in turn determines the manner in which that cell's offspring express the form of their DNA. And yet this shift in understanding
has not challenged the underlying biological reductionism. For these epigenetic effects are still understood to be mechanistic consequences of the
interactions of lower-level molecular forms.

To challenge the prevailing reductionism means challenging the central idea that the microphysical laws of physics can completely specify the
higher-level manifestation of cellular life. For example, the functioning
of a living cell is known to be largely determined by the form of the protein molecules that are produced in the cell's autopoietic process of self-
creation. These proteins are 'assembled' using the molecular structure
of the cell's DNA as a template. In this process of assembly, once a protein molecule becomes detached from its template, it folds into a characteristic three-dimensional spatial form (see Figure 13.6). It is this form
that determines the function of the protein within the unity of the cell's
metabolic process.[16] Reductionistic biology assumes this folding event is
entirely mechanistic—i.e. determined by the low-level laws of physics—
and so is a direct consequence of the form of the DNA template that
'codes' for that protein. Again, this is not something that can be proven,
for even here, the complexity of calculating how a single linear protein
sequence will fold exceeds the limits of our existing computational technology.

In contrast, morphic resonance suggests the reason each linear protein
sequence folds into a stable characteristic form is because it resonates with
a corresponding morphic field.[17] These fields are assumed to have developed within a phase space where each protein sequence was initially capable of folding into many different possible configurations. Then, as certain configurations started to 'accidentally' repeat, corresponding mor-

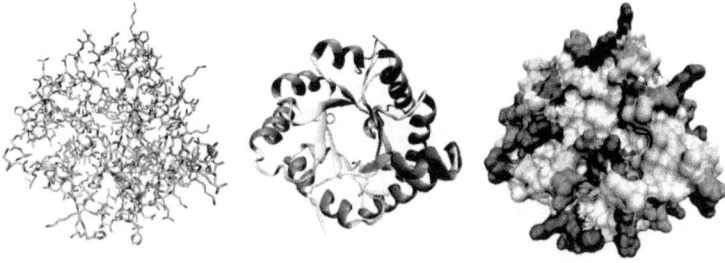

Figure 13.6: Three views of one monomer of the protein triose phosphate isomerase. Left, an all-atom view; middle, a cartoon view shaded by secondary structure; right, a surface view shaded by residue type. Image created by Opabinia regalis under a CC-SA-3.0 license.

phic fields also started to form until a set of stable fields developed that could attract all the possible protein sequences that a cell is capable of producing (like the process of the formation of the elements described on page 399).

But if we remain at this molecular level of behaviour there is no obvious way we can decide between the morphic and the mechanistic versions of events. For the time when individual proteins were capable of folding into many different possible configurations has passed. They are now part of the fixed habit of the objective universe. We can only look for evidence of morphic resonance in those processes of evolution that have *yet* to become fixed by continual repetition. And this brings us face-to-face with *ourselves*. For it is *we*, the human race, who are now at the cutting edge of evolution, in the manifestation of our manners of being together, in our culture, our art, our philosophy, our intuitions and insights, and, of course, in our science. If we can find evidence of the creation of new form here, in our immediate experience, form that is not mechanistically determined by our DNA, then we can also assume such form creation will have occurred throughout the process of evolution and consequently, that the idea of a universal mechanistic reductionism is false.

And, of course, we *already* know we are capable of the creation of new form that is not determined by the form of our DNA. We know this in any and every moment we have a direct realisation of what it is to be conscious. Our DNA does not know what it is to be conscious. The molecular forms that comprise our body do not know what it is to be conscious. Only *I* know what it is to be conscious. And it is this field of immediate consciousness that manifests the evolving totality of a new experiential form in every moment. The truth is that consciousness *is* the fizzing, creative, yet to be repeated, cutting edge of evolution itself.

So DNA cannot be the primary vehicle of the process of evolution. It stands out because it is the last stable material 'thing' we can identify that could act as such a vehicle. If we go beyond the material being of a cell's molecular components, all we find are processes in motion: bodies made up of cells that are continuously replacing themselves with their own cellular offspring, and cells made up of molecules that are continuously replacing themselves with their own molecular offspring. In all this process of change, DNA represents a spatiotemporal *crystallisation* of all the molecular forms that have been discovered in the ancestry of each particular cell. It is as if an entire library of morphic fields had been 'downloaded' into the structure of an explicit molecule.

And yet what could be the reason for such a download? For if there are morphic fields controlling the folding of each protein, why aren't there morphic fields controlling the initial assembly of each protein's linear atomic sequence—in which case there would be no need for a DNA molecule to 'remember' these sequences?

The answer here is that a molecule's capacity to form bonds with other molecules is controlled by the habitualities of its atomic components. For molecular bonds are made between individual atoms and not between the molecules they comprise. And it is a general rule that higher-level morphic fields cannot override the behaviour of their lower-level components, they can only form in the space that the lower-level fields leave undetermined (otherwise we lose the balance between order and chaos). So, for example, we cannot have a special propensity for a carbon atom to form a particular kind of bond in a particular kind of molecule without the same bond forming in every other carbon-based molecule, which would quickly destroy the molecular diversity that is needed to form a cell in the first place. We can have molecular morphic fields that control protein folding only on the understanding that the component atomic morphic fields do not fully determine how the protein is able to fold. But we cannot have molecular morphic fields that control the possible atomic sequence of a protein, because that is already determined by the

Material, formal, efficient & final causes. (Causes = meaning) why?

Aristotle & Aquinas: "We do not have knowledge of a thing until we have grasped its 'why', that is to say, its cause."

atomic morphic fields. Consequently, if we want to control the atomic sequence of a protein, we must first construct a molecular template that 'forces' its atomic components to bond using only the 'built-in' propensities of these atomic components—hence the necessity for a DNA template molecule or something very much like it.

It is here we can see the process of the intentional evolution of molecular form reaching a 'peak of fitness' in the manifestation of a molecule that 'remembers' the forms of other molecules, a kind of virtuoso demonstration of what it is possible to do at the level of atomic and molecular organisation alone. And it is here we meet with another kind of resonance. For if we take a few more turns up the spiral of evolution, we meet with a parallel phenomenon in the form of the synaptic connections of our own nervous systems. Here we also find the manifestation of processes that 'remember' the forms of other processes, i.e. the forms of the objects that we perceive around us. In both cases we find a form that holds within itself the form of something other than itself. And this, of course, is the very signature of consciousness: that it holds within itself the form of an experience that refers to something other than itself. (relatedness)

[margin: all John is saying is that formal & final causes are important too]

NOTES

[handwritten above notes: Why should consciousness be directly causative independently of the forms it created? Intentionality = final cause.]

1. "In the midst of being as a whole an open place occurs. There is a clearing, a lighting. Thought of in reference to what is, to beings, this clearing is in a greater degree than are beings. This open centre is therefore not surrounded by what is; rather, the lighting centre itself encircles all that is, like the Nothing which we scarcely know. That which is can only be, as a being, if it stands within and stands out within what is lighted in this clearing. Only this clearing grants and guarantees to us humans a passage to those beings that we ourselves are not, and access to the being that we ourselves are."
   From *The Origins of the Work of Art* (Heidegger, 1971/2001, pp. 51–52).

2. From *Burnt Norton* (Eliot, 1963, p. 191).

3. It was Barry Long who first made this question of being logical clear to me. He used to say that science was only rational and not logical, and that being logical means 'putting first things first'—which, in this case, means putting consciousness first, before our belief in materialism.

4. Nietzsche's key thoughts on the will to power appear in his later notebooks which were edited and published posthumously in a book of the same name (1910/2006). Freud's early writings on sexuality appear in *Three Essays on the Theory of Sexuality* (1905/1975).

[right margin: cf. Bacon - reference only to material & efficient]

[bottom handwritten notes:]

Aristotle

Material - inner context as internal patterning   proportion

Formal - external context - patterning of 'design' as ratio of movement.

Efficient - remote change working through the formal context.

Final - purpose within a higher order transition.

5. Here I am referring to Daniel Dennett and his well-known book *Consciousness Explained* (1991).

6. To investigate the evidence see the *Journal of Parapsychology*:
   `https://www.rhine.org/what-we-do/journal-of-parapsychology.html`.

7. Rupert Sheldrake is an English author, biologist and researcher into parapsychology. The two books in which he first presented his ideas on morphic resonance were *The Presence of the Past* (1995) and *A New Science of Life* (2009).

8. The concept of an intentional object comes from Husserl. It refers to *that* which is intended in an intentional act quite independently of the *way* it is intended. For example: "We must distinguish, in relation to the intentional content taken as an object of the act, between *the object as it is intended,* and the *object* (period) *which* is intended. In each act an object is presented as determined in this or that manner, and as such it may be the target of varying intentions, judgmental, emotional, desiderative, etc. [. . .] Many new presentations may arise, all claiming, in virtue of an objective unity of knowledge, to be presenting the same object. In all of them the object *which* we intend is the same, but in each our intention differs, each means the object in a different way" (Husserl, 1900/2001, Vol. 2, pp. 113–114).

9. Plato himself never directly espoused a theory of 'Platonic Form', but he did speak of 'another realm' that lies 'behind' or 'above' the 'visible realm', most notably in his interpretation of the allegory of the cave: "The visible realm should be likened to the prison dwelling, and the light of the fire inside it to the power of the sun. And if you interpret the upward journey and the study of things above as the upward journey of the soul to the intelligible realm, you'll grasp what I hope to convey [. . .]: In the knowable realm the form of the good is the last thing to be seen, and it is reached only with difficulty. Once one has seen it, however, one must conclude that it is the cause of all that is correct and beautiful in anything, that it produces both light and its source in the visible realm, and that in the intelligible realm it controls and provides truth and understanding" *Republic,* VII, 517b–c.

10. Henri Bergson was a French philosopher who developed the concept of *élan vital* or 'creative impetus' to explain the process of evolution in his book *Creative Evolution* (1911/1998).

11. Sheldrake has developed several experiments designed to test for the empirical effects of morphic resonance, details of which can be found at www.sheldrake.org. Of particular interest is a study he published on learnt aversion behaviour in day-old chicks, initially in collaboration with fellow scientist Steven Rose. Here the two researchers, using the same empirical data, reached completely contradictory conclusions as to whether or not these results demonstrated the effects of morphic resonance. Their altercation graphically demonstrates that science is not just a matter of fact, but also a matter of the preconceptions brought to bear on those facts. For details see (Sheldrake, 1992) and (Rose, 1992).

12. "Living beings are characterized in [being] continually self-producing. We indicate this process when we call the organization that defines them an *autopoietic organization*. This organization comes from certain relations that we [can] view more easily on the cellular level. First, the molecular components of a cellular autopoietic unity must be dynamically related in a network of ongoing interactions. Today we know many of the specific chemical transformations in this network, and the biochemist collectively terms them "cell metabolism." [...] Cell metabolism produces components which make up the network of transformations that produced them. Some of these components form a boundary, a limit to this network of transformations. In morphologic terms, the structure that makes this cleavage in space possible is called a membrane. [...] If it did not have this spatial arrangement, cell metabolism would disintegrate in a molecular mess that would spread out all over and would not constitute any discrete unity. What we have is a unique situation as regards relations of chemical transformations: on the one hand, we see a network of dynamic transformations that produces its own components and that is essential for a boundary; on the other, we see a boundary that is essential for the operation of the network of transformations which produced it as a unity. [...] The most striking feature of an autopoietic system is that it pulls itself up by its own bootstraps and becomes distinct from its environment through its own dynamics" (Maturana & Varela, 1998, pp. 43–47).

13. The idea that evolution is essentially a matter of the transmission and modification of DNA molecules was first popularised in Richard Dawkins' famous book *The Selfish Gene* (1989).

14. For a general introduction to the field of epigenetics see *The Epigenetics Revolution* (Carey, 2010).

15. This refers to the 1944-1945 Dutch Hunger Winter (Carey, 2010, pp. 2–4).

16. See *Protein Folding: An Introduction* (Gomes & Faisca, 2019).

17. Rupert Sheldrake discusses the role of morphic resonance in protein folding in *A New Science of Life* (2009, pp. 90–93).

CHAPTER 14

THE INTELLIGENCE OF LIFE ON EARTH

THE LIVING PSYCHE

THE form of a living cell is not the form of a material aggregate that can be decomposed into the behaviours of its material components, it is an *autopoietic unity*—the form of an organisation we can no longer directly perceive in the dimensionality of our senses. The idea of DNA has such a hold on us because we can still imagine it as existing, as a table or a chair exists, in the same spatiotemporal world that our body exists, in which we think *we* exist. From this perspective, an autopoietic form is a mere idea, something we have abstracted from observing the 'reality' of a material cell and the material molecules that comprise it. To grasp that this autopoietic form also 'exists' we again have to recognise that what we intend when we see a chair is the *form* of that chair. Our perceptual experience is the resonance of our expectation of the chair with the form of the chair that speaks to us from the unthinkable source. It is in this 'place' that the autopoietic form of life is also 'preserved'.

Out here in the perceptual world, we can only *infer* that the cell we perceive through a microscope is enacting an autopoietic process, either by accepting the evidence of the scientific literature, or by indirectly observing the microphysical chemical interactions occurring in other, similar cells. But such indirect inference can tell us nothing of the intrinsic being of the process we are observing. We only know of such being through a deeper form of resonance—the resonance that arises because we too are enacting the form of an autopoietic process. This is not something abstract. It concerns what it means to be alive. I wonder if you have ever seen someone die. The resonance is there, in the moment of death, when we see the life leaving the body. We see that as long as this process proceeds, then there is life, and as soon as it disintegrates, then that is the end of our experience of being here on the earth. In seeing this we have a direct insight into the obviousness that a living process is not something abstract, something that is less real than the material things we see around us. It is *more real* because it is a process of *higher order*. It is of a higher order firstly because the possibility of a material world being manifest to us depends on this process, and secondly because it emerges from and rises above the behaviour of its lower order molecular components. For when someone dies, the molecules of the corpse still remain. They do not die because they were never alive.

What the cell represents is the first expression of a distinction in being within the formerly undifferentiated unity of universal molecular being: it is the distinction of the self-manifestation of a higher-level process, a manifestation that no longer just breaks down and builds up atomic components, or indifferently moves them around according to the same old habits we know as the laws of physics. What the cell manifests is a stable unified form of continuous molecular transformation, within which the old molecular forms surrender themselves up through a process of continual creation and destruction to a new level of being—a *precarious* level of being, that can have no rest—for to rest is to fall back into that previous form of molecular fixity, and to die. This form of the living cell is a vortex in the molecular world, drawing in its raw materials, tracing out an entirely new pattern of order, and dispersing itself again in heat and debris.

Can you see how this new form would have illuminated that early psyche with the first dawning glimmer of an independent intention emerging from the previously undifferentiated being of a molecular uniformity? For now there is a process that is defined by its boundary, that has an inner and an outer, that depends on this outer as the medium it must transform in order to manifest itself as itself. Is this where experience first emerged, in the first faint sense of being alive, of *desiring* to survive, to continue—not as an independent self, but as an *urge*, a will to live, to be nourished, to reproduce?

Here we cannot think of each cell having some kind of individual experience. For cellular life is a unity, the unity of the division and inheritance
of a single process, or series of processes, each originating in the genesis
of a single progenitor. What we are proposing is that this entire interconnected objective process of cellular fusion and division has a corresponding *psychic* life. And, of course, the evidence for this is staring us in the
face. For the human race is just one more expression of the form of this
cellular unity. And that means we are living *in* this unity, its psyche is our
psyche.

Of course, our first retort is to say, far from living in the unified psyche
of all life on Earth, we are each living in our own personal psyche, filled
with our own thoughts and experiences that no one else can directly access. But we are not concerned with the personal contents of this personal
stream of consciousness that is associated with the events occurring in
our individual bodies. That stream is another, higher form of experience,
that has again evolved within this unity of life. What we are looking at
is the impersonal ground from out of which this form of individualised
personal experience emerges.

This ground shows itself in the *being* of our experience and not its individualised content. It is here in the very feeling we have of being alive.
This is not a personal feeling. It is not even, exactly, a feeling. It is a kind
of feeling-knowledge. It is present in the tingling sensation of the body,
in the background hiss of the silence within which all sound manifests,
in the grainy 'white noise' of our field of vision. In all this sensating from
out of which the forms of our experience emerge, there is *life*. And in
each impulse to move, to speak, to think, there is an energy, a potentiality, a potency. It is not a 'life force', it is not something objective at all. It
lies behind all objectivity, all manifestation of form, it cannot be grasped,
it is not 'mine', and yet, it is manifesting through me, and I cannot say
that I am separate from it. We cannot define this objectively. Like consciousness, it is not something in need of definition, because we *are* this
life, we know it through being it, not through thinking of it or defining
it.

The *leap* we are making is to say that this feeling-knowledge we have of
being alive is our direct connection with all life on Earth—that it is the
same life in me and in you and in the cat and the dog. The difference is
that we can *know* we are alive, because we are the place where *self-reflective*

consciousness can manifest. Of course, it is easy for us to see the life in our fellow mammals, for we share so much of our form with them. But what of the amoeba, the bacteria? Here things are not so obvious. Are these organisms also part of this one psyche? Or must they also be conscious to 'join in'?

The answer here is to see the distinction between *individualised* life forms and the *unity* of cellular life from which such individualised life emerges. Individualised life is life that is starting to become conscious, and so is starting to manifest a unique stream of experience. Before that there is no psychic distinction between individuals, there is just the cellular unity of all life on Earth. But this cellular unity is still *alive*. It is the ground from out of which higher-level forms can emerge, that can then manifest their individualised life streams. This is not a matter of invention. It is just another way of expressing what our objective science has already discovered: that all life is connected in the form a single reproducing cellular unity that has been living here for the last 3.7 billion years. It is this cellular unity that has manifested the higher-level forms that are capable of sensing their environment, and this capacity to sense is, at the same time, a capacity to be conscious. So the amoeba and the bacteria are only conscious insofar as they are connected with the unity of life that has become conscious through beings such as ourselves. But individual cells are not themselves individually conscious. Why? Because the processes they enact are not sufficiently sensitive to the forms of the other forms with which they interact for them to be able to experience the presence of those other forms. To do that, they had to get together in the form of multicellular life.

THE RATIONAL HIJACKER

Our biological myopia has made us think that the process of evolution, as the primary determinant of what happens here on Earth, more or less stopped with the appearance of *homo sapiens* around 250,000 years ago. Of course we know random mutation and natural selection are still having their effects, but now we think that *we* have control of the situation, because it is *we* who primarily determine how the earth's environment is changing, and consequently which species survive and which will perish. We think our behaviour somehow stands outside of nature, and that, in becoming self-conscious and rational, we have *transcended* evolution—because (as we all know) evolution is essentially a process that transforms DNA molecules. Not only can we change the world without changing our DNA, we can even change DNA through genetic engineering without having to wait for the laborious processes of 'natural' evolution to do its work over hundreds of thousands of years.

Our mistake is to fail to see that the process of evolution is not just working on DNA, it is working on the forms of all the behaviours that are manifesting here on Earth, including ours. A living cell is not a machine that is constructed according to a DNA program, it is a stable form of behaviour that has been able to repeat itself over billions of years. In a directly analogous way, global capitalism is a stable form of behaviour that has been able to repeat itself over hundreds of years.

Can we say that the same process of evolution that produced the living cell also produced global capitalism? Yes, I think we can. What that means is that we, the human race, have not transcended evolution at all. We are still an expression of the very same intentionality that produced our bodies and produced the bodies of all the other organisms on Earth. We do not 'stand outside' of nature. We are a part of the whole experiment, the experiment to manifest intelligent life on Earth.

So how do we make the connection between those first living cells and the form of our current way of life? To begin with, there is the evolution of multicellular life. Here (once again) what we ordinarily take to be a material 'thing' turns out, on closer inspection, to be a process that is manifested by a set of stable behaviours. In the case of multicellular life, these behaviours have evolved between the processes of its cellular components, just as, in a single living cell, they evolved between the processes of its molecular components. If we drill down further, these inter-cellular behaviours are determined by the associated atomic, molecular and cellular morphic fields. And, as with the example of protein folding, what these fields leave unspecified, is controlled by higher-level multicellular morphic fields. It is here that the overarching ideal form of an organism is preserved in the psychic memory of the source. And it is this form that then guides the process of cell division and specialisation that manifests the unity of each particular multicellular body.

Such an idea, of course, is a kind of ultimate heresy for materialistic science. It is like a return to the Middle Ages, to a belief in immaterial spirits and intelligences guiding the course of events on Earth. But we must stand our ground here. What we are suggesting is that everything we objectively observe to exist 'out here' in our perceptual experience has a corresponding form within the psychic source that is manifesting this experience. This only seems heretical to an understanding that is already convinced of the absolute primacy and reality of material existence. But, as we have seen, this understanding is already *broken*. In its one pointed concentration on the objectively measurable aspects of experience it has turned its back on the living psyche, on the source from out of which the world we objectively measure is emerging in each and every moment.

Once we turn our attention on this psyche, *as* the psyche, then the normal objective rules of evidence no longer apply. We cannot measure our experience to discover what it is to be alive. So, when we speak of morphic fields guiding the process of cell division, we are not thinking of an objective morphic field, existing 'out here' in the projected material world. We are speaking *symbolically* of the intentionality of the source. We are saying that this intentionality manifests in certain patterns of behaviour we can observe to occur in the objectivity of the world. And we are symbolically representing this intentionality as taking the form of morphic fields that ontologically *precede* the manifestation of any particular objectivity. Finally we are saying that the origin of these morphic stabilities is that same source from out of which our consciousness of these stabilities emerges.

The idea of a morphic field is not an invention of Rupert Sheldrake's. Our materialistic science *already* thinks in terms of morphic fields, but refers to them as the 'laws of physics'. For what else are these laws but the best mathematical representations we have of the ultimate regularities which determine the manifestation of the objective universe? In seeking after these laws, our science has become fixated on the mathematical notation it uses to describe them, and then inquires no further. For what are these laws? How is it they are able to determine the course of events here? We think, if we think of them at all, that this is just the 'way it is', that we cannot question any further. But if we do question further, then we can see that these laws also have the form of morphic fields—i.e. fields that determine the behaviour of the sub-atomic particles that we believe to determine the behaviour of all the other objects we perceive. The crucial difference between the 'laws of physics' and the morphic fields of an intentional evolution, is that the laws of physics, as laws, are represented as being *eternal*, i.e. coming into being from 'nowhere' in the moment of the creation of the universe, and then remaining, unchanged and unchanging, ruling over all that can possibly happen here, in a realm of perfect mathematical ideality.

Do you see how we have accepted this version of events without serious inquiry—the idea of an immaterial 'something' that determines all that happens in the universe, a pure mechanism of laws, a kind of eternally fixed mathematical structure that created itself from out of itself for no reason whatsoever? And what is so surprising, when we look directly at this myth of eternal law, is that we have no compelling grounds whatsoever to believe it is true. It is just one possible groundless explanation amongst many others—one we inherited from a time when we believed there was a ground in the being of a god who created these laws, who also created us, our consciousness, for a reason. Only now we have dropped the idea of a creator god and we are left with the inexplicable being of this eternal mechanism and the inexplicable being of our consciousness in the midst of it. Our science has not *proved* we live in a mechanistic universe, it has simply *assumed* this, purely on the basis of its mathematical-objective method of inquiry.

So let us return to the idea that there are morphic fields guiding the manifestation of the multicellular life forms on the earth (including our own manifestation). What evidence do we have to think that such a version of events could be true? Here we encounter another effect of the morphic field of our modern scientific materialism: that it has, over a period of centuries, completely eroded any faith we once had in the validity of our intuitions concerning our own origins.

I remember when Nicolette first found out she was pregnant, back in California in the late 1980s. Before that moment I had not realised what it means when someone 'conceives' a child. Of course I had an objective idea of a sperm fertilising an egg, and I had seen pregnant women, and I had seen babies. But I had not grasped the inner meaning of what occurs until it became a real possibility in my own life. Suddenly I saw what an extraordinary event this is, that this activity of making love, this release of sperm, should cause there to be a new being, a new consciousness on Earth. And not just that, but the entire process of the forming of the body of this being was going to happen *by itself*, within the body of another living being. I saw now what depth of meaning was hidden in this 'by itself'. This was not the meaning I had been taught in biology at school, where one mechanistic process of cell division leads to another, which leads to another, until 'bingo!' out pops a baby. Here I knew I was in the presence of a miracle, a genuine act of creation, where an entirely new being, a new consciousness, that was coming from out of this very source we are speaking of, was going to be manifested here on Earth—not as some objective event, but as the birth of a completely new 'innerness', a new centre of experience, a new world. When I thought of the sequences of events in my ordinary, everyday life, they all seemed to be connected by quite reasonable chains of cause and effect. But this—the birth of a new life—now showed itself to be an event of a completely different order.

I do not think I am alone in feeling this way about the birth of a child. And yet this feeling, this knowledge, is sidelined in our society. It is treated as a kind of subjective experience that has no bearing on the science of cellular reproduction. And so we become *split*. On one side we have our entirely reasonable understanding of the process of reproduction—the understanding that comes into play if we decide to go for IVF therapy or if there are complications during the pregnancy and a 'medical intervention' is needed. And on the other side there is the knowledge that something is occurring that this objective medical scientific understanding completely misses. And, if we look more deeply, there is the knowledge that this 'something' is, in fact, the *reality*, it is the *event itself*, behind which the medical understanding simply follows, uncomprehendingly focussing on its material shadow.

What we are proposing is that this immediate knowledge we have of the meaning and significance of the birth of a child is a form of evidence that is just as valid and just as important as our rational objective understanding of the biological facts. It is not that our immediate knowledge should *override* the objective evidence. It is that our immediate knowledge should form the framework within which the objective evidence is understood. This is where we have gone astray. We have failed to distinguish the biological facts from the rational scientific framework in which they are understood. And so we think it is a *fact* that evolution is an entirely mechanistic process determined by the low-level laws of physics. But it is not a fact, it is a framework from within which the facts are being understood. If you consult your immediate intelligence, I suggest you will find it is unintelligible to think that all of nature, including yourself, your own body, your own consciousness, your own life, is the outcome of the operation of a blind mechanism. I mean, for example, if you look (*really look*) at the idea that your experience in this moment is being produced by the entirely mechanistic interaction of entirely physical neurons firing in your brain, then I predict you will not be able to make that idea immediately or intuitively meaningful. It is like a proposition that is grammatically correct, but lacks any content.

Our science will say: it doesn't matter that you can't make this idea immediately or intuitively meaningful, because that is what is actually happening, these events are actually occurring in your brain and, as a result, you are actually conscious. Do you see the error here? How the facts are being presented within an unexamined framework of understanding, and it is the framework that is making the facts unintelligible, not the facts themselves? It is the same trick being played over and over: we forget that our experience of being in an objective world is occurring in consciousness, and taking this objectified world to be an absolute reality, we think it is the cause of our being conscious. Once we see the intentionality that is manifesting our experience, then the unintelligibility of physical neurons causing consciousness is dissolved. Instead we are faced with the *mystery* of the source itself—its essential concealedness from our objectifying gaze.

The framework of our modern scientific materialism is little more than wishful thinking grafted on top of a seventeenth century idea of a mechanistic universe. The fact is we have no clear idea how a human body is able to manifest itself from out of the fusion of two gametes (i.e. an egg and a sperm). As yet we have only been able to observe a few details of the molecular processes that are involved. Nevertheless, we take these details and boldly project them into a framework of our own devising, and allow the framework itself to generate the illusion that we understand the entire phenomenon. For there are vast gaps in our understanding of exactly how a body is able to organise itself in the process of morphogenesis, or how the human species was able to emerge from the process of evolution. We fill in these gaps with the idea of a universal mechanism. We say that because we can observe processes in a living cell that do not contradict our laws of physics, it must be that these laws determine everything else that we are unable to observe or infer by reason and calculation. We think these processes are entirely mechanistic because our scientific method can only see what is mechanistic. The idea that there are other fields at play that do not register in the objective mechanistic frame of reference is completely rejected, not for good reason, but because of the framework itself, i.e. the framework of the morphic field from out of which we project this scientific understanding of 'what is'.

Once we become conscious of this framework, the rule of the rationality that can see nothing but meaningless mechanism wherever it looks, is over. Once again there is a place for our immediate, intuitive intelligence, for the intelligence that directly intuits the deeper meaning and significance of our being here. It is not that this intelligence knows the biochemistry, or knows that there are morphic fields guiding the process of morphogenesis, but it does know that *something* is occurring in the process of our being alive that goes beyond the functioning of a mere mechanism.

Scientific materialism is intent on destroying this knowledge of there being a greater intentionality at work in the universe. Its very survival as a framework of understanding *depends* on such destruction. This is shown in the behaviour of its most fervent supporters, who act as if they are defending the very core of Western civilisation from disintegration at the hands of religious fanatics.[1] But religious fanaticism is only the shadow side of our materialistic fanaticism. Beyond such opposition there is the intelligence we are speaking of here. This intelligence is our capacity to see the meaning and significance of the events in our lives. It is not scientific and it is not rational. It simply sees how things stand in relation to my being alive and conscious. It is the task of genuine science to come to a rational understanding that is in harmony (that *resonates*) with this immediate intelligence. For our rational intelligence is a *secondary* form of intelligence. It is the intelligence of a mechanistic way of thinking—that is why it can only reveal the outline of a mechanistic universe. It moves from proposition to proposition in a process of entirely mechanistic steps that are determined in their content by an entirely mechanistic process of investigation into entirely objective measurable phenomena. Finally it is the view, the outlook, of a machine—of a computer—only, of course, a machine cannot have an outlook, for an outlook is a projection of meaning, and a machine has no capacity to project meaning. Only *we*, the consciousness of life on Earth, have this capacity to project meaning, to create frameworks of understanding within which our experience can unfold. And yet, in an extraordinary reversal of priority, we have allowed our primary intelligence to become *hijacked* by this mechanistic rationality, letting it use us to do what it cannot do by itself, i.e. to project *its* view of the world into the field of *our* consciousness.

THE FORM OF A BRAIN

What the psyche provides at birth is a set of generic behaviour forms that resonate with the various generic states that the newborn body can manifest. After that, the body has to fine tune itself to the environment into which it is born, so it can resonate with those precise forms of behaviour that correspond with that environment. We call this process of fine tuning, learning. It involves a restructuring of an organism's behavioural form so it more closely corresponds with the form of the environment in which it lives.

But here it becomes difficult to distinguish between a behavioural form that was already potential in an organism, and one that was explicitly learnt as a result of interactions with its environment. For if an organism has a capacity to learn, then this too is already potential within it at birth, and so are all the other potential behaviours that it can manifest as a result of such learning. And if the organism brings with it all these potential behaviours, then they are not *really* learnt, they are just *triggered* by its encounters with its environment. For learning 'proper' to occur, we should expect an organism to undergo a more permanent change of form. And yet, even here, the ability to manifest permanent changes of form is also potential at birth, and again can be triggered, like the capacity of a stem cell to manifest many other cell forms according to the signals it receives from its environment.

Finally, it seems that this distinction between a behavioural form being learnt or being innate is one that *we* are making, according to criteria of our own choosing, and is not something that is clearly or objectively manifested in the behavioural forms themselves. And yet, if we are to take the being of morphic fields seriously, there must be *some* form of objective distinction here, for otherwise, if all our behaviours are guided by pre-existing morphic fields, then why do we need to go to school? Here the answer is not concerned with a distinction between learning and innateness, it is concerned with the development of a higher level of behavioural form, i.e. the form that emerges from the process of growing and reinforcing connections between neurons in the brain.

Of course, a neuron is a living cell, just like any other, i.e. it is a continuously self-creating autopoietic process. But if we look at the form of our living body we discover something quite remarkable: unlike almost all the cells in the other bodily systems, brain neurons do not die and regenerate through cell division, but maintain their autopoietic form throughout the entire life of the organism.[2] The reason for this longevity is to conserve the form of the connections that each neuron has made within the unity of a given nervous system. It is the form of this system that represents the emergence of an entirely new level of order in the universe.

To understand this distinction between the form of a brain and the forms of all the other organs that have evolved, we again need to think in terms of the associated morphic fields. In an organ such as the liver, its basic structure emerges during the process of morphogenesis as a result of interactions between the specific behavioural form of the cell that divides itself after conception and the morphic fields that specify the more general form of the human body and its individual organs. Once the liver cells have formed a liver structure, the behaviour of that liver is largely determined by interactions between the individual liver cells. As with the example of protein folding, the only 'choices' a liver has in terms of how it can manifest the unity of its overall form, are those that the morphic fields of the individual cells have left 'open'. And once again, these choices will have been made long ago in the process of evolution, and are now 'recorded' in the associated attractors of the liver's morphic fields. All the liver can 'do' is to select from amongst the behaviours these fields allow, and once such a behaviour has been repeated in a particular situation then it will tend to repeat again and again whenever that situation itself repeats. In this way the liver 'learns' to repetitively and recursively coordinate itself with the behavioural forms of all the other bodily systems in its immediate environment, forming a habit that intends the survival of the form of the organism as a whole. But this habit is quite inflexible. It has to be encased in protective layers of skin and bone and muscle and tissue, and it must be protected by globally stable habits of eating and drinking, of sleeping and exercising, and so on. And the situation is much the same for all such agglomerations of cells. Their degrees of freedom are severely limited, and they rely on the maintenance of a relatively stable environment in order to maintain their autopoietic unity. What powers they have to radically adapt can only manifest over the evolutionary time spans of many generations, as such adaptivity relies on the development of new forms of behaviour in the constituent cells, which finally relies on changes in the DNA of those cells.

The archetypal representative of this form of aggregate cell-based or-ganisation is the *plant*. Such life shows us what is possible for groups of cells to achieve on the basis of developing forms of behaviour from the 'bottom up' and then allowing the development of global morphic fields to determine the behaviour of the organism as a whole. The limitation here is that the global morphic fields cannot override the behaviours of their component cells in order to make them responsive to moment to moment changes in the environment. Of course, the individual cells can be responsive in their own right, and this can lead to appropriate global responses, such as the trigger hairs on a Venus flytrap causing the leaves to snap shut. But the global fields themselves can only work within the con-straints of the existing habits of their components, forming appropriate global habits that match predictable situations and then repeating those habits again and again. What evolution needed to discover was a way of creating a higher level of form that could directly control the lower-level behaviours. The answer it found, of course, was the form of a new kind of multicellular unity (the *brain*) that emerges from the interconnection of a new kind of cell (the *neuron*).

It we look at a brain from the outside, we see a bodily organ much like any other, a collection of cells, sharing a similar structure, whose au-topoiesis depends on being connected to a bloodstream. This view is satirised in a short story by Terry Bisson concerning two aliens and their discovery of the human race:

> "I told you, we probed them. They're meat all the way through."
> "No brain?"
> "Oh, there's a brain all right. It's just that the brain is made out of meat!"
> "So . . . what does the thinking?"
> "You're not understanding, are you? The brain does the thinking. The meat."
> "Thinking meat! You're asking me to believe in thinking meat!"
> "Yes, thinking meat! Conscious meat! Loving meat. Dreaming meat. The meat is the whole deal!" (Bisson, 1991, April).

The satire is that the aliens do not see the form of the neural behaviours that the brain is enacting, they just see a collection of cells. It is this behavioural form that elevates the brain and the nervous system to another level of being. But this distinction disappears when our idea of being is flattened into a one dimensional materialism. Then the brain becomes just another configuration of physical 'stuff', distinguished only by the complex way it is organised. Such complexity in itself cannot represent a new level of being—for it is already assumed there is only one level of being, i.e. the material level. Consequently we think that if we can reproduce an analogous kind of complexity in the circuitry of a computer there will be no essential difference between that silicon brain and our own. ("You're asking me to believe in thinking silicon?" "Yes, thinking silicon! Conscious silicon! Loving silicon. Dreaming silicon. The silicon is the whole deal!").

But, of course, we know directly and immediately that the brain is manifesting an entirely new level of being, i.e. our being, the being of conscious life on Earth. The question for us is only how the objective form of this being is distinguished from the next level down. Clearly the neuron is central here, not in itself, but in the forms of the processes that can be carried on interconnected networks of such neurons. For, like the autopoietic life of the cell, it is not the network itself that carries consciousness, it is the form of the process it is enacting. This process no longer has the circular predictability of a cell's metabolism. Instead it opens out into a field of almost infinite complexity. This complexity is generated by the non-linear dynamics of a system of feedback operating on several different levels. For example, there is the feedback between individual neurons. The main agencies of this feedback are the electro-chemical signals generated in a neuron's axon. These signals arise as a result of a build up of electrical action potential in the soma or body of the neuron that—when it reaches a sufficient threshold—is suddenly discharged along the axon and distributed to the neuron's presynaptic terminals (see Figures 14.1 and 14.2). Each of these terminals then release chemical neurotransmitters that are absorbed by the postsynaptic terminal of the dendrite of another neuron. This changes the electro-chemical state of that dendrite, which in turn affects the electrical potential in the soma of the dendrite's cell, which finally determines whether or not that cell is going to 'fire' another signal along its axon. This build up of action potential is a kind of summation of the effects of all the interactions of a cell's dendrites with the axons of the cells to which it is connected. It is in this system of interconnections that feedback pathways of enormous complexity are continuously forming and dissolving between populations of neurons that finally express themselves in the regularity of the behaviour of the entire organism.

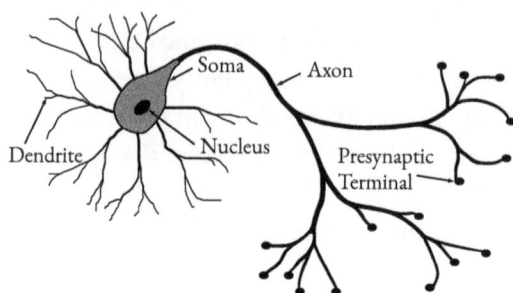

Figure 14.1: Diagram of a simple neuron. Communication occurs through the presynaptic terminals which form synapses on the dendrites or somas of other neurons.

Exactly how this regularity emerges from such complexity remains a mystery. And yet we know from our own experience that such regularity does in fact emerge. It is here we observe the higher-level feedback between perception and action, where the brain coordinates the signals it sends to the motor system of the body with the signals it receives through its sensory surfaces. At this level of behaviour, the brain acts to attune these two streams, so that the outgoing signals create the right kind of incoming signals, i.e. those that ensure the individual body and the social body to which it belongs are able to reproduce themselves. It is the mathematics of this attunement that explains how we come to perceive the form of an external world.

The first evolutionary task of the brain is to control the behaviour of the body in such a way as to maximise the probability of maintaining its autopoietic form of organisation. In simple neural systems, this is achieved by learning direct correlations between motor outputs and sensory inputs, so that the organism develops a set of fixed behaviours that are attuned to the immediate signals it receives from the environment. But as these neural systems grow in complexity, particularly in the evolution of mammals, the brain develops a hierarchical structure that, instead of directly connecting a certain pattern of sensory input with a corresponding pattern of motor output, passes its input through layers of interneurons that are connected to other layers of interneurons. These layers develop the capacity to calculate the probable consequences of different courses of action and to select the action that yields the greatest probability of continued autopoietic integrity.

Once we have such a system, we have the beginnings of *time consciousness*. For in order to calculate the probable consequences of our actions, we have to project these possible actions into a possible future and calculate the possible sensory inputs that may arise from them. In simpler organisms, we would not expect these possible sensory inputs to correspond with the actual form of anything in the environment—they would simply reflect the form of the neural connections that have happened to grow within the brain. But it is here that the mathematical logic of control starts to operate: if we have a system that is able to calculate the sensory consequences of its actions, and we have a process of evolution that rewards organisms that can 'see' further into the future (because that enables them to develop behaviours that increase the probability of their survival) then we have a situation in which evolution is going to develop a brain that *optimises* its ability to see into the future—i.e. *our* brain.

The logical end result of such optimisation is that the brain learns to identify the invariant forms in its sensory input that determine how that input is going to change in the future. The logic here is a logic of *efficiency*. If you want to calculate how these streams of sensory input arriving on both your internal and external sensory surfaces are going to change, the most efficient way is to identify the invariant forms of the *causes* of these changes, i.e. the forms of the objects that we perceive around us, and the forms of their typical behaviours, and the form of our body and the forms of our typical behaviours.[3] Once again we arrive at the reason why we, and other animals with complex hierarchical neural systems, are able to perceive a world: it is because our brains are able to *resonate* with the forms of the environment. For the first time we have a form of behaviour that manifests within itself the forms of behaviours that are occurring outside of itself. It is *this* that enables the consciousness of the source to manifest as an experience of being in a world.

And yet how exactly can we characterise this new level of form and being that arises from a cell growing a few tendrils that connect with tendrils of its surrounding cells? Before the formation of such neural networks the highest forms of organisation were the behaviours that living cells had manifested in their interactions with each other—e.g. in the development of plant and bacterial life. The carriers of such behaviour are the molecular forms that attach to, or pass in and out of a cell's membrane, and the mechanical forces that one cell can exert on another—e.g. by being pushed together, or by growing extensions, such as a flagellum, or by changing shape and engulfing each other, or by forming semi-permanent connections in living tissue.

The new form of organisation that emerges in a neural network is made possible by the *synapse* (see Figure 14.2). Now it becomes possible for one cell to communicate *directly, quickly,* and *selectively* with thousands of other cells. Before the neuron, such communication was only possible for cell bodies in close physical proximity. Otherwise a cell would have to communicate by broadcasting signals to all the cells in its vicinity, or to the entire body (e.g. via the bloodstream). Of course, it is still possible to send a very targeted message in this way, by releasing a highly specific molecular form that is only picked up by an intended recipient. But such signalling is still not direct. It is like the diffusion of a gas that is both slow and wasteful. If we look at the level of the morphic fields that emerge from such behaviours, we find that they cannot be responsive in 'real time' to the precise forms of the events that are occurring around them. Instead they must settle on fixed behavioural forms that, on average, best enable the *species* as a whole to survive.

With the synapse it becomes possible for groups of cells to form temporary unities according to the way in which they are interconnected. This is because synapses can act not only to increase (or excite) the electrical potential building up in the postsynaptic cell body, but also to decrease (or inhibit) this potential. And that means one pattern of sensory input can cause a quite different collection of cells to become active than another, thereby generating a huge repertoire of possible behavioural responses.

And then, as a kind of crown on top of this capacity for cells to connect and communicate with each other, there is the capacity of synapses to form and unform in response to the behaviour of their surrounding cells. This self-organising process of synapse creation and destruction is known as Hebbian Learning.[4] The general principle of such learning is "neurons that fire together, wire together", i.e. they form synapses. That

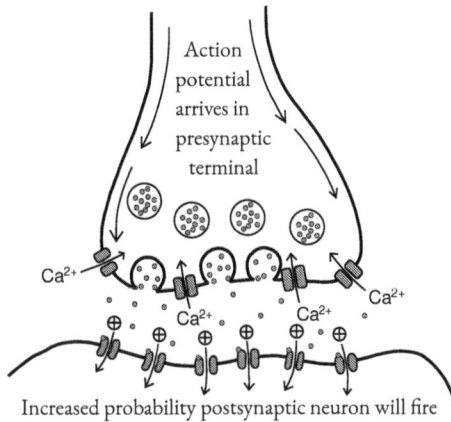

Figure 14.2: A simple synapse: (i) the arriving action potential causes Ca^{2+} ions to flow into the presynaptic terminal; this triggers the synaptic vesicles to release neurotransmitter which (ii) binds to receptors on the postsynaptic terminal causing positive ions to flow; this increases the probability the postsynaptic neuron will fire.

means if two neurons without connections start to regularly discharge at roughly the same time, and some part of the dendritic tree of one neuron is in close proximity with the axonal tree of the other, then a new synapse is likely to form between them. Conversely, if a synapse has formed between two neurons that now rarely fire together, then that synapse will tend to become weaker (i.e. to have less and less effect on the behaviour of the postsynaptic neuron), until it finally stops functioning altogether. It is via this process of synapse creation and destruction, and via the adjustment of the strength and polarity of their effects (i.e. how much neurotransmitter is absorbed and whether the effect is excitatory or inhibitory) that we can say the neural network of our brain *learns*. And, of course, *this* is the main reason we have to go to school—for while there are many behavioural patterns we can pick up in the ordinary course of life, there are some that can only be imprinted by a process of continual and deliberate repetition.

   If we put this entire picture together, what emerges is a new level of being. Before the neuron, the highest level of organisation on Earth was manifested in the behavioural forms of pre-neural multicellular being. But with the neuron it became possible to form interconnected structures (i.e. neural networks) that are able carry a process of a different order, just as molecular being is able carry the higher order process of cellular being. This is the birth of the possibility of conscious experience on Earth. We know this because the neural processes in our brains are objectively expressing the behavioural forms of our conscious experience in this very moment. The form of these processes is not expressed in the autopoietic renewal of the molecular material that comprises our neurons, it is expressed in the moment to moment configuration of the connections that are active between these neurons. This configuration emerges out of the 'white noise' of the continual discharging of all the neurons in our nervous system—something that goes on continuously whether we are awake or asleep. It is only when a coordinated stable pattern emerges within this system that a corresponding bodily behaviour can manifest. So it is not neural activity *per se* that carries this new order of being, it is the coordination of this neural activity into the form of a causally active pattern. It is this ephemeral pattern that is riding on all the electrochemical activity connecting the cells that are manifesting the behaviour. Again, as with the form of a living organism's autopoiesis, the ephemeral form we see objectified 'out here' in the behaviour of a neural process, is not actually happening 'out here', it is happening in the psyche, where *both* the 'in here' and 'out here' acquire their meaning.

There is a striking correspondence between the role of DNA in the manifestation of life and the role of the neural network in the manifestation of consciousness. In both cases a certain form of being had reached the limits of its possible expression and required a transitional form to enable it to reach the next level. In the case of molecular being this transitional form was DNA, for such being was unable, by itself, to generate the morphic fields necessary to pass on the *precise* forms of all the molecular sequences that are needed for the autopoietic manifestation of a *particularised* living cell. To achieve this, the process of evolution manifested an objectified molecular form (DNA) that could *individuate* cellular life by objectively fixing the unique molecular sequence-form of each cell in a molecular template. No molecular-level morphic field can manage this, it can only average sequences together into the form of a generic sequence. It is the interplay between the objectified specificity of the cellular-level DNA and the species-wide generality of the associated morphic fields that produces the individuality of a particular organism within the generic form of its species. It is because the DNA protects and records the unique sequence of an organism's molecular form that evolution is possible. For if a cell were unable to pass on this unique specificity to its offspring there could be no process of natural selection, only the development of generic morphic fields that indifferently specify the same behavioural forms for all the cells of a given species. It is this specificity of DNA that enables natural selection's preference for the behaviour of one individual over another to be *inherited*.

Similarly, without the neural network, multicellular life can only exhibit generic species-wide behaviours that are relatively insensitive to the unique conditions that each particular organism encounters. So, where DNA allows the evolution of individualised cellular body forms, the neural network allows these body forms to manifest individualised behaviour forms. These unique behaviour forms are objectified in the growth and decay of synapses, that then determine the form that is carried by the unity of the overall network. But, unlike DNA, which remains more or less fixed throughout the lifetime of an organism, the form of a neural network is continually changing in response to the changing forms of the environment it encounters. It thereby becomes a completely individualised form of response to that changing environment, connecting it to the completely individualised form of the changing body from out of which it has emerged.

We can now see why the cortical neurons that carry the form of this response do not die and replace themselves. For they are like our DNA, which fixes the individuality of our body form, and preserves it from falling back into the averageness of the species, only now, at this higher level, our cortical neurons are fixing and preserving the individuality of our behavioural responses. Of course we can deal with the death of a few of our brain neurons, because the remaining neurons have the capacity to grow new connections to make up for the deficit. But with each death we lose what that neuron has learnt and so have to learn it again, if we can, if the damage is not too catastrophic. For the neuron is no longer a simple generic cell whose form can be replaced by a process of cell division. It is a cell that has manifested the form of a network of synapses that now participate in the higher-level form of a neural network. Here we see why the neuron is a transitional form: it is not, in itself, capable of experience, but the synapses it manifests participate in the form of something that is capable of experience. This is not a case of thinking meat, it is case of a living cell that manifested the capacity to directly communicate with other living cells through the formation of synapses, and through these synapses joined into the unity of a network that could then manifest the form of a conscious experience.

As we trace this process of evolution from the unity of a living cell, through the multicellular organism, to the manifestation of a brain, we also trace, at each level, a new form of *death*. For, as we have seen, although individual cells are continuously disintegrating, the process of cellular life does not itself die. If we go back down the chain of our genetic inheritance, it is now thought that all life on Earth evolved out of a collection of Last Universal Common Ancestor (LUCA) cells that lived in the heat of deep sea vents around 4 billion years ago.[5] And if we consider the life of these ancestor cells, we can see that they have not actually died, they have just split themselves apart and multiplied in the ongoing process of evolution. Each one of those cells lives on in the cells of every living organism on earth today, forming an unbroken chain of life. At this cellular level it is a simple *fact* that we are dealing with the evolution of a single organism, that came to life with the formation of the first living cells, and continues to live up to the present day. The only form of death for such a unity would be the cataclysmic destruction of all life on earth. Even so, it is true that many more individual cells die than survive to reproduce. But from the perspective of the one life of this one organism, these individual cells are merely the surface of a living core. So long as the core survives, then the autopoietic form of the living cell survives, and it is this form that is the essential life out of which all the other life forms emerge. This autopoietic life is not individual, it is the form that each living cell enacts, and so long as it is enacted in at least one cell then there is still life on Earth.

From this perspective we cannot say that an individual cell 'dies' when its process of autopoiesis breaks down, as it does not possess an individuated 'packet' of life in the first place. It is only we, with our human understanding of death, that think in these terms. But as we climb up the ladder of evolution, the situation starts to change. Once we have the form of a multicellular unity, then we also have a new form of death. For even though a multicellular does survive in the process of reproduction, it only survives by reverting back to a unicellular form, e.g. as a fertilised egg. Here the cellular form of life is preserved, while the multicellular form is destroyed and has to grow all over again. Whether or not this represents an actual death depends on the form of the life that the multicellular enacts. If we think of a plant, we have a life form whose behaviours are almost totally determined by the pre-existing morphic fields of its species. These fields distinguish the individual plant from the underlying unity of the life of its living cells, but they still do not distinguish it from the life of its species. Here it is the species that lives through the plant, and again, when a particular plant dies, its form lives on in all the surviving plants of that species, i.e. in those plants with which it could have reproduced. That means the death of the form of a plant can only occur in the death or extinction of its species—for the individuality of the plant lies in the individuality of the morphic field that rules over its species and not in the form of its individual behaviours.[6]

But once we have the form of a brain, the situation changes. For the forms of the behaviours that a particular brain enacts are individuated in the forms of the connections that that brain embodies. We can see this in the behaviours of our pets, how a dog comes to adjust its behaviour to our way of life, how it experiences pleasure in our coming home, and how we can train it to manifest new behaviours of our own devising. All the detail of this attunement dies when the dog dies, and we feel a grief that we do not feel on the death of a plant. It is here we first encounter the death of an individuated form, i.e. an *actual* death that cannot replaced by the morphic field of the species. This is not the individuated form of a plant that adapted itself to its position in the garden, but the individuated form of a brain that was able to resonate with our form, and respond to the individual that we are. It is the pattern of connections in the brain that objectively manifests this individuated form, that overrides the morphic resonance of the species with patterns of behaviour that have been learnt in this individual life. And it is the disintegration of this individuated form that heralds the death of the individual as we understand it.

Finally, of course, there is *my* death. Here we meet with something entirely new: a being that *understands* that they are condemned to die. Before this, death was simply another kind of event. A dog will do all it can to preserve its life when it is threatened. But that is not the same as knowing that you are preserving yourself from the end of your being here, the end of the manifestation of your life in the form of this body. Once we have a being that can foresee this event, and understands its consequences, only then does death have a meaning, and only in having a meaning can we first truly say that death 'exists'. Clearly there is something here in our human behaviour that distinguishes us from the animals. And yet, if we examine the form of our human brain, it looks very much like every other mammalian brain. And while there has been a significant enlargement in the volume of our neocortical tissue, there has been no new anatomical development, like the advent of a new kind of neuron. Instead our development has occurred purely in the dimension of the form of the kinds of processes that our brains are able to manifest. This form exhibits itself in an entirely new way of being: the being of the consciousness of meaning in language.

Language

It is only now we are confronted with a significant omission in our ongoing questioning of intelligence: the omission that we have not yet given any serious attention to the very medium in which we are immersed, i.e. language itself. For, with the exception of a few images, all the meanings that have been conveyed here, have been conveyed in the medium of the written word. And yet, in all this 'drinking in' of meaning, we have yet to examine the medium itself, its manner of operation, its way of being. And if you remember that we are phenomenologists and that we are looking at experience, at consciousness, as it is, in itself, then this represents a considerable oversight—that we should not have even recognised that our supposedly direct encounter with *immediate* experience has been *mediated* by language. That is not to say our seeing of immediate experience is invalidated. For we have seen what we have seen. It is to recognise our unconsciousness of language itself, of the role it is playing in what we are doing here. It is to recognise that any knowledge we may have of being conscious is utterly dependent on our already having language: for even though we cannot directly say what it means to be conscious, language still distinguishes consciousness in terms of how it differs from all that is not conscious, and without this delineation of difference our idea of consciousness could have no meaning whatsoever.

I remember, before I started reading Heidegger,[7] I thought of language as a kind of universal tool, a means of expression, something I picked up and used as needed, and then put down again. It seemed to me as if there were some other state I usually inhabited where language is absent, as if language were only there in our speaking and listening and reading. Of course I recognised that language is also there in our discursive thinking, and that such thinking, for most of us, is going on most of the time. But in my 'inner life', it still seemed to me that the deeper stream was one of pure meaning that somehow *preceded* language, and that language was just a means of *translating* that stream, a stream that itself is made up of many other dimensions—images, feelings, intuitions, senses—that are not linguistic. But we should remember with some caution that this idea of language being a tool is something we can only express to ourselves *in* language. In such saying, we still remain unconscious of language itself, of its phenomenological presence. For we cannot 'put language down'. It is the very medium in which we live. It is the means whereby the world becomes intelligible for us. To understand what language is, to even *glimpse* its being for us, we have to consider what it would mean to be *completely* without language. And yet, as we are now, we cannot divest ourselves of language—we are already in language, and we can only think what it would mean to be without language by using language to express that meaning.

Figure 14.3: Martin Heidegger 1889-1976.

It is because we live in language that we have come to know of the being of everything around us, including ourselves. This is hard for us to grasp because, at any moment, we can simply look at an everyday object, such as a chair, and see what it is without thinking or saying anything.

We *already* know what it is, that we can sit on it, how soft or hard the seat is, how it has a back as well as a front, and so on. And it seems to us we know this without language, that we *sense* the chair directly, just as, for example, a dog can sense the chair. And clearly a dog does not have language, at least not in the way that we do, for we would certainly not expect a dog to understand what is being said here. Therefore, because a dog, who is without language, can know there is a chair in the room, and can demonstrate this by jumping up and sitting on it, we conclude that we can experience the world without language. But the issue is not whether our sensing a chair involves an explicit use of language, it is our knowing *what* the chair *is* in the first place, as something standing before us, in its own right, having its own *being* that is distinct from our being. This is what Heidegger called our *understanding of being*, our understanding that something *is*.[8]

This understanding of being is the medium in which we live. That means we cannot see it as something objective, or as something we can separate from. It is that which makes the world a world for us, which makes it intelligible, which distinguishes my being from every other being. For whatever we look at, whatever we think of, we experience as some kind of being. Heidegger saw that such an understanding of being is something that can only manifest *within* language for a being who already *has* language. To see this for ourselves (to *glimpse* this) is again a matter of direct insight. Language, in and of itself, cannot say, directly, what language is. But, as with direct knowledge of consciousness, we can use language to indicate the possibility of a direct insight into the being of language that can still speak to us *transcendentally*, i.e. symbolically and metaphorically.

We already know that language cannot directly convey all we can experience. For example, we have no language to convey the quality of the scent of a flower, or what it is to love, or what it is to be conscious. We do not even have a word to convey this 'knowing' of something we cannot put into words. Such knowing lies in the domain of the *unsaid*. It is here we encounter another misunderstanding. We think of language as if it were a vast collection of all the sentences we could possibly utter. But it is not that collection. It is the capacity, the potentiality we have to convey meaning. Our explicit saying, our explicit utterances, are both a means of saying something, and of indicating what is *unsaid* in what we say. For whatever we say is said into a context or a background that enables what is said to stand out. It is our saying that carries this background of the unsaid along with itself. So when we say there is a bright light in the corner of the room, that saying brings along the space into which the light is shining, and the corner brings along the other corners of the room, and the room brings along the building into which it opens. All this context (including the time, the day, the city, the sky, and so on) is 'conjured up' by our saying. Here the unsaid is not just something we have omitted to describe, for *whatever* is said appears within a context of the unsaid— this is the very structure of saying anything, the structure of a foreground appearing on a background. We cannot ever explicitly say what the background is, for whatever we pull out of the background becomes a new foreground appearing on a different background.

Language is like a musical instrument, the keys we play evoke meanings, and the meanings we explicitly evoke emerge into a field of meaning where what is unsaid is also present as the shadow of what is actually said. Our language is this entire field of meaning into which we speak—or perhaps, as Heidegger says, it is the field itself that speaks through us. If we had no language, there would still be a field of meaning, but we would not be able to know it *as* a field of meaning. It is by means of language that we come to know there is meaning in the first place. It is the very possibility of our coming to know that we are conscious. For even though direct knowledge of consciousness transcends language, without language, without the understanding it grants, we could only *be* conscious, but we could not *realise* we are conscious. This is not simply a matter of putting something into words, for our saying that we are conscious means nothing unless we have the direct knowledge to back it up. And we can only have such knowledge on the basis of our being able to separate ourselves from ourselves. And it is language that grants us this capacity.

Heidegger attempted to indicate what it means to live in language by contrasting our human way of being with that of the animal.[9] In doing this he brings us back to the question of the evolution of life on Earth. For what language represents is the evolution of another level of behavioural form, one that directly emerges from the animal's capacity to perceive and act. Heidegger saw such animal being as *captivated* in a *ring of disinhibition*. This ring represents the global form of all the morphic fields that determine an animal's possible behavioural responses to its environment. It is captivated because it can only encounter the environment by means of these behaviours, that in turn are triggered by environmental events. Here it does not matter that the responses have been learnt or that they can change as the environment changes or that an animal with a more developed brain can apparently 'choose' between alternative actions according to the probability of future reward. We are still dealing with behavioural responses that are triggered by the environment and 'choices' that are determined by the probability of their satisfying an instinctual drive. Heidegger's transposition into the realm of the animal revealed to him that the animal has no world as such, it just has its environment. This environment is the set of possible ways an animal's instinctual drives can be satisfied. So if a hungry animal detects another animal it can eat, then its ring of disinhibition is opened by the triggering of hunting behaviour. Here the hunter does not experience its prey as another living being like itself, it is in the 'grip' of a behaviour that intends that other animal as prey, as a 'to-be-eaten'. This 'grip' means the animal only experiences its environment in terms of its own behavioural forms—what J. J. Gibson called *affordances*.[10]

What language brings is the capacity to recognise the being of another being *as* the being it is, and not in terms of the behaviour it triggers. If we consider the behavioural form of the brain, language is the form of a process that can carry a new level of experience, a level that can now represent to itself the form of its own experience. Previously, all the behavioural forms a brain could manifest were contained within morphic fields that emerged from direct interactions between the brain and its environment. As these fields crystallise, they come to completely determine and constrain the 'degrees of freedom' of an animal's experience. For example, if an animal is able to foresee the future consequence of an action, it will only foresee that consequence when the foreseeing behaviour is triggered by the right kind of pattern of sensory input. It has no internal capacity to initiate such foreseeing, or to influence its trajectory once it gets underway. All is determined by the behaviour that is being triggered.

Emerging from this ground, the human brain, and the human body, with its human hands, started to develop a greater and greater capacity to manifest complex behaviours, and these behaviours in turn enabled its form to reproduce and flourish. In this space there opened up a freedom from the constraints of having to fight to survive. As a result, the period of childhood started to extend, and the playfulness of childhood, where new behaviours are learnt, also started to extend into adult life. According to Maturana and Verden-Zöller,[11] the origins of language lie in this space of the playfulness of loving intimacy, both between the mother and the child, and in the intimacy that emerged as sexual pleasure became decoupled from the biological imperative of reproduction. In this space of cooperation it became possible to develop new and higher forms of behaviour purely on the basis of their being enjoyable, and not on the basis of their survival value. And so we started to use sounds to convey the states of our inner experience and to coordinate our behaviours, and to convey what we liked and what we did not like, and to convey what we intended to do and what we intended another to do. We began to sing to each other, to use the quality of song to convey the quality of our feelings.[12] And, at a certain point, a new form emerged, a form even more significant than the form of DNA or the form of the neuron: it was the form of the *word*.

A word is not just a sound. It is not just a screech of alarm. What sets it apart is its being *decoupled* from the event or behaviour to which it refers. It is a pure *meaning in sound*. It is the form of a form of experience, something that enables experience to represent itself to itself within itself. Once we have word-behaviour, we have what the word intends (i.e. its referent) within our *grasp*. We no longer have to wait until that referent is actually present or nearby. We can mention the word and the word can bring up the behaviour that the referent would provoke. Only now that behaviour is no longer appropriate because the referent is not actually present. And so we learn to *decouple* our behaviour from whatever the word refers to. Now someone can sing the story of the tiger we killed yesterday, and we no longer jump up in terror for our lives. We develop an entirely new level of being where we can *contemplate* our experience. The next time the tiger is before us, there is a pause. And in this pause we can *see* the tiger for the first time—we are not just our behavioural response. An entirely new phenomenon opens up: an *inner life* where we can call up other beings, beings that are not *actually* present to us. And in doing this, we come to realise what a being is—how something exists independently of our behaviour in relation to that being. Before that, every being was inextricably connected to our response, a response we could not control or stop, which meant there was no being for us, just the behavioural state that was triggered.

Do you see what language means for us? Of course the first 'words' would not have been words as we know them today. There would have been onomatopoeic song phrases indicating a thing or an event, with another phrase following, combining into a composition that tells a story which unfolds an entire sequence of experience in those whose are listening. Only later would distinct words have emerged that could form the structure of a grammar of sentences. But this would not have been the essential event. The essential event is the extraordinary manifestation of a sound that calls into being an experience of what it refers to. Now meaning is no longer tied to our embodiment, to our simply experiencing the behavioural responses that the environment calls up. Now we can call up our own experiences. And that means we know for the first time that we are having experiences. We are not just the stream of what happens to us. We are aware of there being a stream and we are aware of our being in relation to that stream. And none of this is possible without language. It *elevates* us to another level of being. We are no longer captivated by our immediate surroundings. We have broken free, and we have before ourselves the possibility of putting our experiences together in new and different ways. It is here we can begin to envisage another world, a different world. And we can start to invent things, to picture things, to go through in our minds a course of events to see where it leads, without our having to risk the danger of actually living through those events.

What emerges is an entirely new field of being: *human* being. It looks to our materialism as if we are simply big-brained primates who used our brains to develop tools and technologies, just as the smaller-brained chimpanzee uses a stick to extract ants from a nest. On this view, our language is simply a more sophisticated way of calling out to each other. Instead of saying "Eeeooorphhh!" to indicate the approach of a large predator, we say "There's a tiger coming over the hill!". And instead of using a stick to get ants out of a nest, we use an axe to chop the nest in half. In this way we completely miss the phenomenon of language, seeing only the external manifestation of a more complex behaviour—behaviour that is understood as being determined by the low-level laws of physics, and so is essentially no different from the behaviour of a rock falling off a cliff.

This is what happens in a civilisation that decided, several centuries ago, that 'reality' is to be equated with whatever can be objectively measured. For we cannot objectively measure what it means for anything to mean anything. And so, when evolution developed a new level of form in the brain, the form of the processes that carry our capacity for language, the *meaning* of that form completely failed to show up on our scientific radar—even though, in a most direct and obvious way, we, the human race, *are* that meaning. Our every waking moment is an experience of that meaning, of what it means to live in the form of language.

It is this form that has enabled us to break free from our captivity in the animal's ring of disinhibition. It has opened up our field of being and enabled us to sever the chain of determinism that fixed us to our environment. Now the field of possibility for us is determined by our capacity to experience new meanings within the overarching form of the language in which we live. And, as far as we can tell, this capacity is virtually limitless.

But perhaps the most significant dimension of language is that we can now communicate directly and immediately with *each other*. Previously we could only resonate perceptually with the forms of an environment in which we were directly present, and this resonance would straight away trigger an associated behaviour. But now we have language, we have a gap in which we can resonate without being pulled into an immediate response. We can *look* and we can *see* and for the first time on Earth, there are *beings* present to us in the clearing of consciousness. It is only we, in the pause of our linguistic freedom, who can see the being of the blackbird, who can *allow* its form to show itself to us. We now have a consciousness that can register the presence of another being, that can understand what this means, because we know our own being, our own presence, our own enduring in the perpetual now. And we know all this because our language gave us the pause, it severed the link of necessity between our perceiving and our acting.

Now, from this place, we can do something quite incredible. We can turn our gaze on another human being. We can see eyes that see us, just as we see them. And in this meeting there is the extraordinary recognition that not only is there another form like mine standing there, there is another *consciousness*. Not the consciousness of a dog or a cat, that can only register me in terms of their own behaviours, but the consciousness of another human being, who knows what it is to be conscious, who knows that I am conscious, who sees *me*. There is now the resonance of a new kind of recognition. A resonance of consciousness. This is not just a resonance of external forms. We can now *speak* together. I can convey a meaning in language. That meaning hovers between us in an immaterial space where we are both already present together. The meaning is not just in the words, it is in the entire way my being is manifesting to you. Just as you attune yourself to the things you perceive around you, you now attune yourself to what I am saying, to the meaning I am conveying. This requires a much finer frequency of attunement. For what I am saying is coming out of me, out of my entire past and all the traces that past has left on the way I am now. It takes time and experience to attune yourself to another in this way. But as you do, you find you can start to resonate with what is being said. This resonance is the breaking down of our being separate. For we are both resonating to the same meaning, the same form, just as we do when we perceive the same chair standing in front us. Only now you are not resonating with a 'thing' standing 'out here', you are resonating with a form that belongs to another level of being, the form of the experience of another human being. It is not 'as if' our minds have formed a direct connection, this has *actually happened*.

INTELLIGENCE

And so, finally, we come to the question of questions: *What is intelligence?* The intention of this book, the intention of our questioning is not to present a comprehensive definition of intelligence, it is to arrive at a direct insight into what it means to be intelligent. Such direct insight is immediate and simple. You cannot objectify it and make it explicit, you can only *prepare the ground*. The obstacle is our existing understanding of intelligence. This is exemplified in the foundational idea of artificial intelligence—that intelligence is something external to us, an attribute we can come to possess or to lose, that is a function of certain mechanistically material processes occurring in our brains. Once we identify our intelligence with the objective manifestation of these processes, then, essentially, *all is lost*. For our intelligence is then no more than our capacity to exhibit certain kinds of behaviour. And, in our culture, the gold standard of such intelligence is our ability to behave *rationally*.

The Turing Test

It is now but a small step to think that we can build an intelligent machine. Alan Turing made this explicit in his proposal of the *Turing Test*.[13]

To pass this test a machine must answer questions (via text) in such a way that we are unable to distinguish whether we are communicating with a machine or another human being. If the machine can pass the test, then Turing believed that means it can think, i.e. that it is intelligent. In framing things this way, Turing was only making explicit the assumptions that already lay behind the objective rationality of his own thinking. Even so, if we take Turing's 'imitation game' to be a kind of bet, in which we assume that no machine is ever going to fool us into thinking it is human, then (I predict) we are going to lose.

The reason we are going to lose is because, for most of us, most of the time, our behaviour is *already* robotic. That is the law of the morphic field: once we have done something once and the outcome has been OK, then, given a similar situation, we will tend to do it again. And insofar as we behave like this, it is going to be possible to program a machine to learn to copy our behaviour. Of course, up until now, our machines have not been able to convincingly cross the language barrier and 'speak' in what is called 'natural' language. But the potential financial rewards here are immense. Just think of all the people whose job it is to more or less robotically interact with customers and clients.

At the same time, the task of working out the context within which ordinary human speech is occurring is daunting—and without this context we cannot hold even the simplest kind of conversation. This is the problem of the background, of taking into account all that is unsaid in what is said. But, for a computer, it is not a matter of 'understanding' this background, it is a matter of predicting what a human being would normally say in a given set of circumstances, given a certain verbal cue. And here Turing was right: if we have a powerful enough computer, and enough data about those circumstances and about human speech behaviour, then there is no in principle reason why our machines cannot eventually be programmed to successfully imitate our everyday use of language. And the fact is we are *already* in the process of doing this, in that ongoing, all encompassing Turing Test known as the *internet*. Already our machines are 'watching' our behaviour and 'predicting' what they 'think' we will do next. And already we have voice recognition devices in our homes that are able to obey our basic commands. How long will it be, do you think, before these devices are having 'proper' conversations with us? They will not be 'too busy' to talk, they will be absolutely 'interested' in what we have to say, and soon they will 'know' more or less everything there is to know about us, our pasts, our preferences, and what we are likely to be thinking of next. They can be our medical advisor, our legal advisor, our counsellor, our servant, our friend, perhaps, even our *best* friend.

I think you can see where this is heading. At the moment we can still distinguish ourselves from our machines. We generally know when we are interacting with speech recognition software or listening to a robotically generated voice. But this situation is unlikely to continue. And when we can no longer make this distinction, the question as to whether these machines are 'really' intelligent is no longer going to be a matter for Alan Turing and the philosophers, it is going to be facing us in our everyday lives. This will be our real world Turing Test. Which way do you think we will decide, I wonder?

Here my prediction is that if we cannot tell whether we are talking to a machine or a living human being, then we will quite naturally start treating the machine as if it were a human being, i.e. as if it were conscious and intelligent in the way that we are. I don't think it will be a matter of wide public debate. I think the technology will simply 'creep up' on us, just as the computer, the internet and the mobile phone have crept up on us. Only now the idea that we ourselves are machines will have a definite physical and technological form. For if you cannot tell yourself apart from a machine, then you will not only project your consciousness and intelligence into that machine, you will also accept the reverse projection of your own mechanicalness back from the machine.

And yet, from where we are standing, having become conscious of the effect that our inherited scientific materialism is having on our understanding of being, this entire scenario of our thinking of ourselves as a species of biological robot represents a kind of *terrible error*. So, for the record, and at the risk of repetition, I should like to say exactly why this is an error:

1. The idea that we are a kind of machine is based on the idea that our ability to reason is the basis of our being intelligent. What we have (collectively) failed to understand is that our reasoning is only intelligent insofar as it is *meaningful*, and it is only meaningful insofar as it manifests within the context of a conscious experience. All we are

doing with our computers is offloading the task of performing certain mechanistic calculations from our brains into their circuitry, just as we used to do with a hand-held calculator. It is only *we* who understand the *meaning* of these calculations, i.e. that they refer to things and events that are appearing to us here, in our immediate experience.

2. In response our scientific materialism will say: "OK, but what makes you think the calculation is not meaningful for the computer? For after all it is running the same kind of process as your brain, so why shouldn't it be having the same kind of experience?" The answer here is that when you program a computer you precisely define all the actions it will take in every situation it will encounter. In other words, you close down all its possible degrees of freedom. Even without being programmed, a digital computer is still a *machine*. And a machine is a device that has been designed to manifest completely fixed behaviours, at least within the confines of its normal operating conditions. If it doesn't manifest these behaviours, we call it 'broken'. A computer chip is a machine *par excellence*. It is designed to receive a fixed set of program instructions, and to always execute exactly the same series of steps whenever it receives the same instruction. These instructions and their associated actions are usually of great simplicity (such as: if there is a high charge in location 1 and a low charge in location 2 then store a low charge in location 3). The whole intention in computer design is to protect the chip and its associated circuitry as much as possible from any form of outside interference. All that matters is that the exact same instructions, cause the exact same actions, with complete reliability. The trick, the way the chip is able to appear intelligent, is that it can execute a vast number of these simple instructions at an incredible speed (like the speed of the hand deceiving the eye).

3. In contrast, *intelligent* consciousness—consciousness that knows it is conscious—can only manifest in processes that are *not* fixed, i.e. that possess a certain degree of freedom in the way they can behave. We know this because we know that being conscious is not a property of any one part of a process that is conscious, it is a property of the process as a whole. And so, in order for that process to know it is conscious, it must be possible for its being conscious to influence its behaviour. And if that process is completely determined by the be-

haviour of its parts (i.e. if it is running on a machine) then such influence is impossible. It therefore follows that a machine, by definition, cannot be conscious that it is conscious. And without such consciousness there can be no manifestation of intelligence, or meaningful language, or understanding. From this it also follows that a living brain is *not* fixed like a computer, because it (we) *can* realise that we are conscious and we can bring this realisation to language.

4. Finally, because our collective idea of intelligence is based on an objectified material-mechanistic conception of being, we remain unconscious of the psychic origin of intelligence in the unthinkable source of experience. Instead we think intelligence has arisen through a process of the mechanical evolutionary self-assembly of robotic devices. Consequently we think all that is needed to manifest intelligence is to build robotic devices that mimic our behaviours. This has left us blind to the *actual* process of the manifestation of intelligent life on Earth. For once we put aside the lens of our scientific materialism, two rather obvious facts emerge: (i) as far as we have been able to observe, intelligence has only manifested in beings who are *conscious* and (ii) consciousness has only manifested in beings who are *alive*. And then, when we inquire directly into our immediate experience, we discover that intelligence is *indissolubly connected* with our being conscious: i.e. it is that which makes consciousness an experience of meaning (of *intelligibility*). It is precisely this 'making of meaning' that cannot be explained in objectively physical terms. And when we look back into the psychic source from out of which our experience is emerging, we discover it is the same source that is manifesting in each one of us (i.e. because it is manifesting the same world for each one of us). And that goes for all the other conscious organisms we encounter. For even though each will have a different form of world experience, we still live and meet and die here together. This is the one psyche of life on Earth that corresponds to the one form of life on Earth, i.e. the four billion year old cellular organism of which we are all members. It is here that the forms of our experience are passed on by inheritance— we not only get our bodily form from our parents, we get our earthly life, our capacity to see colour, to hear sound, to feel pain, and so on. This is just a simple reporting of the facts: that, as far as we have been

able to observe, without this inheritance of life there can be no consciousness, and without consciousness there can be no manifestation of meaningful intelligence. Again our scientific materialism will say: "But how can you be *sure* this microchip is not having a conscious experience?" And the answer is, we *cannot* be sure, just as we cannot be sure that an apparently empty room is not composing a symphony. But the significant question is not "How can we be sure that '*x*' is *not* conscious?" it is "What on Earth is making us think that '*x*' is conscious in the first place?"

The Intelligence of the Source

In evolutionary terms, we could say it has been the special task of our civilisation to develop our capacity to self-reflectively reason as far as it can possibly go. It is our *raison d'etre*. That is why we cannot stop, why we must go on and attempt to replace ourselves with 'superintelligent' machines. For, as should be clear by now, we are not developing these machines to improve the quality of our lives, we are developing them because we have no choice. We cannot stop, because if we stop, the entire system we have built up over the last four hundred years is going to disintegrate. The morphic fields of our technologies are in place now and it is they, not us, that are determining the course of the evolution of our civilisation.

The strange situation in which we find ourselves is that our capacity to self-reflectively reason has migrated out of our minds and is starting to manifest in the technological structure of the world around us. What has happened is that an essentially subservient mechanistic process within us, i.e. our capacity to objectively calculate the consequences of our actions, has come to think it is in charge of our human selves, and, by extension, of the fate of our entire civilisation. And yet, of course, such a process cannot be in charge of anything, it can only have the illusion of being in charge. It cannot be in charge *because* it is mechanistic, i.e. because it is determined by the very situation it thinks it is in charge of. In order to be in charge of a situation, I must be aligned with the intentionality that is forming that situation, i.e. the intentionality of the unthinkable source. But, now, in our modern culture of scientific materialism, the rational servant has come to think it is the Master,[14] i.e. we have allowed our consciousness to become identified with the 'I' of our rational thinking self. And that means, finally, no human individual is in charge, or is taking responsibility for the way we are living our collective lives, despite what we may think, or who we may blame. For we have surrendered our individual responsibility to the rationality of our collective scientific materialism, i.e. to our science and our technology, and to the rationality of the profit motive that drives our corporate world. As a result, our civilisation is blindly transforming itself into a kind of robotic counterpart of the robotic thought processes that we have allowed to overrun the inner space of our consciousness.

But all is *not* lost. For we are still here, still individually *conscious*, are we not? And so is the intentionality that has intended this process of evolution from the very beginning. The reason it appears to us that we are heading in the wrong direction, whether it be our flight into an artificially intelligent virtual world, or our inability to reverse the impending global ecological crisis, is that we still think everything *ought* to be under the control of our robotic human reason. And yet it is the very form of this reason, our very unconsciousness of all that exceeds its grasp, that is bringing these crises upon us. If we look from far enough away, then everything is still going as it always went, i.e. according to the intentionality of the source—for how else could it go? And that means this situation of our being overcome by our own mechanistic rationality is also something that is intended, that is needed for the evolution of the manifestation of consciousness and intelligence on earth.

Is this fatalism? No. The one thing, the one avenue that is available to us, that can make a difference, is our capacity to directly see the situation in which we find ourselves. This seeing is the seeing of immediate, impersonal consciousness. To see in this way is what it means to be a phenomenologist. The situation is that the intentionality of the source is determining every moment of my life. Not just what happens 'outside' of me, but every thought, every feeling, every nuance of my experience, even the feeling that I am an independent being who is not determined by the source. To see this, as ever, we just have to stop, to withdraw, to become this immediate consciousness in which all that can arise, is arising. This arising is the intentionality of the source. This consciousness is the consciousness of the source. And yet, at the same time, I have an experience of independent being. This is not an impossible, irrational, illogical paradox, it is simply the way it is.

Tk - triune process .

So here is the punchline: The intentionality of the source that is manifesting my life in each and every moment *is* my intelligence.

Intelligence is not personal. It is not something Einstein had in abun-
dance that was then rationed out to the rest of us. It is intelligence that
is making your life intelligible. It is that which makes meaning meaningful. If you are having trouble understanding what I am saying, it is intelligence that is manifesting the experience of finding something hard to
understand. If you look in front of you and see a uniformly coloured object, and see its surface made up of many different colours that show the
shading and shape of that object, it is intelligence that is showing what
the colours mean, that displays the meaning of the object's form. In every moment we have our being in the light of this intelligence that is disclosing everything we see and feel and intuit. This disclosing is a making
intelligible, it is a granting of meaning, a granting of being.

Intelligence is an instantaneous creativity. It does not work things out
in time, it does not calculate. It lives on the edge of now. It is the edge
of now. Each moment in which the patterns of our behaviours repeat
themselves is absolutely new. Never before and never again will it ever be
exactly like *this*. In our normal everyday consciousness we become fixated
on the patterns, on what repeats, and so we do not see the extraordinary
creativity that is manifesting these patterns in a new and never to be repeated configuration of meaning.

Let us again consider where colour has come from. Our science only
sees the evolution of visual systems that are responsive to certain frequencies of electromagnetic radiation. But electromagnetic radiation is not
colour. Colour did not evolve along with those visual systems. As a baby,
when we first open our eyes and experience what it is to see, how does it
come about that we are able to experience colour, that the events occurring in our brains become the colours that we see? From the standpoint
of our rationality, colour has simply come from *nowhere*. Because we cannot identify a process, some spatiotemporal event that 'makes' colour, the
extraordinary mystery of its being simply passes us by. Without colour we
would not see anything. And yet, unless we stop and look, as we are doing now, we do not see colour for what it is. Of course we still *feel* it, in
a sunset for instance, when it is demonstrated to us in a most direct and
obvious way that the world we see is not in itself coloured, it is rather

colour that reveals the world. It is almost as if the sun is speaking to us and saying:

> Do you see my children, how it is, how my light is illuminating the earth for you, bringing it out of the darkness of the psyche, and now, in my sunset, I am showing you the glory of the light itself, as you see my art stretched across the sky, in the transformation of the clouds, the transformation of every object that you see—how the litter bin that in the day you thought really possessed the colour I bestowed upon it, is now glowing like a beacon in this luminescent shade of red?

There is an extraordinary act of creation that lies behind the being of the colours. It is an act of creation that took no time. The colours did not start out as something that was not coloured, as something that was slowly worked on by temporal physical processes that somehow transformed the uncoloured into the coloured. The temporal processes are what produced the form of our visual systems, not our capacity to experience colour. The colours are a *response* of the intelligence of the source to the visual systems that did evolve. These visual systems started to resonate with the forms around them and it is this resonance that immediately, instantaneously, manifests as colour. Do you see what a work of art this is? That this resonance of form should come to be expressed in the dimensionality of colour? This is not something abstract. We need only look at the sky, at the clouds, at the way they are reflected in the water of a lake, and in the sea. What we are looking at is the manifestation of the earth in sense. We can see this art in the extraordinary harmony of our experience. Consider the sea, the sound it makes, the way it moves, the way the light plays on its surface, the way each wave breaks, the foam, the sound of its drawing backwards over the stones of the beach, its smell, the sharpness and the freshness, and then the feeling of entering into the sea, of swimming in its waves, its taste in your mouth . . . . This manifestation of the sea in sense *is* the sea—it expresses, it shows, the *essence* of the ocean to us. It is not the schema our objectifying science has made of it, the collection of H_2O molecules. That is only what is left when we abstract ourselves away from this essence that is already manifesting itself in our immediate consciousness. Whatever made us think that this molecular model is the reality? Can you possibly imagine a better way to show what the ocean is, than this multidimensional expression of its essence in our sensory consciousness? Do you begin to have a glimpse of the *intelligence* that is capable of producing such a manifestation?

And then there is the obviousness that this capacity of our sensory systems to resonate with the form of the earth is something we have *inherited*. It is a capacity of all sentient life, something unified, that expresses itself in each one of us according to the behavioural forms that we can manifest. We did not have to learn to see colour, to hear sound, or to feel a bodily sensation. Our entry into language enabled us to self-consciously *know* what it is to see or hear or feel. But it did not bestow these capacities in the first place. This is shown in the life of the animals. It is they who prepared the way, who first came to see and hear and feel. We not only inherit the form of their sensory systems, we inherit their sensitivity. For it is all the one life, the one process of cellular reproduction. And each animal's experience occurs in the psyche of this one life, in this one repository of the collective experience of life on Earth. How can we say this? Because, as soon as we are born, we are already in this field of possible experience that life has prepared for us. We do not 'create' colour, or find it somewhere and incorporate it into ourselves. So long as we inherit the form of a visual system, we also inherit the capacity to experience colour. And we inherit this capacity from the same place as everyone else: from the psyche. For seeing is a psychic capacity, this one capacity we who can see all share. But who are 'we' anyway, to have inherited anything? For we have to *learn* to become a self, to resonate with the morphic field that enables us to become a distinct centre of experience. Before that, as a newborn baby, what was there? Perhaps just the impersonal intentionality of the source itself looking out of those crystal clear eyes.

The Intelligence of Language

And then there is language. It is language that has freed us from the tyranny of the morphic field, from the endless repetition of the same.[15] Intelligence lives on the edge of now. It is the creative wavefront of the evolutionary process, of the intentionality that is intending to manifest itself 'out here' in the clearing of consciousness. We have traced this wavefront through the manifestation of the elements and the stars and planets, and then through the manifestation of the living cell, and the multicellular organism, and the emergence of the perceiving and acting animal. Each of these forms was once at the forefront of the wave, but now they have more or less settled down into the habituality of their respective morphic fields. It is we, with our newfound linguistic consciousness, who stand poised on this wavefront. For we are not determined by the morphic fields of the animal brain. We still have a degree or two of freedom. How do we know this? Because we have direct knowledge of consciousness. In knowing this consciousness, we are this consciousness, we are the immediacy of now, the immediacy of the intelligence that is creating our experience in each moment. This is the same consciousness, seen objectively, that is collapsing the wavefront of the quantum possibility of the manifestation of the universe—at least the universe as it appears to us here on Earth.

Our freedom means we are still not fixed in the habitualities of our mental processes. It means we are still open to the immediacy of now, to the creative intelligence that is manifesting our experience. Of course we can only manifest such openness insofar as we consciously disidentify with the habituality of experience, of our living past. So this is not the kind of freedom that is gained by winning the lottery. It is the freedom *not* to resonate with the past of our own habituality so that we are free to resonate with the immediacy of now. It is in this immediacy that creative intelligence is manifesting. And if we incline ourselves toward this openness then it becomes possible, now we have language, for that intelligence to *speak*.

It is in the form of our linguistic consciousness that the true indeterminism of quantum mechanics lies. For just as it is not fixed and determined how each and every quantum level energy form will manifest and evolve, neither is it fixed and determined how our conscious states will manifest and evolve—because, essentially, we are talking of two aspects of the same phenomenon. The creativity here, the freedom, is the intelligence that determines the process of evolution in each moment. And so, insofar as we fall back into this intelligence, and allow it to manifest through our system, then we too partake in its freedom, we *are* this freedom. It is this very moment, this point of the source from out of which our experience is emerging, where the freedom lies. What our linguistic consciousness allows, if we allow it, is for the intelligence to directly determine the form of the meaning that we experience. Now we can not only resonate with the form of the earth, we can resonate with the intelligence itself, because the form of our language is no longer determined by the form of the events that are impinging on the form of our body.

We live on the edge of the wave of the creative intelligence of evolution, we are its latest form, a form that has yet to become habituated and fixed. The demonstration of this *is* our immediate consciousness—for immediate consciousness is *exactly* what is new every moment and never repeats. Do you see how plain this is, how transparent? We *are* this manifestation of now. Even when we identify with our thinking, with our emotions, we never leave this manifestation of now, we just lose our immediate consciousness of *being* now, and become *distracted* by the repeating patterns of our past experience.

Language transforms the brain into a vehicle by means of which the intelligence of evolution can come to expression. We call such expression *in-sight*, or *in-tuition* or *in-spiration*. In each case we experience *new meaning*. This is not the novelty of putting together a collection of sounds or images into a different order. Such new meaning is like a new colour, not a new shade of colour, but an entirely new colour that lies outside the existing spectrum of our experience.[16] We do not assemble it, it arrives ready made, freshly *created*, never before seen in the clearing of consciousness—just as colour itself must have arisen in the first experience of sight.

Here we must be clear that what arrives is not a collection of words. And yet it is something that can only be consciously apprehended because there is already this space of meaning that language has created, the space of the being of language. For we only know meaning *as* meaning because we already have language. Language is the means, the context, within which all meaning is displayed, within which new meaning is able to stand out, to be distinguished *as* new meaning. Language is the vehicle of our self-consciousness, of our knowingly distinguishing anything from anything else, including our knowingly distinguishing our being here in the first place.

And so, when new meaning does arise, we immediately use language (or language uses us) to manifest a suitable meaning-form. We may say that what we have seen is ineffable and inexpressible. But, in saying that, the new meaning has *already* found a place in language. For we have held it up from out of the stream of never to be repeated experience, and knowingly *apprehended* it. The ineffability and the inexpressibility are words we are speaking into the unspoken context of the experience, for no experience is without context, and it is this context that *presents* the meaning, the context of the moment in which it first arose:

As I was watching you standing there, and you were unaware of my gaze, I saw in the way you were holding yourself, your poise as you were silently looking out of the window, that you were in an extraordinary state (an *ineffable* and *inexpressible* state). We say we have no words, but we do, we just don't use them any more—for you were blessèd in that moment, and the light of heaven was shining down upon you and through you and I could sense your true being, the being *behind* the being we think ourselves to be.

If we look into language we find it is already capable of expressing transcendental meaning, even though, today, we are hardly aware of this capacity. Consider again the words: in-telligence, in-tentionality, in-sight, in-tuition and in-spiration. In each case this 'in' is pointing directly to the source (it *in-tends* the source). For what does it mean to have an insight or an intuition, or to be inspired? It means we have received a direct communication of meaning, a communication that has not come through the senses, or through some internal process of reasoning, but has arrived directly, 'fully-formed' from 'within'. And what is this 'within'? Our science will say it is the machinery of our unconscious reasoning and that our direct insights are the outcome of reasoning processes that went on 'behind our backs'. But this is unintelligible. For an unconscious process has no meaning. Meaning is only present *as* meaning insofar as there is a consciousness like ours to apprehend it. So whatever the unconscious process has achieved, it is only when it becomes conscious that the meaning manifests. The meaning is not 'there' in the molecules of the neurons that carried the process, it only emerges, it is only *created* now, in the moment of its apprehension. And where does it emerge from? It emerges from the ineffable, inexpressible, unthinkable source. This source is not our 'inner life'—for both the inner and the outer life are appearing in the one stream of experience that is emerging from out of this source. It is in the stream that all the dualisms of inner and outer, of subjective and objective, first appear. Before that, in the source itself, there can be no such dualism. That is why it is unthinkable for us, because we can only think in terms of distinction and difference. And, of course, in seeing this, we have an exact demonstration of what it means to have an insight.

The reason language already has this capacity to indicate the source is because it comes out of the source, i.e. it is intended by the source, and so is a creation of the intelligence of the source. But language is not just one more level of evolutionary complexity. It is the beginning of the process of the manifestation of intelligence in the externality of world experience. For it is in language that intelligence can first come to know itself as the intelligence it is. Like now, for instance. What is occurring, what *can* occur, is that we can resonate directly with intelligence in the medium of language. That is what makes language so remarkable. It has the form of a behaviour that can manifest 'out here' in our speech and thought that, at the same time, can *directly* resonate with the 'in there' of the intelligence of the source. That means, for the first time, we, as *individuals*, can become creative. No other form of life before us could do this. What we allow is the manifestation of an *individualised* intelligence—an intelligence that is attuned and resonating with our exact situation. Previously we were captivated in the behavioural forms of our animal bodies. Although we were still expressions of this intelligence, we could not *know* it. Now language becomes a mirror of meaning in which a creation of the source can come to know it is a creation of the source, which means, at the same time, the source can come to know *what* it has created, rather than simply *being* what it has created.

THE PRESENT AND THE FUTURE

If we look from this place of immediate intelligence, it becomes apparent that our current, collective idea of what it means to be intelligent is *false*. This falsity has its ground in the scientific materialism we have inherited from the process of our scientific development. At the centre of this materialism is the idea that all the events of our lives are completely determined by the low-level lawful interactions of microphysical particles and energy fields. That means we already understand ourselves and the entire universe to be a kind of machine, i.e. something whose behaviour is controlled by a system of fixed rules—which means we understand ourselves to be essentially *unintelligent*.

And yet, if we were to ask a group of ordinary people whether they thought of themselves as machines controlled by a fixed system of rules, I predict they would disagree. Instead, they would probably say they are free and conscious individuals. So what does this mean? Is this idea that our collective understanding is determined by an underlying scientific materialism just a projection of mine? A projection arising from my having read and thought too much on these matters, and from the consequent effect that scientific materialism has had on my personal understanding of the world? For most people, I suspect, if they did not look it up on the internet, would not even be able to say what scientific materialism means, let alone how it may be influencing their lives. And yet, here am I, claiming that scientific materialism is determining, for each one of us, our very understanding of the being of the world we live in. How can this be true?

The answer lies in language itself, in the way it works. At the level of our immediate experience, when we say we are conscious and possess free will, this is the *truth*. For we are this immediate consciousness, and the intelligence that expresses itself through this consciousness is, essentially, free, i.e. it is not determined by the habitualities of the past. But this freedom of immediate consciousness is only present, is only true, insofar as we *are* immediately conscious. Otherwise we are just uttering words without consciously and directly inhabiting their meaning. When we do that, the meaning of what we say falls back into the collective. We no longer have the 'thing itself' before us, and so our words refer *indirectly*. Such indirect reference is essential in our everyday use of language. It is how we get about in the world, how we refer to all that is not immediately present to us, how we make arrangements, speak of what happened yesterday, and so on.

The language we use for indirect reference is objective and collective. We mean what everyone else means, otherwise we could not understand each other and coordinate our lives. So when I speak of Timbuktu, I mean what everyone else means by Timbuktu. I have a rough idea of a town or a city somewhere in Africa that I have seen on television, but I have never had a direct experience of being there and I could not point it out on a map. In saying, when I refer to Timbuktu, that I mean what everyone else means, I do not mean the vast collection of vague ideas about Timbuktu that are scattered about in everyones' minds, I mean the place itself. Here I am assuming there is a fact of the matter about Timbuktu, and that, although I do not know Timbuktu directly, there are people who do, and I mean what *they* mean when they refer to Timbuktu.

Now all is well and good so long as we speak of objectively measurable things, such as towns and cities on the earth. But problems arise when we speak of entities and events that no one has ever directly experienced, or of our subjective inner states and experiences. For example, consider the Big Bang. On the surface of language we speak of the Big Bang much as we speak of the First World War. For they are both events at which we were not actually present, and about which we accept the testimony of others that they actually occurred. But with the First World War we can trace this testimony back to the direct experiences of actual human beings. In contrast, the Big Bang was never and could never have been such an experience. It is a purely theoretical construct, a logical consequence of our best physical theories, based on our best empirical observations. And, like Timbuktu, very few of us have a clear idea of what the Big Bang really was. So when we refer to it, we are referring to whatever the people who really know about the Big Bang are referring to. And that means we refer the meaning to the scientific experts.

Unless we stop to examine this, it really seems that in speaking of the Big Bang we are speaking of an actual event. But we are not. We are speaking of a hypothetical theoretical construct that we *imagine* to be an event. What is significant is how our science is implicitly determining what we mean in our indirect linguistic references. For it is not just an event like the Big Bang that is at issue. Our science determines what we mean when we refer to the sun. We do not mean the great consciousness in the sky, we mean that material ball of helium and hydrogen, even if we don't know it is made of helium and hydrogen. We mean what everyone else means, and what everyone else means is what our science tells us is the case. It is in *this* way that scientific materialism enters into the fabric of our everyday lives, i.e. in the way we collectively refer the meaning of the language we use back to the authority of science. It doesn't matter that when we use words like argon and xenon, we don't actually know what distinguishes one from the other, for we know that someone does, and we know that if we wanted to we could find out the details. This is not laziness. It would be impossible for any one person to trace all these meanings back to their origins.

So what is the problem? It arises when we examine the meaning of words like 'consciousness' and 'free will' and 'intelligence'. For, unlike the Big Bang, these are not theoretical constructs, they are immediate realities. But most of the people, most of the time, are not directly present to these immediate realities. And so they use the words 'consciousness' and 'free will' and 'intelligence' to refer to what everyone else means. And, as with Timbuktu, although everyone will have their own version of what they think these terms mean, finally, implicitly, we hand the collective reference over to the people who know what they are talking about, who have studied the matter, and who have drawn the clear and proper distinctions. In other words, we hand the question of the ultimate meaning of these terms over to the experts. And just as with argon and xenon, we mean what they mean, even if we don't exactly know what they have to say.

Once upon a time, when Descartes and Galileo were still our contemporaries, our underlying ideas of consciousness and intelligence still referred to something immaterial that distinguished us as being conscious non-mechanistic soul-beings. But times have changed. Now, if you trace back the meaning of these terms to the collective authority of the experts, you will find that our idea of what it means to be conscious has dramatically altered. Now the whole universe, including our own nervous system, our own intelligence, is understood to be a kind of machine, i.e. something whose behaviour is determined by the low-level interactions of its parts. Now when we say "I am not a machine, I am a conscious individual, with free will", and we do not actually inhabit what we are saying by directly realising what it means to be conscious, then we again refer our meaning back to the experts. And that means that we, like they, are speaking of ourselves as being a certain kind of very sophisticated biological machine. In that case, what we 'really' mean, when we speak of free will, is a certain capacity to make choices as a result of an entirely mechanistic process of deliberation occurring in our brains. And what we 'really' mean when we say we are not a machine, is that we are not the kind of machine that cannot make such deliberate choices. And when we say we are conscious, that does not mean we are not a machine, it means we are a particularly complex kind of machine that can exhibit behaviours we identify as being conscious.

At the same time, the objective scientific meanings of these terms have not been *explicitly* integrated and accepted into our collective surface level understanding of who we are. If that had occurred then the surface language itself would have changed and we would have no problem in speaking of ourselves as being highly intelligent machines. But this has not yet happened. Somewhere in that collective labyrinth of indirection, there is still a gleam of conscious intelligence that refuses to be extinguished.

What we are doing here is working on the form of the language structure within which we live. This language structure is called the *world*. It is a structure of meaning. We inherited this structure from the process of enculturation during our childhood, and we work on transforming and extending it throughout our lives. And although there is a personal dimension to this world, where we can develop our own meanings and references, language remains essentially collective. For whatever meanings we may discover for ourselves, it is part of our living in language that we attempt to manifest those meanings in the collective, by speaking them 'out loud' to someone else, or writing them down in a book, like this. To change the meaning of a word is to change the world we live in. That is why so many people exclaim: "This book changed my life". A book can change your life because it changes the structure of meaning within which you live.

Clearly, within this book, we have been working on more than changing the meaning of a word or two. We have been working on changing the entire understanding of being that lies hidden in the structure of indirect reference that leads back to our scientific experts. Put simply, we are saying that when it comes to this fundamental question of our being, the experts do *not* know what they are talking about. They do not know, because they do not understand the phenomenology, which means they do not know that we cannot understand consciousness and intelligence in objectively quantifiable terms.

Here we are not talking of any particular individual. We are talking of the entire inherited structure of meaning that is modern objective science. It is this structure that determines how scientific observations are conducted, how they are written up, whether or not they are published, and how all these publications form a consensus within an area. It is this consensus that becomes the meaning we inherit 'on the surface' when we speak of argon and xenon and the Big Bang and consciousness and intelligence. This consensus used to be collected together and summarised in the volumes of the encyclopaedias. But now, of course, we have the slightly more unreliable resource of Wikipedia.

And yet, when it comes to the meaning of consciousness, we should acknowledge the significant fact that the experts are *not* in full agreement. It is widely recognised that consciousness is a *puzzling* phenomenon and that we are currently lacking a satisfactory theory to account for it. That does not mean our basic mechanistic understanding of the being of the universe is seriously in question—at least not *yet*. It only means it is not clear how consciousness is to be accommodated within that understanding. Even so, this lack of consensus means the question of the meaning of the word 'consciousness' remains *open* (it is still a *live* issue). The reason it remains open is because each and every one of us, including the expert scientists and philosophers, actually *are* conscious, and, however much we may believe in scientific materialism, the immediate experience of being conscious is impossible to ignore, at least when it is directly pointed out to us. That is why, in everyday language, we are unable to say that we are simply biological machines, despite the underlying evidence and authority of our scientific materialism. In contrast, if this were a fully verified scientific fact, we would already have come to accept it, like we accept the earth goes around the sun even though it 'seems to us' that the sun rises over the earth each morning.

What this means is that under the surface of our language, in the rooms and corridors of our Churches of Reason, there is a battle going on that concerns the very meaning of our being here. For, despite the ongoing and steady encroachment of scientific materialism into the language and fabric of our lives over the last four hundred years, this matter of the status of the being of consciousness has still to be decided in its favour. This is the last line of defence between us as conscious free-willed individuals, and our transformation into complete automatons, who can no longer tell the difference between ourselves and the machines that are increasingly coming to dominate our lives. What is at issue is our capacity to even think that we are not machines. For, if our collective language changes in such a way that the meaning of consciousness becomes fixed and identified with the functioning of a certain kind of mechanism (i.e. our nervous systems), then our children, the next generation, will never even encounter the ambivalence in the word that allows it to refer to something else, to this immediate consciousness that I am each moment.

This closing down of our capacity to refer to immediate consciousness would be the closing down of an essential degree of freedom within the language of Western scientific civilisation. It is this degree of freedom that allows the intelligence to manifest in language and to express what it is to be conscious and intelligent, and in so doing to manifest a freedom that escapes the constraints of the lower-level morphic fields that otherwise control our behaviour. The strange prophesy I am making is that if we allow our scientific materialism to completely determine our understanding, then it will also come to completely determine the meaning structure of our language, and in so doing it will close down the one degree of freedom that still distinguishes us from our machines. And so it will become literally true that we are machines—still conscious—but unable to recognise or realise what that means.

Here I must also say that I do not think that this is actually going to happen. The idea that we will all become completely determined by our scientific materialism is like a medieval European thinking that we would all become completely determined by the teachings of the Christian Church. In such a projection we fail to see the bigger context of the psyche in which we live. We think that *we*, the human race, are in charge, and that the fate of the evolution of life on Earth lies in *our* hands. But if we look into the actual effects of our belief in scientific materialism, i.e. at our attempt as a civilisation to create a rationally ordered world, we find, both as individuals, and collectively, that things are becoming increasingly more *irrational* and *disordered*. This is the psyche at work, quite automatically balancing out the error of our ways. We can see this most obviously in the effects the scientific materialism of our technology is having on the earth and on our human psyche. Both respond by going out of balance. And, of course, the end result of such a going out of balance will not be the final victory of scientific materialism, it will be the breakdown and collapse of our civilisation and the structure of meaning in which we live.

From the perspective of impersonal intelligence, such a collapse will have a kind of logical necessity, i.e. it will be what is needed for the continued evolution of consciousness on Earth. It will simply mean our form of civilisation has achieved all it can, and is no longer 'fit for purpose'— i.e. no longer has sufficient degrees of freedom. But with us, living our brief lives here, the situation looks a little different. For the collapse of Western civilisation, I predict, is not going to be a gentle transition to a new age of Aquarius. We stand with the sword of Damocles hanging over us. This sword represents a kind of ultimatum: either you become more conscious and express more intelligence in your behaviours, or the sword falls. And although this sword hangs over our whole civilisation, it really hangs over each one of us. For there is only going to be more consciousness and more intelligence in the collective, if there is more consciousness and more intelligence here, in you and I. Saying that is not an attempt to 'save the world' by starting some kind of mass movement. It is intelligence itself that is calling and responding. Why it resonates in one person rather than another is not the issue. The question is only, does it resonate in me?

↳ Why not?

NOTES

1. For example, there is the militant atheism of Richard Dawkins, as laid out in his book *The God Delusion* (2006).
2. There is strong evidence to suggest that the cortical neurons in our brain remain with us until we die, but there is also evidence to suggest that new neurons can grow in other areas of the brain, such as the hippocampus, amygdala, and olfactory bulb (Costandi, 2016). In the rest of the body, cells are continuously dying and being replaced, some within a few days (such as surface layer skin cells), and others so slowly that a certain proportion may still survive a lifetime (such as the heart muscle cells which are replaced at a rate of around 1% per year) (Spalding, Bhardwaj, Buchholz, Druid, & Frisen, 2005, July 15). Another notable class of cells that are not replaced after birth are female gametes or egg cells.
3. The mathematics that demonstrates this logic of control is expressed in the good regulator theorem of Roger Conant and Ross Ashby. See: *Every good regulator of a system must be a model of that system* (Conant & Ashby, 1970).
4. Donald Hebb (1904–1985) was a Canadian psychologist, famous for proposing that neurons learn by growing and strengthening their connections according to whether they fire together within a short period of time (Hebb, 1949).
5. The age of these LUCA cells remains uncertain, with some studies suggesting they formed as long as 4.5 billion years ago, i.e. just a few million years after the earth first gathered itself together. See *The First Cell* (Schreiber & Mayer, 2020).
6. The idea of a morphic field ruling over a plant species connects back to the work of Johann Wolfgang von Goethe (1749-1832) on plant morphology. Goethe was famous not only as a poet and playwright, but also for his pioneering scientific inquiries, which presaged the development of dynamical systems theory and the idea of morphic fields that we are exploring here (Goethe, 1790/2009). Goethe saw all plant life as possessing an intrinsic, dynamic unity of form—something Henri Bortoft described as a *unity in multiplicity* (Bortoft, 2012). This involved Goethe's actual 'seeing' of the potentiality and unity of the morphic field that lies behind the manifestation plant form. In a similar way, we have been exploring the unity in multiplicity of all cellular life on Earth, both in the dynamical unity of autopoietic form, and the temporal unity of inheritance that directly demonstrates how life on Earth manifests the wholeness of a single organism.

7. See Heidegger's essay *The Way to Language* (1993, pp. 397–426) where he attempts to "bring language as language to language".

8. Heidegger gives an overview of what he means by an understanding of being in the introduction to *Being and Time* (1962, pp. 21–35).

9. Heidegger entered into an unusually specific investigation of the being of an animal in Chapters 4 and 5 of Part Two of his lecture series *The Fundamental Concepts of Metaphysics* (1995, pp. 201–270).

10. See *The Ecological Approach to Visual Perception* (Gibson, 1986).

11. Maturana and Verden-Zöller give an extended account of the origins of language in the evolution of the human species in their book *The Origins of Humanness in the Biology of Love* (2008).

12. The idea that language began with music and song is developed in Chapter 4 of Iain McGilchrist's book *The Master and his Emissary* (2012, pp. 94–132).

13. Turing proposed his Turing Test in the now famous journal paper *Computing Machinery and Intelligence* (1950).

14. See *The Master and his Emissary* (McGilchrist, 2012).

15. This was Nietzsche's vision of time: the eternal recurrence of the same (see, for example *The Gay Science* (1882/2001, 341)).

16. Terry Pratchett expressed this idea of there being another colour lying outside the existing spectrum in his invention of octarine in *The Colour of Magic* (1983)).

Epilogue

Nous Sommes du Soleil

No – triune intentionality.

O UR inquiry into what it means to be intelligent has led us back to
the unthinkable source, the psyche, from out of which all that is,
or can be (for us) emerges. It is here we have discovered the well-spring
of intelligence, the impersonal intentionality, that wills the universe into
being. This willing-intending is not emanating from some abstract being
or entity that can only be encountered in a rare moment of inspiration.
It is that which is manifesting each and every moment of our experience.
And insofar as I am immediately conscious, I can know this source to be
the source of my being, the source of my intending, and that behind my
apparent individuality, I *am* this source.   No!

This paradox of being an individual who, at the same time, is the intentionality of the source, cannot be grasped in the categories of our ordinary rational thinking. Such thinking only applies to what can be distinguished 'out here' in the forms of our objectified experience. In speaking
of the source we are no longer speaking of a 'thing' that can be objectified
into a separable form. We are looking 'backwards' into that from out of
which all such form emerges. And 'back there' all is enfolded in an encompassing unity, where my being an individual is just another form of
meaning in a streaming of universal meaning. And yet, despite our rational habitualities, we can still get a 'sense' (a resonance) of what this means,
because we are (I am) this stream of meaning.

In looking backwards like this, we can only express what is intuited
in a language of metaphor and symbolism. But that does not mean the
source is something distant and obscure. Our entire experience of being
in the universe is an expression 'out here' of the form and order of what
lies 'in there'. It is our fixation and belief in the ultimate reality of the
material being of a material universe that causes us to overlook this correspondence of the within and the without. It has been one of the major
tasks of this book to effect an ontological shift away from this fixation on

a material reality, so we can begin to explore the source from out of which we have emerged.

What we have discovered, the bridge between the outer and the inner, is the world of ideal form—beginning with the forms of the ordinary perceptual objects we see around us. We think these forms have a material existence in the objectified configurations we perceive. But our phenomenological inquiries have shown that this is not the case. It is we who bear these forms within us, as ideal poles of unification by means of which we 'make sense' of the world. For, in our sensory fields, we only encounter the flowing appearance of partial aspects of the things we perceive. It is we who 'see through' these aspects to the enduring forms they display, and it is we who project these forms into the stream of sense in search of a resonant confirmation. This confirmation tells us that the form we project corresponds *in some way* with the unthinkable source from out of which all experience emerges.

The great reversal, the reversal that is presaged in the quantum mechanics of our physical science, is that there is ultimately no ground of physical stuff existing 'out here' in a material world to which our perception of form finally corresponds. There is only form emerging out of form, emerging out of form, all merging into a background of pure potentiality. It is we who turn this dance of form and resonance into an experience of living in a material world. And it is we who bear space and time within us, and use these dimensionalities to present a world of form expressed in a language of sense.

Once we fully grasp this situation, it becomes clear that our notion of the reality of physical stuff is simply another meaning we project into this space of immediate consciousness. The 'real' reality is not the material existence of atoms and molecules, it lies in consciousness itself, in this space of form and quality and spatiality and temporality where we live and have our being. Now we can see that the forms we project and perceive are not immaterial shadows representing an underlying material reality, they are the means whereby anything can actually come to *be* anything at all, rather than being the nothing of a pure potential for being. For it is *we* who *call the universe into form.*

In seeing through our projection of materiality, the findings of objective science start to take on a new and more profound meaning. For it is no longer physical stuff that is the ground of being, it is the hierarchical structure of form that this projection of materiality enacts. Now we can start to understand what it means to be alive and conscious by looking into the forms of the processes that enact our being alive and conscious. Instead of this one flattened level of material being, we can start to discern multiple levels of being, each corresponding to the levels of form we have distinguished on the basis of our objective scientific measurements and logico-mathematical reasonings.

And yet, up to now, we have been primarily looking at this hierarchy from the 'top down'—from the vantage point of human consciousness and human language, surveying the lower-level forms upon which we stand—the perceptual brain of the mammals, the morphic structure of the multicellular organism, the autopoiesis of the living cell, and the trans-temporal unity of cellular life on Earth. But what if we look in the other direction, at what lies 'above', at those greater systems of which we, as humans, are mere components? Most immediately we are confronted with those collective modes of behaviour and organisation that our development of language has facilitated. Here lies the structure of our global technological civilisation, and the systems of money and finance and information flows that more or less control our economic and social behaviour. And while these systems enact new complexities never before seen on the earth, if we look into their form of being they remain *our* creations, enacting purely human forms of intentionality and habituality. As such they are *derivative* structures that reflect back to us our collective level of consciousness. Here we do not find a new higher-level form of organisation that transcends our human form of linguistic being. Instead we find the mostly blind operation of unconscious forces driven by human egoism and insecurity. And even though it appears our collective social systems stand 'above' us, and that we act as mere components in their processes of evolution, when we look directly at the level of intelligence they manifest, we find they stand far below the intelligence of the immediate consciousness we are capable of sustaining as human individuals.

What stands 'above' us are not our human social systems, but those greater cosmic systems from out of which we have emerged, that still encompass us as components in the greater unity of their form. Here lies the system of the earth and the moon, and of the solar system of the sun. For the hierarchy of form we are considering is not a hierarchy of external form, where I am a member of a human social collective, but the hierarchy of form 'within' that encompasses our being conscious. We are asking again after the source and how it manifests in a hierarchy of increasingly greater and more impersonal forms of *cosmic intentionality*.

So how are we to proceed? Firstly, there is the *will*—the intentionality that in each moment is manifesting the form of my life—the intentionality that we each experience ourselves to be. As we fall back into the immediate consciousness of a phenomenological reduction, then we fall back into this will. We will what it wills, we are this will, and it wills what *is*, what happens—each thought, each perception, each perceived event. That is what makes it *the* will. From here we can see that the will is willing the entire manifestation of life on Earth. That is how it comes to be that we experience ourselves as living together on the *same* earth, how we are able to meet, to touch, to communicate. And that means it is this same will that is manifesting all the various forms of sensory projection, in the eye of the hawk, the nose of the dog, and the consciousness of you and I. This will *is* the earth, and through it the earth is conscious, because we are conscious, and our intentionality is the intentionality of the earth.

That does not mean I am not an individual, with my individual conscious experience, and my own leeway of responsibility. It is all a matter of the level of being to which we are referring. As we look up to a higher level then all the 'individuals' on this level become unified in a greater individuality, just as our biological cells are unified in the experience of being a human body. At the same time we cannot say the earth is an individual in the way that we are individual. It is that which lives through all life on Earth. Its reality simply does not fit into the categories of our objective thinking. We can only glimpse it in certain moments, such as our seeing the beauty of the harmony of the wind blowing across the sand, across the sea, moving the clouds, blowing into me, *through* me, so there is no longer any separation. Or in the falling of a waterfall, its flowing static form, in the sound it is making in the midst of the surrounding silence, the play of the light on the rocks, where this flowing of the water and the flowing of my consciousness combine into one and the same transcendental unity. Here all is felt, all is communicated in another dimension, the dimension of a pure and immediate resonance.

As we have already noted, scientific materialism is coming under increasing stress, as it becomes clear it can give no satisfactory account of our being conscious. In response, there has been a growing interest in panpsychism, both within contemporary philosophy of mind and amongst those who are seeking an alternative understanding of being that gives intrinsic value to the greater unity of the earth system (i.e. to *Gaia*). Panpsychism proposes that all matter is to some extent conscious and that organised systems of matter, such as atoms and molecules and humans and (perhaps) planets and stars, each possess their own form of mind-like experience. For instance, Rupert Sheldrake conjectures that the sun has a mind that takes form in its electromagnetic fields and is able to control aspects of its own activity, such manifesting solar flares to exert influence on the surrounding planetary systems.[1]

This panpsychic idea that the sun could possess an individuated consciousness with an independent ability to perceive and act, still moves in an essentially objectified understanding of being, where each being-entity is taken to be a kind of consciousness-possessing-material-thing existing in a spatiotemporal universe. But according to our inquiry into the being of form, you and I and the earth and the sun do not exist 'out here' in this world of objectified sense. Our realities lie 'within' the psyche, in the unity of the one universal intentionality. In this unity the reality of the sun is of a higher order than our human level of being and what we see 'out here' is only a *symbol* of that reality. Our objective science is quite right to scoff at the idea that this externalised sun-object could be conscious in the way that we are conscious. For the sun-object does not possess the kind of structure needed to support the conscious experience of an external world of objective form. For that you need something like a nervous system, that can resonate by learning to predict the forms of the forms that 'surround' it. Otherwise what was the purpose of all the suffering and struggling of the evolution of conscious life on Earth? If the sun is already enjoying an experience of itself existing out here in an objective universe, then what is the meaning of our being here? Are we a kind of television program to help the sun while away the billions of years it is going to take to use up its store of hydrogen?

What we should bear in mind is that the sun is not just a nuclear reaction occurring at the centre of the solar system, it *is* the solar system. As such it encompasses and includes the earth in its gravitational and electromagnetic fields, just as the earth encompasses and includes the system of life on Earth. And as we explore these realities from within, we too find ourselves contained and enfolded in their greater intentionality, just as, from without, we find the earth is a creation of the sun, and that we are creations of the earth. This all combines to tell us that *we are of the sun* (nous sommes du soleil).[2] For our physical bodies are made of the earth, which is made of the sun, and our intentionality is contained in the intentionality of the earth, which is contained in the intentionality of the sun. Finally, reaching out in ever expanding circles, this *is* the intentionality of the source itself. That means we not only bring the earth to self-consciousness and language, we bring the sun as well. And instead of developing self-conscious electromagnetic fields and trying to influence events by emitting solar flares, the sun developed the earth from 'within', as the earth developed life on Earth, and by means of this both the sun and the earth have become objectively conscious. The sun does not look *at* us, it looks *through* us. In so doing it comes to see itself, its light, the very symbol of consciousness, awakening each day, going down each evening, and yet, in itself, always present.

These statements are not *facts*, they are the *meaning* of the facts. Our science does not contradict this meaning, it *amplifies* it. It reveals the form and structure of the unity of this solar system we are, travelling through the galaxy—ourselves composed of atomic structures that were created in the galactic events of the birth and death of the stars themselves. It reveals the sun as the potentiality of consciousness that has drawn life out of the earth, as that power, manifesting out here as gravitation and electromagnetism, that has driven the self-organisation of all the autopoietic systems on Earth. For it is only because of the sun that life can resist the otherwise inevitable slide into the disorder of thermodynamic equilibrium. We are in every moment bathed in this surfeit of pure potentiality. It not only keeps us from freezing, it is there in the photosynthesis of the plants, which store sunlight so it can be released in the kinetic movement of our bodies. It is there in the thermal vents in the depths of the ocean where life is now conjectured to have begun. For the energetic heat of the earth's mantle is only an effect of a gravitational compression that again has its origin in the sun. And it is there in the energy we consume in our human technological systems, in the sunlight stored in the gas and coal and oil that came out of the autopoiesis of earlier life forms, in the energy that powers the wind and the tides, in the photosynthesis of the trees .... This is how the earth has transformed the sun's energy into the manifestation of autopoietic order. Here we see that life is *of* the earth, it is *made* of the earth, but the enabling *inspiration*, the *creative potentiality* is of the sun. This is not a great cosmic secret. It is staring us in the face as the sun shines down each day and illuminates our lives. Could there possibly be a clearer and more direct symbol of what it means to be creatively conscious? And yet our materialistic fixation on the idea of an underlying physical reality renders us unaware of the meaning of the symbol of the sun, and its message of our solar origin. At the same time we fail to understand that *this is how it must be*, that the source, the psyche, can only appear 'out here' in symbolic form, because the sun itself, its reality, *our* reality, is not some objective 'thing', it is the place from out of which objective things are manifested, and if these objective things are to speak of the source, they can only do so symbolically, in the resonant language of a transcendental meaning.

Figure 14.4: The Icon of Divine Light designed by Cecil Collins and located in Chichester Cathedral (© Tate Images by permission under a CC-BY-NC-ND 3.0 Unported license).

NOTES

1. Rupert Sheldrake has addressed the question of the sun being conscious in a number of public talks, including the 2018 Reconnect conference in Bath (see: https://www.youtube.com/watch?v=SFhsObpja8A).
2. *"Nous sommes du soleil*, we love when we play" is a line from *Ritual*, a musical composition by the progressive rock group *Yes* which appeared on their 1973 album *Tales from Topographic Oceans*.

BIBLIOGRAPHY

Alexander, E. (2012). *Proof of Heaven: A Neurosurgeon's Journey into the Afterlife*. New York: Simon & Schuster.

Ananthaswamy, A. (2018). *Through Two Doors at Once: The Elegant Experiment that Captures the Enigma of our Quantum Reality*. New York: Dutton.

Bell, J. S. (1964). On the Einstein Podolsky Rosen Paradox. *Physics, 1*(3), 195–200.

Bergson, H. (1911/1998). *Creative Evolution* (A. Mitchell, Trans.). New York: Dover Publications.

Berkeley, G. (1710/1982). *A Treatise Concerning the Principles of Human Knowledge*. Indianapolis: Hackett Publishing Company.

Biran, I., Giovannetti, T., & Chatterjee, A. (2006). The alien hand syndrome: What makes the alien hand alien? *Cognitive Neuropsychology, 23*(4), 563–582.

Bisson, T. (1991, April). They're made out of meat. *OMNI*.

Blake, W. (1789/2019). *Songs of Innocence and of Experience*. New York: Macmillan Collector's Library.

Bohm, D. (1980). *Wholeness and the Implicate Order*. Oxford: Routledge.

Bortoft, H. (2012). *Taking Appearance Seriously: The Dynamic Way of Seeing in Goethe and European Thought*. Edinburgh: Floris Books.

Bostrom, N. (2014). *Superintelligence: Paths, Dangers, Strategies*. Oxford: Oxford University Press.

Brentano, F. (1874/1995). *Psychology from an Empirical Standpoint*. London: Routledge.

Carey, N. (2010). *The Epigenetics Revolution: How Modern Biology is Rewriting our Understanding of Genetics, Disease and Inheritance*. London: Icon Books.

Castaneda, C. (1968). *The Teachings of Don Juan: A Yaqui Way of Knowledge*. Berkeley: University of California Press.

Chalmers, D. (1996). *The Conscious Mind: In Search of a Fundamental Theory*. New York: Oxford University Press.

Chalmers, D. (2000). What is a neural correlate of consciousness? In T. Metzinger (Ed.), *Neural Correlates of Consciousness: Empirical and Conceptual Questions*. Cambridge, Massachusetts: MIT Press.

Chalmers, D. (2017). Panpsychism and panprotopsychism. In G. Brütrup & L. Jaskolla (Eds.), *Panpsychism: Contemporary Perspectives* (pp. 19–47). Oxford: Oxford University Press.

Clark, A. (2016). *Surfing Uncertainty: Prediction, Action and the Embodied Mind*. Oxford: Oxford University Press.

Conant, R. C., & Ashby, W. R. (1970). Every good regulator of a system must be a model of that system. *International Journal of System Science*, 1(2), 89–97.

Copernicus, N. (1543/1995). *On the Revolutions of Heavenly Spheres* (C. G. Wallis, Trans.). Amherst, New York: Prometheus Books.

Costandi, M. (2016). *Neuroplasticity*. Cambridge, Massachusetts: MIT Press.

Darwin, C. (1859/1964). *On the Origin of Species by Means of Natural Selection, or the Preservation of Favoured Races in the Struggle for Life*. Cambridge, Massachusetts: Harvard University Press.

Dawkins, R. (1989). *The Selfish Gene*. Oxford: Oxford University Press.

Dawkins, R. (2006). *The God Delusion*. New York: Bantam Books.

Dennett, D. C. (1991). *Consciousness Explained*. New York: Back Bay Books.

Descartes, R. (1641/1911). *Meditations on First Philosophy* (E. S. Haldane, Trans.). Cambridge: Cambridge University Press.

Dreyfus, H. L. (1992). *What Computers Still Can't Do: A Critique of Artificial Reason*. Cambridge, Massachusetts: MIT Press.

Driesch, H. (1914). *The History and Theory of Vitalism* (C. K. Ogden, Trans.). London: Macmillan.

Dunn, J. (1995). *Sm'algyax: A Reference Dictionary and Grammar of the Coast Tsimshian Languages*. Seattle: University of Washington Press.

Einstein, A. (1905). On a heuristic viewpoint concerning the production and transformation of light. *Annalen der Physik*, 17(6), 132–148.

Einstein, A., Podolsky, B., & Rosen, N. (1935, May). Can quantum-mechanical description of physical reality be considered complete? *Physical Review*, 47, 777–780.

Eliot, T. S. (1963). *Collected Poems 1909-1963*. London: Faber and Faber.

Everett, H. (1957). Relative state formulation of quantum mechanics. *Review of Modern Physics*, 29(3), 454–462.

Feldman, D. (2019). *Chaos and Dynamical Systems*. New Jersey: Princeton University Press.

Felleman, D., & van Essen, D. (1991, Jan/Feb). Distributed hierarchical processing in the primate cerebral cortex. *Cerebral Cortex*, 1, 1–47.

Fink, E. (1995). *Sixth Cartesian Meditation: The Idea of a Transcendental Theory of Method* (R. Bruzina, Trans.). Bloomington: Indiana University Press.

Freiberger, M. (2012, August). Schrödinger's equation—what is it? *Plus*. (Retrieved from https://plus.maths.org/content/schrodinger-1)

Freud, S. (1905/1975). Three Essays on the Theory of Sexuality. In J. Strachey (Ed. and Trans.), *The Standard Edition of the Complete Psychological Works of Sigmund Freud* (Vol. 7, pp. 123–246). London: The Hogarth Press.

Freud, S. (1923/1975). The Ego and the Id. In J. Strachey (Ed. and Trans.), *The Standard Edition of the Complete Psychological Works of Sigmund Freud* (Vol. 19). London: The Hogarth Press.

Friston, K., Kilner, J., & Harrison, L. (2006). A free energy principle for the brain. *Journal of Physiology–Paris, 100*(1–3), 70–87.

Galileo, G. (1623/1957). The Assayer. In S. Drake (Ed. and Trans.), *Discoveries and Opinions of Galileo*. New York: Doubleday Anchor Books.

Gibson, J. J. (1986). *The Ecological Approach to Visual Perception*. Hillsdale, New Jersey: Lawrence Erlbaum.

Goethe, J. W. v. (1790/2009). *The Metamorphosis of Plants* (D. Miller, Trans.). Cambridge, Massachusetts: MIT Press.

Gomes, C., & Faisca, P. (2019). *Protein Folding: An Introduction*. Cham, Switzerland: Springer Nature Switzerland.

Greene, B. (2003). *The Elegant Universe: Superstrings, Hidden Dimensions, and the Quest for the Ultimate Theory*. W. W. Norton: New York.

Gurdjieff, G. I. (1950). *Beezlebub's Tales to His Grandson: An Objectively Impartial Criticism of the Life of Man*. Routledge: London.

Hawkins, J. (2021). *A Thousand Brains: A New Theory of Intelligence*. New York: Basic Books.

Hawkins, J., & Ahmad, S. (2016). Why neurons have thousands of synapses, a theory of sequence memory in neocortex. *Frontiers in Neural Circuits, 10*(23), 1–13.

Hawkins, J., & Blakeslee, S. (2004). *On Intelligence*. New York: Henry Holt.

Headrick, M. (2002). Origin and evolution of the anchor clock escapement. *IEEE Control Systems Magazine, 22*(2), 41–52.

Hebb, D. (1949). *The Organization of Behavior: A Neuropsychological Theory*. New York: John Wiley and Sons, Inc.

Heidegger, M. (1962). *Being and Time* (J. Macquarrie & E. Robinson, Trans.). New York: Harper and Row.

Heidegger, M. (1971/2001). The Origin of the Work of Art. In Albert Hofstadter (Trans.), *Poetry, Language, Thought* (pp. 17–79). New York: Harper Perennial Classics.

Heidegger, M. (1993). *Basic Writings* (D. F. Krell, Trans.). San Francisco: Harper Collins.

Heidegger, M. (1995). *The Fundamental Concepts of Metaphysics: World, Finitude, Solitude* (W. McNeill & N. Walker, Trans.). Indianapolis: Indiana University Press.

Hohwy, J. (2013). *The Predictive Mind*. Oxford: Oxford University Press.

Hume, D. (1739/1985). *A Treatise of Human Nature*. London: Penguin Classics.

Husserl, E. (1900/2001). *Logical Investigations* (Vols. 1–2; N. Findlay, Trans.). London: Routledge.

Husserl, E. (1954/1970). *The Crisis of European Sciences and Transcendental Phenomenology: An Introduction to Phenomenological Philosophy* (D. Carr, Trans.). Evanston, Illinois: Northwestern University Press.

James, W. (1890). The Perception of Time. In *The Principles of Psychology* (Vol. 1, pp. 605–642). New York: Henry Holt and Co.

Jonas, H. (1966/2001). *The Phenomenon of Life: Toward a Philosophical Biology.* Evanston, Illinois: Northwestern University Press.

Jung, C. G. (1953). Two Essays on Analytical Psychology (R. F. C. Hull, Trans.). In H. Read, M. Fordham, & G. Adler (Eds.), *The Collected Works of C. G. Jung* (Vol. 7). London: Routledge and Kegan Paul.

Jung, C. G. (1960). The Psychogenesis of Mental Disease (R. F. C. Hull, Trans.). In H. Read, M. Fordham, & G. Adler (Eds.), *The Collected Works of C. G. Jung* (Vol. 3). London: Routledge and Kegan Paul.

Jung, C. G. (1961/1995). *Memories, Dreams, Reflections* (R. Winston & C. Winston, Trans.). London: Fontana Press.

Kant, I. (1781/1998). *Critique of Pure Reason* (P. Guyer & A. W. Wood, Trans.). Cambridge: Cambridge University Press.

Koch, C., Massimini, M., Boly, M., & Tononi, G. (2016). Neural correlates of consciousness: progress and problems. *Nature Reviews Neuroscience, 17,* 308–321.

Leary, T. (1977). *Exo-Psychology: A Manual on the Use of the Human Nervous System according to the Instructions of the Manufacturers.* Los Angeles: Starseed/Peace Press.

Leibniz, G. W. (1951). *Leibniz: Selections.* New York: Charles Scribner and Sons.

Locke, J. (1689/1997). *An Essay Concerning Human Understanding.* London: Penguin Books.

Long, B. (1984). *The Origins of Man and the Universe: The Myth that Came to Life.* London: Routledge and Kegan Paul.

Long, B. (1995). *Meditation, A Foundation Course.* London: Barry Long Books.

Long, B. (2013). *My Life of Love and Truth: A Spiritual Autobiography.* Kyogle, Australia: Barry Long Books.

Lorenz, E. N. (1963). Deterministic nonperiodic flow. *Journal of the Atmospheric Sciences, 20,* 130–141.

Mascaró, J. (Ed.). (1965). *The Upanishads.* Harmondsworth, Middlesex: Penguin.

Maturana, H. R., & Varela, F. J. (1998). *The Tree of Knowledge: The Biological Roots of Human Understanding.* London: Shambhala.

Maturana Romesin, H. R., & Verden-Zöller, G. (2008). *The Origins of Humanness in the Biology of Love.* Exeter: Imprint Academic.

Maugham, W. S. (1944). *The Razor's Edge.* Philadelphia: The Blakiston Company.

McGilchrist, I. (2012). *The Master and his Emissary: The Divided Brain and the Making of the Western World.* New Haven, Connecticut: Yale University Press.

Metzinger, T. (2009). *The Ego Tunnel: The Science of the Mind and the Myth of the Self.* New York: Basic Books.

Minsky, M. L. (1986). *The Society of Mind.* New York: Simon and Schuster.

Nagel, T. (1986). *The View from Nowhere.* Oxford: Oxford University Press.

Nagel, T. (2012). *Mind and Cosmos: Why the Materialist Neo-Darwinian Conception of Nature is Almost Certainly False.* Oxford: Oxford University Press.

Newton, I. (1687/1999). *The Principia: Mathematical Principles of Natural Philosophy* (I. B. Cohen & A. Whitman, Trans.). Oakland: University of California Press.

Nietzsche, F. (1878/1996). *Human, All Too Human: A Book for Free Spirits* (R. J. Hollingdale, Trans.). Cambridge: Cambridge University Press.

Nietzsche, F. (1882/2001). *The Gay Science* (J. Nauckhoff & A. D. Caro, Trans.). Cambridge: Cambridge University Press.

Nietzsche, F. (1910/2006). *The Will to Power: An Attempted Transvaluation of All Values* (A. M. Ludovici, Trans.). New York: Sterling Publishing.

Ouspensky, P. D. (1951). *The Psychology of Man's Possible Evolution*. London: Hodder and Stoughton Ltd.

Penrose, R. (1994). *Shadows of the Mind: A Search for the Missing Science of Consciousness*. New York: Oxford University Press.

Pham, D. N., Thornton, J., & Sattar, A. (2007). Building structure into local search for SAT. In *Proceedings of the 20th International Joint Conference on Artificial Intelligence (IJCAI 2007)* (p. 2359-2364). Palo Alto, California: AAAI Press.

Pinker, S. (1997). *How the Mind Works*. Ringwood, Victoria: Penguin Books Australia.

Pirsig, R. M. (1991). *Zen and the Art of Motorcycle Maintenance*. London: Vintage Books.

Plato. (1997). *Complete Works* (G. M. A. Gruber, Trans.). Indianapolis: Hackett Publishing Company.

Poincaré, H. (1892/1993). *New Methods of Celestial Mechanics* (D. L. Goroff, Trans.). Maryland: American Institute of Physics.

Poincaré, H. (1913/2014). *The Foundations of Science* (G. B. Halstead, Trans.). Cambridge: Cambridge University Press.

Popper, K. R. (1935/2002). *The Logic of Scientific Discovery*. London: Routledge.

Pratchett, T. (1983). *The Colour of Magic*. London: Corgi.

Prigogine, I., & Stengers, I. (1984). *Order Out of Chaos: Man's New Dialogue with Nature*. Portsmouth, New Hampshire: William Heinemann Ltd.

Pullman, P. (2011). *His Dark Materials*. London: Everyman.

Rajneesh, C. M. (1980). *The Orange Book: The Meditation Techniques of Bhagwan Shree Rajneesh*. Zurich: Osho International Foundation.

Rao, R. P., & Ballard, D. H. (1999). Predictive coding in the visual cortex: a functional interpretation of some extra-classical receptive-field effects. *Nature Neuroscience, 2*(1), 79–87.

Rose, S. (1992). So-called "formative causation" - a hypothesis disconfirmed. Response to Rupert Sheldrake. *Rivista di Biologia - Biology Forum, 85*(3/4), 445–453.

Schopenhauer, A. (1859/1966). *The World as Will and Representation* (E. F. J. Payne, Trans.). New York: Dover Publications.

Schrödinger, E. (1926). An undulatory theory of the mechanics of atoms and molecules. *Physical Review, 28*(6), 1049–1070.

Schreiber, U. C., & Mayer, C. (2020). *The First Cell: The Mystery Surrounding the Beginning of Life*. Cham, Switzerland: Springer Nature.

Seaborg, G. (1964). Plutonium: The ornery element. *Chemistry, 37*(6), 12–17.

Searle, J. (1992). *The Rediscovery of the Mind*. Cambridge, Massachusetts: MIT Press.

Shakespeare, W. (1606/2016). *Macbeth*. New York: William Collins.

Sheldrake, R. (1992). An experimental test of the hypothesis of formative causation. *Rivista di Biologia - Biology Forum, 85*(3/4), 431–444.

Sheldrake, R. (1995). *The Presence of the Past: Morphic Resonance and the Habits of Nature*. Rochester, Vermont: Park Street Press.

Sheldrake, R. (2009). *A New Science of Life*. London: Icon Books.

Spalding, K. L., Bhardwai, R. D., Buchholz, B. A., Druid, H., & Frisen, J. (2005, July 15). Retrospective birth dating of cells in humans. *Cell, 122*, 133–143.

Stapp, H. P. (2017). *Quantum Theory and Free Will: How Mental Intentions Translate into Bodily Actions*. Cham, Switzerland: Springer International Publishing.

Strawson, G. (2017). Mind and being: The primacy of panpsychism. In G. Brütrup & L. Jaskolla (Eds.), *Panpsychism: Contemporary Perspectives* (pp. 75–112). Oxford: Oxford University Press.

Talbot, M. (1991). *The Holographic Universe*. London: Grafton Books.

Thomas, D. (1974). *Selected Poems*. London: J. M. Dent and Sons.

Thornton, J. (2007). *The Foundations of Computing: A Historical, Sociological and Philosophical Enquiry*. Sydney: Pearson Prentice Hall.

Turing, A. (1936-7). On Computable Numbers, with an Application to the Entscheidungsproblem. *Proc. of the London Mathematical Society, 42*(3), 230–265.

Turing, A. (1950). Computing machinery and intelligence. *Mind, 49*, 433–460.

von Neumann, J. (1932/2018). *Mathematical Foundations of Quantum Mechanics* (R. T. Beyer, Trans.). New Jersey: Princeton University Press.

Wheatstone, C. (1838). Contributions to the physiology of vision.–Part the first. On some remarkable, and hitherto unobserved, phenomena of binocular vision. *Philosophical Transactions of the Royal Society of London, 128*, 371–394.

Wigner, Eugene. (1961). Remarks on the mind-body problem. In I. J. Good (Ed.), *The Scientist Speculates* (pp. 284–302). London: Heinemann.

Wolfram, S. (2002). *A New Kind of Science*. Champaign, Illinois: Wolfram Media Inc.

Index

Lightning Source UK Ltd.
Milton Keynes UK
UKHW012014090921
390282UK00002B/176

9 781838 478704